MATRIX-BASED MULTIGRID

Numerical Methods and Algorithms

VOLUME 2

Series Editor:

Claude Brezinski
Université des Sciences et Technologies de Lille, France

MATRIX-BASED MULTIGRID

Theory and Applications

Second Edition

YAIR SHAPIRA
Technion-Israel Institute of Technology

 Springer

Yair Shapira
Technion – Israel Institute of Technology
Department of Computer Science
32000 Haifa, Israel
yairs@cs.technion.ac.il

ISBN: 978-1-4419-4321-7   e-ISBN: 978-0-387-49765-5
DOI: 10.1007/978-0-387-49765-5

ISSN: 1571-5698

AMS Subject Classifications: 65N55, 65F10

Printed on acid-free paper

9 8 7 6 5 4 3 2 1

springer.com

In memory of Prof. Moshe Israeli, my advisor and mentor

Contents

Part I Concepts and Preliminaries

List of Figures

List of Tables

Preface

Many important problems in applied science and engineering, such as the Navier–Stokes equations in fluid dynamics, the primitive equations in global climate modeling, the strain-stress equations in mechanical and material engineering, and the neutron diffusion equation in nuclear engineering contain complicated systems of nonlinear partial differential equations (PDEs). When approximated numerically on a discrete grid or mesh, such problems produce large systems of algebraic nonlinear equations, whose numerical solution may be prohibitively expensive in terms of time and storage. High-performance (parallel) computers and efficient (parallelizable) algorithms are clearly necessary.

Three classical approaches to the solution of such systems are: Newton's method, preconditioned conjugate gradients (and related Krylov-subspace acceleration techniques), and multigrid. The first two approaches require the solution of large sparse linear systems at each iteration, which are themselves often solved by multigrid. Developing robust and efficient multigrid algorithms is thus of great importance.

The original multigrid algorithm was developed for the Poisson equation in a square, discretized by finite differences on a uniform grid. For this model problem, multigrid converges rapidly, and actually solves the problem in the minimal possible time (Poisson rate).

The original multigrid algorithm uses rediscretization of the original PDE on each grid in the hierarchy of coarse grids (geometric multigrid). Unfortunately, this approach doesn't work well for more complicated problems with nonrectangular domains, nonuniform grids, variable coefficients, or nonsymmetric or indefinite coefficient matrices. In these cases, matrix-based multigrid methods are required.

Matrix-based (or matrix-dependent) multigrid is a family of methods that use the information contained in the discrete system of equations (rather than the original PDE) to construct the operators used in the multigrid linear system solver. This way, a computer code can be written such that it accepts the coefficient matrix and right-hand side of the discrete system as input and produces the numerical solution as output. The method is automatic in the sense that the above code is independent of the particular application under consideration.

Because the elements in the coefficient matrix contain all the information about the properties of the boundary-value problem and its discretization, matrix-based multigrid methods are efficient even for PDEs with variable coefficients and complicated domains. In fact, matrix-based multigrid methods are the only multigrid methods that converge well even for diffusion problems with discontinuous coefficients, even when the discontinuity lines do not align with the coarse mesh.

This book offers a new approach towards the introduction and analysis of multigrid methods from an algebraic point of view. This approach is independent of the traditional, geometric approach, which is based on rediscretizing the original PDE. Instead, it uses only the algebraic properties of the original linear system to define, analyze, and apply the multigrid iterative method. This way, multigrid methods are well embedded in the family of iterative methods for the numerical solution of large, sparse linear systems. Indeed, as is shown below, multigrid methods can actually be viewed as special cases of domain-decomposition methods.

The present edition of this book introduces the multigrid methods from a unified domain-decomposition point of view. In particular, advanced multigrid versions (such as black-box multigrid, algebraic multigrid, and semicoarsening) can all be interpreted as domain-decomposition methods. Furthermore, it introduces a new semi algebraic approach for systems of PDEs. Each chapter ends with relevant exercises.

The first three parts are introductory. The first part introduces the concept of multilevel/multiscale in many different branches in mathematics and computer science. The second part gives the required background in discretization methods, including finite differences, finite volumes, and finite elements. The third part describes iterative methods for solving the linear system resulting from this discretization. In particular, it introduces multigrid methods from a domain-decomposition point of view.

The next three parts contain the heart of the book. The discussion starts from the simplified but common case of uniform grids, proceeds to the more complicated case of locally refined grids, and concludes with the most general and difficult case of completely unstructured grids. In each of these three parts, we concentrate on a particular multigrid version that fits our framework and method of analysis, and study it in detail. We believe that this study may shed light not only on this particular version but also on other multigrid versions as well.

These three parts are ordered from simple to complex: from the simple case of rectangular uniform grids, where spectral analysis may be used to predict the convergence factors (Part IV), through the more complex case of locally refined (and, in particular, semistructured) grids, where upper bounds for the condition number are available (Part V), to the most general case of completely unstructured grids, where the notions of stability and local anisotropy motivate and guide the actual design of algebraic multilevel methods (Part VI).

The second edition also contains the mathematical background required to make the book self-contained and suitable not only for experienced researchers but also for beginners in applied science and engineering. Thanks to the introductory parts, no background in numerical analysis or multigrid methods is needed. The only prerequisites are linear algebra and calculus. The book can thus serve as a textbook in courses in numerical analysis, numerical linear algebra, scientific computing, and numerical solution of PDEs at the advanced undergraduate and graduate levels.

Acknowledgments

I wish to thank the referees for their thorough and helpful reports, and the Wingate foundation for their kind support. I wish to thank Prof. Moshe Israeli and Prof. Avram Sidi for their valuable advice in the development of the AutoMUG method, and Prof. Irad Yavneh for his valuable comments. I also wish to thank Dr. Joel Dendy and Dr. Dan Quinlan for useful discussions about semistructured grids, and Dr. Mac Hyman for his valuable support. I wish to thank Dr. Dhavide Aruliah and Prof. Uri Ascher for supplying the coefficient matrix for the Maxwell equations. Finally, I wish to thank my sons Roy and Amir for their constant patience and support.

Yair Shapira
August 2006

Concepts and Preliminaries

Multigrid is a numerical method to solve large linear systems of algebraic equations arising from mathematical modeling of physical phenomena. *Multiscale* is a more general concept that helps to distinguish between different scales in a given mathematical or physical problem. *Multilevel* is a yet more general concept, which helps to define and understand abstract mathematical hierarchies and use them to solve practical problems in different branches in mathematics and applied mathematics.

In the first chapter, we introduce the concepts of multiscale and multilevel and illustrate them in different sorts of problems. In the second chapter, we list some basic definitions and prove some standard lemmas, which are used later throughout the book. In particular, we use these lemmas in the multilevel hierarchy known as the Fourier transform.

Note that the Table of Contents sometimes uses short versions of the titles of chapters and sections, as used at the head of the pages.

The Multilevel-Multiscale Approach

In this chapter, we describe the concept of multilevel and multiscale, and use them in several elementary examples: from basic algorithms in arithmetics of integer numbers, through mathematical induction and recursion, to data structures and parallel algorithms.

1.1 The Multilevel-Multiscale Concept

Multilevel is an abstract concept that stands for a hierarchy of objects. Most often, the objects in the hierarchy are just different forms of the same object. In this case, the object may take a more developed form in the high levels in the hierarchy than in the low levels. The high levels may be more delicate and fine, and contain more details and information about the object.

The objects that form the levels in the multilevel hierarchy are usually abstract rather than physical. In fact, they may well be just ideas, processes, operations, or points of view. Therefore, the multilevel concept may be very helpful not only as a tool for organizing thoughts, but also as a framework for gaining insight, developing ideas, and solving problems.

In some cases, the fine details that are contained in the high levels in the multilevel hierarchy can be viewed as fine scales that can measure or notice delicate features, whereas the low levels deal with the large-scale and global nature of the phenomenon or object under consideration. The multilevel hierarchy may then be viewed as a multiscale hierarchy, with fine scales corresponding to high levels and coarse scales corresponding to low levels. The concept of multiscale is thus a little more concrete than the concept of multilevel, as it specifically considers the growing number of details that can be observed in finer and finer scales. Still, the fine features in the fine scales are not necessarily physical; they may well be abstract. Nevertheless, for simplicity we start with examples that concern physical features.

Almost every measurement made in day-to-day life is based on scales. Consider, for example, the measurement of distances. Although short distances can be measured in meters or feet, these are not very useful for distances between cities, where kilometers or miles are more suitable. The latter units are nevertheless useless when distances between stars are considered, where light years become handy.

On the other hand, in the microprocessor industry, a unit greater than a micron is never to be mentioned.

Thus, measurements can be made in many different scales, each of which is useful in its own context. A less obvious observation is that the very representation of numbers is also based on multiscale. By *number* we mean the abstract concept of number, not the way it is represented or written. In fact, we will see below that there may be many legitimate ways to represent, or write, one and the same number. We will see below that all these representations are actually based on multiscale.

1.2 The Integer Number

From the dawn of civilization, people have had the need to count objects; most often, property objects. Trade was based on counting goods such as cattle, sheep, and, of course, valuables such as gold and silver coins. The need to have an economical counting method that would use a small number of symbols to represent arbitrarily large numbers became most important. The decimal method that is based on hierarchy and regrouping fulfills this need.

In the decimal counting method, groups of ten are used, presumably because people have ten fingers. Thus, only ten different symbols, or digits, are needed; these symbols are used to represent numbers from zero to nine. The next number, ten, is no longer represented by a new symbol, but rather as one group of ten units. The counting goes on in groups of ten, plus a residual number (remainder) that is smaller than ten. The number of groups of ten is represented by the more significant, decimal digit, whereas the residual is represented by the less significant, unit digit. When ten groups of ten have been collected, they are regrouped in a group of ten groups of ten, or a group of one hundred, and so on. The hierarchy of more and more significant digits allows the representation of arbitrarily large integer numbers. For example,

$$1000 + 500 + 20 + 3 = 10^3 + 5 \cdot 10^2 + 2 \cdot 10^1 + 3 \cdot 10^0 = 1523.$$

The coefficient of the next larger power of ten is represented by the next more significant digit in the decimal representation. The next more significant digit represents the next larger scale or the next higher level in the representation of the number.

When digital computers were introduced, the binary method for representing integer numbers became more important. The computer has only two "fingers," "on" and "off." Therefore, it prefers to count integer numbers in groups of two rather than ten, and then regroup them in groups of $2 \cdot 2 = 4$ rather than 100, and so on. The result is the following algorithm for transforming a number represented in the decimal method to the same number in the binary method.

Algorithm 1.1

1. *Input the number k that is written in the decimal method.*
2. *Let s denote the current power of 2 in the binary representation. Initialize s by $s \leftarrow 0$.*
3. *If $k = 1$, then the digit at scale s in the binary representation is also 1, and the algorithm is complete. If $k > 1$, then proceed as follows.*

4. *Divide k by 2 with integer residual:*

$$k = 2m + r,$$

where $r = 0$ if k is even, and $r = 1$ if k is odd.
5. *The residual r calculated in the previous step is the digit in scale s in the binary representation.*
6. *Replace k by m:*

$$k \leftarrow m.$$

7. *Replace the scale s by the next larger scale:*

$$s \leftarrow 2s.$$

8. *If $k \neq 0$, then go to step 3.*

Let us illustrate this simple algorithm for the input number 77 (in the decimal method). The last digit in the binary representation is 1, because

$$77 = 2 \cdot 38 + 1.$$

The next, more significant digit is 0, because

$$38 = 2 \cdot 19 + 0.$$

The next two more significant digits are 1's, because

$$19 = 2 \cdot 9 + 1$$

and

$$9 = 2 \cdot 4 + 1.$$

The next two more significant digits are 0's, because

$$4 = 2 \cdot 2 + 0$$

and

$$2 = 2 \cdot 1 + 0.$$

Finally, the most significant digit is, as always, 1. The output of the algorithm is, thus,

$$1001101,$$

which indeed is the binary representation of

$$77 = 2^6 + 2^3 + 2^2 + 2^0.$$

The multiscale approach is also helpful in simplifying lengthy algorithms by using recursion. This is because multiscale is based on repetition: what is good for a certain scale, must be equally good for the next, larger scale. Therefore, an algorithm can often be written in a short and compact way by initializing it for the first scale and applying it recursively to the next, larger scale as well. In our case, the above binary-representation algorithm takes the form:

Algorithm 1.2 Binary(k):

1. *Input the number k that is written in the decimal method.*
2. *If $k = 1$, then the output is also 1. If $k > 1$, then proceed as follows.*
3. *Divide k by 2 with integer residual:*

$$k = 2m + r,$$

where $r = 0$ if k is even, and $r = 1$ if k is odd.
4. *The output is the following sequence of digits:*

$$\textbf{Binary}(m)r.$$

The recursive call in the last step produces the binary representation of the "coarse-scale" approximation to k, namely, m. By adding r to it as a last digit, the algorithm produces the entire binary representation of k.

The multiscale approach is thus useful not only in representing mathematical objects such as the above integer numbers, but also in manipulating them as in the above algorithm. This is illustrated next in another elementary algorithm.

1.3 Division of Integers

Most often, the user is unaware of using different scales or levels in the algorithm. Consider, for example, the standard algorithm for division of integers. Let us look at the following example: find an integer x that satisfies the equation

$$8x = 192.$$

The basic algorithm splits the problem into different scales: 1's, 10's, 100's, and so on. The solution is then obtained by summing the contributions from the different scales.

The first step in the algorithm is to confine or restrict the right-hand side 192 to a coarse scale, namely, the scale of 10's. The restriction operation is just to divide the integer by 10 (with residual), resulting in the number 19. (Indeed, the original right-hand side 192 contains 19 10's.) Then, the problem is solved at the coarse scale of 10's, giving the result $19/8 = 2$ (division of integers with residual). This contribution from the coarse level is then prolonged back to the fine level by simply multiplying it by 10, resulting in the number 20. The residual (remainder) at the approximate solution $\tilde{x} = 20$ is

$$192 - 8\tilde{x} = 192 - 8 \cdot 20 = 32.$$

This residual is then used to obtain the contribution from the fine scale of 1's, namely, $32/8 = 4$. The solution is, thus, the sum of contributions from coarse and fine levels, namely,

$$x = 20 + 4 = 24.$$

This algorithm is described schematically in Figure 1.1. The illustration in this figure is similar to the Latin letter F, hence the name "F-cycle." (Compare to the V-cycle in Figure 6.1 below.)

In the next section, we show how multiscale can be used in a more complicated algorithm.

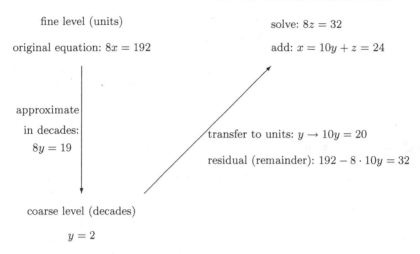

fine level (units)   solve: $8z = 32$

original equation: $8x = 192$ add: $x = 10y + z = 24$

approximate
in decades: transfer to units: $y \to 10y = 20$
$8y = 19$
                                               residual (remainder): $192 - 8 \cdot 10y = 32$

coarse level (decades)

$y = 2$

Fig. 1.1. The so-called F-cycle for division of integers: the present example is $192/8$.

1.4 The Greatest Common Divisor

In the above algorithm, the different scales are predetermined independently of the particular problem: they are always the decimal scales of units, tens, hundreds, and so on. Other multiscale algorithms may be nonlinear in the sense that the scales depend on the data supplied in the particular instance to be solved. Consider, for example, Euclid's algorithm for computing the greatest common divisor of two integers. Let us compute, for instance, the greatest common divisor of the numbers 100 and 36. First, we divide (with residual):

$$100/36 = 2 \text{ with residual } 28.$$

This division may be interpreted as using the scale obtained from multiples of 36 as the "coarsest" scale for locating the number 100 in the number line. The residual (remainder) in the above division, 28, will serve as the next finer scale. Indeed, in the next step we divide with residual:

$$36/28 = 1 \text{ with residual } 8.$$

The residual 8 will serve as a divisor to calculate the next finer scale:

$$28/8 = 3 \text{ with residual } 4.$$

Since 8 is divided by 4 with no residual, there is no need to use a further finer scale. The scale consisting of multiples of 4 is sufficiently fine to locate both 100 and 36 on the number line. Thus, 4 is the greatest common divisor of 100 and 36.

Thus, Euclid's algorithm uses variable scales: 36, 28, 8, and so on. These scales are obtained from the algorithm itself, and depend on the particular application. The algorithm that divides integer numbers in Section 1.3, on the other hand, uses fixed scales: 100, 10, 1, and so on, which are independent of the particular application.

This distinction is analogous to the distinction made later in this book between structured and unstructured grids: in structured (uniform) grids the coarse grids are always constructed in the same way (independently of the particular application), whereas in unstructured grids the coarse grids may be defined during the multigrid algorithm, using information from the particular problem that is currently being solved.

1.5 Multilevel Refinement

In the early days of the science of applied mathematics in the 19th century and the first part of the 20th century, boundary-value problems were recognized as a powerful tool in modeling many physical phenomena. In a boundary-value problem, one needs to find a function that is defined in a given spatial domain, satisfies a given differential equation in the interior of the domain, and agrees with some given function on its boundary. Unfortunately, only few model boundary-value problems can be solved analytically in closed form.

The invention of the digital computer in the middle of the 20th century changed the science of applied mathematics completely. Complicated boundary-value problems that model complex physical phenomena such as weather and compressible flow suddenly became solvable. The solution process uses fine grids that are embedded in the domain. The discretization of the original problem on these grids produces large systems of difference equations that approximate the original differential equations. The science of numerical analysis that studies the quality of these approximations and the science of scientific computing that studies the performance of computational methods to solve these discrete models have been born.

In many realistic boundary-value problems, the domain is so complicated that it cannot be approximated well by a standard rectangular grid. The original grid should, therefore, be further refined near the boundary of the domain and also at locations where the solution is expected to have large variation. This process can be repeated to provide further and further high-resolution refinement where it is needed. The final grid that is obtained from this process of multilevel refinement has a better chance to approximate well the complicated domain and the delicate features in the solution function.

1.6 Example in Computer Science

The notion of multilevel is also useful in computer architectures and, in particular, storage strategies. The cache memory hierarchy is based on the idea that often accessed data should be stored in the primary memory, where access is fast and efficient, whereas rarely accessed data should be stored in the secondary memory, where access is slow and expensive [47]. This principle is used in the entire cache hierarchy, which may contain many levels: the highest level that is accessed most easily contains data that are accessed most often, and lower and lower levels that are accessed less and less easily contain data that are accessed less and less often. This way, frequently accessed data are stored in the easily accessed top of the hierarchy, and rarely accessed data are stored at the hard-to-access bottom of the hierarchy.

1.7 Multilevel in Mathematical Logics

A mathematical system contains objects and relations between them. For example, the system of Euclidean geometry contains points, lines, angles, and so on, with the possible relations that a particular point may lie on a particular line, a particular line may pass through a particular point, a particular line may be parallel to some other line, and so on.

The mathematical system also contains some assumptions, known as axioms. For example, two axioms in Euclidean geometry state that through two distinct points passes one and only one line, and that through a given point passes one and only one line that is parallel to another given line that doesn't pass through the point, etc.

A theorem in the mathematical system says that, under some given assumptions, a particular statement (or assertion) is true. The proof of the theorem uses the concept of multilevel in the context of mathematical logic. Indeed, the proof starts from the assumptions in the theorem and some of the axioms in the mathematical system in the lowest (or finest) level. Then, some items from the lowest level are combined to conclude some properties, facts, or assertions in the next higher level. Similarly, the items in these two levels are combined again to conclude more advanced properties, facts, or assertions in the next higher level. By repeating this process, the proof "climbs" from level to level, resulting in more and more advanced assertions, concluded from the assertions that have been proved before in the lower levels. Finally, in the top level there should be only one item: the assertion of the original theorem. This item has thus been proved successfully by the entire multilevel proof.

1.8 Multilevel in Language

Multilevel is also a powerful tool in human language, in both text and speech. Indeed, the letters, which are the most elementary bricks that form the lowest level, are combined with each other to produce syllables, which are in turn combined to produce words in the next higher level. Words, in turn, are combined to produce subsentences, which are then combined according to the laws of grammar to produce sentences. Finally, sentences are combined to form paragraphs, each of which sheds light on one of the aspects of the subject discussed in the text.

We refer to the above multilevel hierarchy as the syntactic structure, because it concerns the formal structure of the text. The original plan of the contents of the text, however, as well as its interpretation by the reader, must use a different structure: the analytic structure.

Like the syntactic structure, the analytic structure also uses a multilevel hierarchical logics. However, this multilevel logics may be totally different from the one used in the syntactic structure.

Indeed, in the analytic multilevel logics, the main idea in the text lies in the highest level. The motives, reasons, aspects, or consequences of this idea lie in the next lower level. The facts, factors, and processes behind these motives, reasons, aspects, and consequences form the next lower level. Finally, the details of these facts, factors, and processes form the lowest level.

When one wants to make a short summary of the text, it is advisable to parse (open) the above analytic hierarchy. First, the main (often abstract) idea in the text should be introduced. Then, the main motives, reasons, aspects, and consequences of this idea should be listed, each with its main factors. The details can be omitted.

Although the analytic structure is essential in planning the text, the syntactic structure is no less important in the process of writing it in a good and readable style. In fact, the syntactic structure is relevant not only in natural (human) languages but also in formal languages, such as computer languages. Indeed, in good and elegant programming, one uses the elementary variables and arithmetic operations as low-level bricks to write elementary functions in the next higher level. These functions are then used to write yet more complex functions in the next higher level. This process eventually produces the hierarchy of functions required to write elegant, modular, and easily debugged computer programs.

1.9 Multilevel Programming

Multilevel is a useful concept also in computer programming. Indeed, the computer actually uses the most elementary programming language: the machine language. This low-level language tells the computer in detail how to fetch a datum from a particular address in the memory, update it by some arithmetic operation, and then return it to its original address in the memory. Fortunately, the programmer is free from bothering with these technical details: he/she is writing in a high-level programming language such as Fortran, C, or C++, where fetching data from the memory and updating it is done implicitly by the assignment operator, and arithmetic operations are supported by the standard arithmetic symbols. This way, the programmer can focus on implementing the mathematical algorithm to solve his/her particular application.

The high-level programming language may thus be viewed as an interface for the programmer to access the computer. The programmer can use this interface to write various functions, and let other people use them by giving them access to his/her well-documented code. This way, the programmer actually creates a yet higher-level interface for users who don't really need to program but only want to use the functions that he/she has written.

1.10 Object-Oriented Programming

Object-oriented programming is a programming approach, in which the programmer can design and introduce new objects along with the functions that characterize them. This way, the programmer practically extends the standard high-level programming language by introducing new words which refer to the new objects and functions.

In fact, the programmer can place his/her code in a library, so anyone who has access to this library can include it in his/her code and use the names of the new objects and functions as if they were standard keywords in the high-level programming language.

Object-oriented programming can be viewed as a kind of multilevel programming. Indeed, the above objects may be viewed as elementary or low-level objects. Now, the users of the above library can design more complicated "higher-level" objects, which may contain or use the low-level objects. Again, these higher-level objects can be placed in a higher-level library for further use in yet more complicated objects. By repeating this process, a hierarchy of more and more complicated objects is created, each of which uses the more elementary objects in the "lower" levels below it.

The above approach actually enriches the high-level programming language substantially by introducing into it new words which refer to the new objects and the functions that characterize them. In [103], it is shown how this approach can be used successfully in the numerical solution of PDEs.

1.11 Example in Data Structures

The concept of a "tree" in data structures is another example of a multilevel structure that is most useful in graph theory, data structures, and algorithms. The "tree" structure is suitable for recursion, and is thus often used to implement data structures such as connected lists and algebras. Mathematical objects such as two-sided connected lists and arithmetical expressions are best implemented in binary trees (Figure 1.2).

1.12 The Sorting Problem

Some mathematical problems also find their best solution by using virtual trees. For example, the problem of sorting an unsorted list of elements according to a given order is best solved by recursion. The original list is divided into two sublists. On each sublist, the sorting procedure is applied recursively. Once the two sublists are sorted, they are merged into one sorted list. The complexity, or cost, of this method is linear, that is, it is linearly proportional to the number of elements in the original list.

1.13 Parallelism

The notion of multilevel is also helpful in parallel computing, where one often uses a "divide-and-conquer" strategy. When a problem is encountered, one first identifies certain problem components that can be solved independently of each other. These

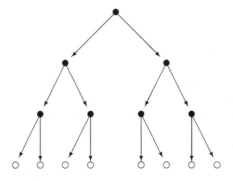

Fig. 1.2. The binary tree is also based on the multilevel concept.

components, which may be thought of as the "fine level" of the problem, are solved simultaneously in parallel. Other problem components that cannot be divided because they are inherently sequential and deal with "global" features of the problem, must be solved sequentially in a "coarse level" phase and, thus, may create a bottleneck in the process that slows it down.

1.14 Self-Similarity

The notion of self-similarity used in the science of chaos is another example for the usefulness of the multiscale approach. In this notion, certain physical and other kinds of phenomena, albeit chaotic, may still be characterized by the property that small elements in the system behave in much the same way as the system itself. In this way, tiny pieces of a leaf may be similar to the leaf itself, small portions of the seashore may have the same pattern as the global seashore, and the weather in a small area may have similar properties to the global weather. The individual elements in the system may be viewed as a fine scale, whereas the global system forms the coarse scale. Because of the self-similarity, these elements contain further smaller elements, which form a yet finer scale in the multiscale hierarchy, and so on.

The mathematical concept known as "fractal" is characterized by the self-similarity property, and can thus model such phenomena. The self-similarity property may also be used in numerical simulations, where different scales are handled separately in the solution process.

1.15 The Wavelet Transform

The wavelet transform of one-dimensional functions is another example of a useful multilevel hierarchy. In the wavelet transform, the function is repeatedly "averaged" at each point, using values at adjacent points whose distance from the current point is successively doubled when moving from level to level. The "averaging" process can use both positive and negative weights, and the regularity of the transformed function (the number of continuous derivatives) grows linearly with the number of adjacent points used in the "average." The wavelet transform is useful in the discretization and numerical solution of integro-differential equations in rectangular domains. The discrete analogue of the wavelet transform is often used in image processing and, in particular, in coding-decoding and compression.

1.16 Mathematical Induction and Recursion

A most important mathematical concept that can also be viewed as a multilevel process is the concept of mathematical induction. In logic, *induction* means the extension of an argument, which is already known for certain individual objects, to the entire family of objects. Mathematical induction also attempts to extend a property, which is already known for a particular mathematical object, to an infinite, well-ordered set of such objects. In order to have such an extension, one must prove the induction step, that is, that the fact that the property holds for a particular object implies that it also holds for the next object. Once we know that the property holds for the first object and the induction step is valid, we actually know that the property also holds for every object in the entire set, because we can always start from the first object and step (or "climb") from object to object using the induction step until the desired object is reached.

Mathematical induction is also useful in the definition of infinite sequences of mathematical objects. In fact, the very definition of integer numbers is done in Peano theory by mathematical induction. Indeed, assume that '1' denotes the first integer, and the fact that n is an integer implies that $n+1$ is an integer as well (the induction step). Then, we immediately have the entire infinite sequence of positive integer numbers that have just been defined by mathematical induction. Indeed, every positive integer m can now be constructed by starting from 1 and applying the induction step $m-1$ times.

Mathematical induction is also useful in solving more complicated problems. Assume that the problem is characterized by a positive integer n, i.e., it concerns n distinct elements. Assume that the solution of smaller problems that are associated with $1, 2, \ldots, n-1$ implies the solution of the problem associated with n (the induction step). Assume also that we know how to solve the simplest problem that is associated with the integer 1. Then the problem is solved by solving the problem associated with 1, then using it in the induction step to solve the problem associated with 2, then using the solution of both these problems in the induction step to solve the problem associated with 3, and so on, until the problem associated with n is solved.

The above process is also known as *recursion*. In recursion, however, the process is viewed the other way around. The induction step is viewed as a reduction step, in which the problem associated with n is reduced to subproblems associated with integers smaller than n. These subproblems are in turn reduced further into subproblems associated with yet smaller integers, until every subproblem has been reduced to the simplest subproblems associated with 1, which are easy to solve.

1.17 The Tower Problem

A good example for using mathematical induction and recursion is the tower problem. Consider three columns denoted by "A", "B", and "C". Assume that a tower of n rings is built around column A, in which the rings are ordered from the largest at the bottom of the tower to the smallest at the top of the tower. In other words, the largest ring (the ring of largest radius) lies at the bottom of the tower, the next smaller ring sits on top of it, and so on, until the smallest, nth ring that sits on top of all the $n-1$ larger rings at the top of the tower. The task is to transfer the entire tower of n rings from column A to column C (possibly by using also column B if necessary), while preserving the following rules.

1. A sequence of moves should be used.
2. In each move, one ring is moved from one column to the top of either of the other two columns.
3. A ring cannot be moved if another ring lies on top of it.
4. A ring cannot be placed on a smaller ring.

The solution of this complex problem by mathematical induction is straightforward. Clearly, the solution of the problem associated with $n = 1$ is trivial, because it only requires the transfer of one ring from column A to column C. Let us now show how the problem is solved for $n > 1$. For this purpose, we need to construct the induction step. Assume that we already know how to transfer towers of $n - 1$ rings from column to column, while preserving the above rules. Let us use this knowledge also to transfer a tower of n rings from column A to column C. First, let us transfer the top $n - 1$ rings from column A to column B, using our assumed knowledge. Then, let us move the remaining, largest ring from column A to column C. Finally, let us use our assumed knowledge once again to transfer the entire tower of $n - 1$ rings from column B to column C and put it on top of the largest ring that already lies there. This completes the transfer of the entire tower of n rings from column A to column C, as required.

In practice, the problem is solved by recursion (Figure 1.3). The original problem to transfer a tower of n rings is reduced to the problem of transferring a tower of $n - 1$ rings. The reduced problem should be solved twice in the algorithm. In order to solve it, the reduced problem is further reduced to the problems of transferring a yet smaller tower of $n - 2$ rings. This problem should be solved four times in the algorithm: twice in each "solve" (solution process) of the previous problem. The reduction is repeated in the same way, until only trivial problems of transferring towers of one ring need to be solved. In fact, the above recursion produces a virtual binary tree of reduced problems that are to be solved. Each leaf at the bottom of this tree contains a trivial problem. Mathematical induction can then be used to prove that the algorithm indeed solves the original problem.

Because each reduction doubles the number of times that a subproblem has to be solved, the total number of moves that are required in the entire algorithm

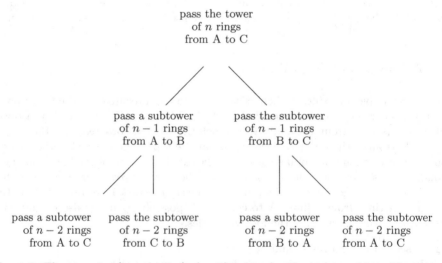

Fig. 1.3. The recursive (or inductive) algorithm to solve the tower problem. The entire task is placed in the upper level. This task is carried out by two subtasks in the second level. Each of these subtasks is carried out recursively, as can be seen in the third level, and so on.

is $2^n - 1$. If each move requires a time unit, then the solution of the problem requires exponentially large time. Even when implemented efficiently on a strong computer, the above algorithm requires a prohibitively large amount of time for large n. Unfortunately, this amount of time cannot be reduced; the problem cannot be solved more rapidly by any algorithm.

In this book, we are more interested in problems that can be solved in polynomial time, that is, in n^p time units, for some constant p that is independent of n. In the next section, we consider such a problem, and show how it can be solved efficiently on a parallel computer.

1.18 The Parallel Product Algorithm

In this section, we describe a parallel algorithm for computing the products of elements in a group. We show that it can actually be viewed as a multilevel algorithm and also as a domain-decomposition algorithm.

Let $k > 1$ and $p > 1$ be given integers. Consider the following problem. Use $p-1$ processors to calculate efficiently the $kp - 1$ products

$$g_1 g_2$$
$$g_1 g_2 g_3$$
$$g_1 g_2 g_3 g_4$$
$$\cdots$$
$$g_1 g_2 g_3 \cdots g_{kp},$$

where $g_1, g_2, g_3, \ldots, g_{kp}$ are elements in some group or semi-group.

The algorithm starts with a "fine scale," parallel step that computes the $p-1$ products

$$G_1 = g_1 g_2 g_3 \cdots g_k$$
$$G_2 = g_{k+1} g_{k+2} g_{k+3} \cdots g_{2k}$$
$$G_3 = g_{2k+1} g_{2k+2} g_{2k+3} \cdots g_{3k}$$
$$\cdots$$

Then, a "coarse scale," sequential step is used to calculate the $p-1$ products

$$H_1 = G_1$$
$$H_2 = G_1 G_2$$
$$H_3 = G_1 G_2 G_3$$
$$\cdots$$

Finally, another "fine scale," parallel step is used to compute the products

$$H_r g_{rk+1}$$
$$H_r g_{rk+1} g_{rk+2}$$
$$H_r g_{rk+1} g_{rk+2} g_{rk+3}$$
$$\cdots$$
$$H_r g_{rk+1} g_{rk+2} \cdots g_{(r+1)k}.$$

where r, $1 \le r \le p - 1$, is the index of the task assigned to the rth processor. This algorithm illustrates the usefulness of the multilevel approach in parallel computing. As a matter of fact, the "coarse level" part in the algorithm may be further

Fig. 1.4. The parallel product algorithm from a domain-decomposition point of view. Each circle or bullet represents an element from the group. The algebraic "subdomains" consist of the circles, and are separated by the bullets.

parallelized by using the same algorithm itself recursively, although not all the processors can be used in the recursion.

The parallel product algorithm can also be viewed as a domain-decomposition algorithm. The kp elements that form the product are ordered in a long line as in Figure 1.4. In this figure, the bullets stand for g_k, g_{2k}, g_{3k}, and so on, and divide the line into algebraic "subdomains." In the first step in the above algorithm, local tasks are carried out in the individual subdomains. In the second step, a global task is carried out over the bullets. In the third and final step, more local tasks are carried out in the individual subdomains. This relation between domain decomposition and multilevel is highlighted throughout the book.

1.19 Multilevel in Statistics

Multilevel is also used often in quantitative research in medicine and social sciences. Suppose that we have a phenomenon in, for example, economy, sociology, or psychology that we want to analyze. Of course, we cannot study the entire population, so we consider a sufficiently large sample (say, 1000 people) and ask them about the phenomenon. Then, we place their answers in a 1000-dimensional vector. For example, in economy, the phenomenon might be their annual salary. In this case, the annual salaries of the 1000 people are the components in the 1000-dimensional vector.

Now, suppose that we want to study the effect of gender on the salary. For example, we may suspect that there may be discrimination in our society in terms of gender, and we want to know if this is indeed true. To do this, however, it is not enough to check whether there is any correlation between gender and salary; after all, there may be inherent differences between the genders. For example, women may work fewer hours than men, so they may earn less not because they are women but merely because they work less. We must therefore design our research carefully to avoid misleading results.

The above vector is called the dependent vector, because it is assumed that salary depends on some other phenomena that are more obvious to us, say gender and amount of work per day. To quantify this dependence, we form three 1000-dimensional independent vectors. The first vector is the constant vector, all of whose components are equal to 1. This vector explains (approximates) the average salary in the entire sample. The second vector is a binary vector, whose components are either 1 (if the corresponding interviewee is a woman) or 0 (if the corresponding interviewee is a man) This independent vector may be viewed as a "coarse-level" vector, whose purpose is to explain the variance between the salaries of women as a group and the salaries of men as a group. The third 1000-dimensional independent vector contains the number of hours that one works per day. This "fine-level" vector is supposed to explain the fine-scale variance among the individuals in the sample in terms of the amount of time they work per day.

Now, we approximate the dependent vector by a linear combination of the three independent vectors in the best way we can (say, in the l_2-norm). The coefficients of the independent vectors in the linear combination then tell us the results. In fact, the first and third independent vectors are only used for control purposes. What really interests us is the coefficient of the second independent vector, the "coarse-level" vector. Indeed, this coefficient tells us whether gender really affects salary: if the coefficient is negative, then it would mean that the salary of a woman may indeed decrease only because she is a woman. If, on the other hand, it is not statistically significant, then it would mean that we are unable to conclude any gender discrimination from our experiment.

Of course, more realistic research must include many more "fine-level" independent vectors to account for many more kinds of variance in the level of individuals. For example, one must also control variance in terms of education, marital status, number of children, and so on. Only when every relevant "fine-level" independent vector is included in the model (namely, it is controlled), can one use the coefficient of the "coarse-level" independent vector to draw any indication about any discrimination against any particular group in the society.

This procedure is called *linear regression* or *analysis of variance*. Indeed, it attempts to study the effect of the "coarse-level" variance between different groups, after the effect of the "fine-level" variance among individuals has been filtered out.

In more thorough research, one might want to introduce also "intermediate-level" independent variables to control variance among subgroups in the population. For example, one might want to control the profession of the interviewee, as it might have a significant effect on his/her salary. To do this, one should introduce many more binary independent vectors with the value 1 if one is in a particular profession and 0 otherwise. These independent vectors will filter out the effect of profession on the salary, leaving the pure effect of gender in the coefficient of the gender vector.

1.20 Multilevel in Music

Multilevel is also often used in music. The fine level in the musical piece is determined by the basic motif, which consists of a small number of notes that contain the basic musical idea. In the coarser level, this motif is repeated with slight variations, while preserving its original spirit, to form the complete musical piece.

In other words, the motif in the fine level is the heart of the musical piece. In the coarse level, it is varied by adding to it extra notes and harmonies. Thus, the coarse level is where the musical piece is actually designed and tied from the elementary motif.

1.21 Exercises

1. Consider the (naive) inherently sequential algorithm to solve the problem in Section 1.18:

$$L_1 = g_1$$
$$L_{i+1} = L_i g_{i+1}, \qquad i = 1, 2, 3, \ldots, kp - 1.$$

Compute the operation count (number of multiplications) required in this algorithm. (The answer is a function of kp.)

2. Assume that a multiplication requires α seconds, where α is a small positive parameter. Calculate the computation time required in the above algorithm. (This is called the *serial time*.)

3. Calculate the number of multiplications required in the parallel algorithm in Section 1.18. Is it larger than the number of multiplications in the above naive algorithm? By what factor?

4. Assume that a parallel computer with p processors is available. Assume that a multiplication in each processor requires α seconds. Calculate the *parallel computation time*: the time required to perform the multiplications required in the parallel algorithm on the parallel computer with p processors. (You may assume that the coarse-level step in the parallel algorithm is carried out on one of the processors.)

5. Assume also that delivering data from one processor to another requires $\gamma + \delta B$ seconds, where γ and δ are small positive parameters, and B is the number of bytes delivered in the message. Calculate the *communication time*: the time required to deliver the messages required in the above parallel algorithm on the above parallel computer. Note that communication is required mainly in the coarse-level step in the parallel algorithm in Section 1.18. (You may assume that this step is performed by first sending all the required data from all the processors to a single processor, and then carrying out the calculation in this processor only.)

6. Calculate the *parallel time*: the parallel computation time plus the communication time.

7. Calculate the *speedup*: the serial time divided by the parallel time.

8. Calculate the *average speedup*: the speedup divided by p.

9. Assume the typical values $\alpha = 10^{-6}$, $\gamma = 10^{-3}$, $\delta = 10^{-7}$, and $kp = 10^9$. How do the speedup and average speedup behave as functions of p?

10. Repeat the above exercises, only this time assume that the coarse-level step in the parallel algorithm in Section 1.18 is carried out recursively in parallel by the same parallel algorithm itself. How does this affect the speedup and average speedup?

11. Write the computer code required for the recursive algorithm in the previous exercise. You may assume that $k = 2$. (You must use a programming language that supports recursion, such as C or C++.)

12. Write the computer code that implements the algorithm in Section 1.3 that divides two integer numbers (without using the '/' operation available in the computer language.)

13. Repeat the above exercise, only this time use recursion.

14. Write the computer function that implements the algorithm in Section 1.4 to find the greatest common divisor of two positive integer numbers m and n ($m > n$).

15. Repeat the previous exercise, only this time use recursion. For example:

Algorithm 1.3 *GCD(m, n): if m is divisible by n, then return n; otherwise, return GCD(n, m mod n).*

16. Use induction on m to show that the above code indeed works.

17. Show that the least common multiple of two integer numbers i and j is ij/m, where m is the greatest common divisor of i and j.

18. Use your answers to the previous two exercises to write a code that adds two simple fractions (without using the arithmetic operations available in the programming language) by finding the least common multiple of the two denominators and using it as a common denominator.

19. Write the computer code that implements the algorithm in Section 1.2 that provides the binary representation of an integer number. The solution can be found in Section 1.18 in [103].

20. Repeat the previous exercise, only this time use recursion. The solution can be found in Section 1.18 in [103].

21. Write the computer code that implements the algorithm in Section 1.12 to sort a list of elements according to a given order (e.g., a list of unordered integers that should be ordered in increasing order). The solution can be found in Chapter 3 in [103].

22. Write the computer code that implements the algorithm in Section 1.17 to solve the tower problem. Implement the three towers of rings in a list of three well-ordered connected lists of integer numbers, as in Chapter 3 in [103]. (Use an extra fictitious item to make sure that each connected list is never completely empty.)

2

Preliminaries

In this chapter, we list some basic definitions and prove some standard lemmas used throughout the book. In particular, the lemmas are used to obtain some useful properties of the one- and two-dimensional Fourier (sine) transform, which are used often in the book.

2.1 Preliminary Notation and Definitions

Here we present some elementary notation and definitions from linear algebra that are useful throughout the book.

For every integer i, define

$$i \bmod 2 = \begin{cases} 0 & \text{if } i \text{ is even} \\ 1 & \text{if } i \text{ is odd.} \end{cases}$$

Also, for every two integers i and j, we say that

$$i \equiv j \bmod 2$$

if $i - j$ is even.

Let $A = (a_{i,j})_{1 \leq i,j \leq K}$ be a square matrix. Then K is called the *order* of A. The reverse direction is also true: saying that A is of order K implies that A is a square matrix.

The lower- (respectively, strictly lower-) triangular part of A is a matrix $L = (l_{i,j})_{1 \leq i,j \leq K}$ with the elements

$$l_{i,j} = \begin{cases} a_{i,j} & \text{if } i \geq j \text{ (respectively, } i > j) \\ 0 & \text{if } i < j \text{ (respectively, } i \leq j). \end{cases}$$

Similarly, the upper- (respectively, strictly upper-) triangular part of A is a matrix $U = (u_{i,j})_{1 \leq i,j \leq K}$ with the elements

$$u_{i,j} = \begin{cases} a_{i,j} & \text{if } j \geq i \text{ (respectively, } j > i) \\ 0 & \text{if } j < i \text{ (respectively, } j \leq i). \end{cases}$$

We say that A is lower- (respectively, strictly lower-) triangular if it is equal to its lower- (respectively, strictly lower-) triangular part. Similarly, we say that A is upper- (respectively, strictly upper-) triangular if it is equal to its upper- (respectively, strictly upper-) triangular part.

Let R denote the real field, and let C denote the complex field. For every $z \in C$, $\Re(z)$ denotes the real part of z, $\Im(z)$ denotes the imaginary part of z, and $\bar{z}$ denotes the complex conjugate of z. We also denote the k-dimensional vector space over C by

$$C^k \equiv \{(z_1, z_2, \ldots, z_k) \mid z_i \in C,\ 1 \leq i \leq k\}.$$

Similarly, we denote the 2-D Euclidean plane by

$$R^2 \equiv \{(x, y) \mid x \text{ and } y \text{ are real numbers}\},$$

and the 2-D infinite grid of pairs of integer numbers by

$$Z^2 \equiv \{(k, l) \mid k \text{ and } l \text{ are integer numbers}\}.$$

If a nonzero vector $v \in C^K$ and a complex or real number λ satisfy

$$Av = \lambda v,$$

then we say that v is an eigenvector of A and λ is an eigenvalue of A that corresponds to v. The set of all eigenvalues of A is called the *spectrum* of A. The maximal magnitude of a point in the spectrum of A is called the *spectral radius* of A and is denoted by $\rho(A)$.

The following theorem is the celebrated Gersgorin's Theorem.

Theorem 2.1 *The spectrum of A is contained in the following set.*

$$\bigcup_{i=1}^{K} \left\{ z \text{ is a complex number} \ \middle| \ |z - a_{i,i}| \leq \sum_{1 \leq j \leq K,\ j \neq i} |a_{i,j}| \right\}.$$

Proof. Let v be an eigenvector of A with the corresponding eigenvalue λ. In other words,

$$v \neq \mathbf{0} \quad \text{and} \quad Av = \lambda v. \tag{2.1}$$

Let i be the index of the component in v of maximum modulus; that is,

$$|v_i| \geq |v_j|, \qquad 1 \leq j \leq K,\ j \neq i. \tag{2.2}$$

Consider the ith equation in (2.1):

$$\sum_{j=1}^{K} a_{i,j} v_j = \lambda v_i,$$

or

$$(a_{i,i} - \lambda) v_i = - \sum_{1 \leq j \leq K,\ j \neq i} a_{i,j} v_j.$$

By taking absolute values in both sides of the above equation and using also (2.2), one obtains

$$|a_{i,i} - \lambda|\,|v_i| = \left| \sum_{1 \le j \le K,\ j \neq i} a_{i,j} v_j \right|$$

$$\le \sum_{1 \le j \le K,\ j \neq i} |a_{i,j}|\,|v_j|$$

$$\le \sum_{1 \le j \le K,\ j \neq i} |a_{i,j}|\,|v_i|.$$

From (2.1) and (2.2), we also have that $|v_i| > 0$. Therefore, one can divide both sides of the above inequality by $|v_i|$, from which the theorem follows.

Furthermore, we say that A is *positive definite* if all its eigenvalues are positive. We say that A is *positive semidefinite* if all its eigenvalues are positive or zero. We say that A is *indefinite* if it has both eigenvalues with positive real parts and eigenvalues with negative real parts.

The square root of a positive-semidefinite matrix is a matrix with the same eigenvectors as the original matrix but with corresponding eigenvalues that are the square roots of the corresponding eigenvalues of the original matrix.

We say that A is diagonal if $a_{i,j} \neq 0$ implies $i = j$, that is, only main-diagonal elements can be nonzero, and all off-diagonal elements vanish.

Let I denote the identity matrix of order K, that is, the diagonal matrix whose main-diagonal elements are all equal to 1. Define the standard unit vector $e^{(i)}$ by

$$e_j^{(i)} = \begin{cases} 1 & \text{if } j = i \\ 0 & \text{if } j \neq i. \end{cases}$$

In other words, $e^{(i)}$ is the ith column in the identity matrix I.

We say that A is *tridiagonal* if $a_{i,j} \neq 0$ implies $|i - j| \le 1$. That is, A is a matrix with only three nonzero diagonals: the main diagonal and the two diagonals that are adjacent to it. All other diagonals vanish. In this case, we write

$$A = tridiag(b_i, c_i, d_i);$$

that is, the only nonzero elements in the ith row are b_i, c_i, and d_i, in this order. Of course, the first row contains only two nonzero elements c_1 and d_1, and the last row contains only two nonzero elements b_K and c_K.

The diagonal part of A, $diag(A)$, is the diagonal matrix with the same main-diagonal elements as A:

$$diag(A) = diag(a_{i,i})_{1 \le i \le K} = diag(a_{1,1}, a_{2,2}, \ldots, a_{K,K}).$$

Similarly, $bidiag(A)$, $tridiag(A)$, $pentadiag(A)$, and $blockdiag(A)$ contain, respectively, two diagonals, three diagonals, five diagonals, and a diagonal of blocks with the same elements as in A, and zeroes elsewhere.

We say that A is *nonnegative* if all its elements are positive or zero. We say that A is an *L-matrix* if all its off-diagonal elements are either negative or zero. We say that A is an *M-matrix* if it is an L-matrix with positive main-diagonal elements and a nonnegative inverse A^{-1} [118].

The M-matrix property is particularly important in discrete approximations to differential operators: because the Green function that represents the inverse of the differential operator is usually nonnegative, so should also be the inverse of the matrix that contains the discrete approximation to that differential operator.

Finally, A is diagonally dominant if, for every $1 \leq i \leq K$,

$$a_{i,i} \geq \sum_{1 \leq j \leq K,\ j \neq i} |a_{i,j}|.$$

The following standard lemma is useful in the analysis in Parts III and IV below.

Lemma 2.1 *If A is diagonally dominant, then its spectrum is contained in the set*

$$\bigcup_{i=1}^{K} \{z \text{ is a complex number } | \ |z - a_{i,i}| \leq a_{i,i}\}.$$

As a result,

$$\rho(A) \leq 2\rho(diag(A)).$$

If, in addition, A has a real spectrum, then it is also positive semidefinite.

Proof. The lemma follows from Theorem 2.1.

The transpose of A, A^t, is defined by

$$\left(A^t\right)_{i,j} = a_{j,i}.$$

We say that A is symmetric if A is real and $A = A^t$. If A is symmetric and positive definite, then we say that A is symmetric positive definite or SPD.

The adjoint to A is defined by

$$\left(A^*\right)_{i,j} = \bar{a}_{j,i}.$$

In other words,

$$A^* = \bar{A}^t.$$

We say that A is hermitian if $A = A^*$.

In this book, however, we deal with real matrices only, so one can assume $A^* = A^t$. The only place where complex matrices are used is in the numerical examples with indefinite matrices. Therefore, we use "$*$" only when we would like to emphasize that, for complex matrices, the adjoint should be used rather than the transpose. In fact, it would be possible to replace "t" by "$*$" and "symmetric" by "hermitian" throughout the book, with only few additional adjustments.

The inner product of two vectors $u, v \in C^K$ is defined by

$$(u, v)_2 \equiv u \cdot \bar{v} \equiv \sum_{i=1}^{K} u_i \bar{v}_i.$$

The l_2 norm of a vector $v \in C^K$ is defined by

$$\|v\|_2 \equiv \sqrt{(v, v)_2}.$$

The l_1 norm of a vector $v \in C^K$ is

$$|v|_1 = \sum_{i=1}^{K} |v_i|.$$

The l_∞ norm of a vector $v \in C^K$ is

$$|v|_\infty = \max_{1 \leq i \leq K} |v_i|.$$

The l_2 norm of the matrix A is defined by

$$\|A\|_2 = \max_{v \in C^K,\ v \neq 0} \frac{\|Av\|_2}{\|v\|_2} = \max_{v \in C^K,\ \|v\|_2 = 1} \|Av\|_2.$$

The l_1 norm of the matrix A is defined by

$$\|A\|_1 = \max_{v \in C^K,\ v \neq 0} \frac{\|Av\|_1}{\|v\|_1} = \max_{v \in C^K,\ \|v\|_1 = 1} \|Av\|_1.$$

The l_∞ norm of the matrix A is defined by

$$\|A\|_\infty = \max_{v \in C^K,\ v \neq 0} \frac{\|Av\|_\infty}{\|v\|_\infty} = \max_{v \in C^K,\ \|v\|_\infty = 1} \|Av\|_\infty.$$

The following lemma states useful matrix-norm inequalities, including the triangle inequality.

Lemma 2.2 *Let* $\| \cdot \|$ *denote the* l_1, l_2, *or* l_∞ *norm. Then we have*

$$\rho(A) \leq \|A\|.$$

Moreover,

$$\|Av\| \leq \|A\| \cdot \|v\|$$

for every K-*dimensional vector* v. *Furthermore, if* B *is another matrix of order* K, *then*

$$\|AB\| \leq \|A\| \cdot \|B\|$$

and

$$\|A + B\| \leq \|A\| + \|B\|.$$

Proof. Let λ be the maximal eigenvalue of A (in terms of magnitude), and let u be the corresponding eigenvector. Then we have

$$\|A\| = \max_{v \in C^K,\ v \neq 0} \frac{\|Av\|}{\|v\|} \geq \frac{\|Au\|}{\|u\|} = |\lambda| = \rho(A).$$

Next, for every nonzero vector $v \in C^K$,

$$\|Av\| = \|A(v/\|v\|)\| \cdot \|v\| \leq \max_{w \in C^K,\ \|w\|=1} \|Aw\| \cdot \|v\| = \|A\| \cdot \|v\|.$$

As a result, we have

$$\|ABv\| \leq \|A\| \cdot \|Bv\| \leq \|A\| \cdot \|B\| \cdot \|v\|,$$

which implies that

$$\|AB\| \max_{v\in C^K,\ v\neq 0} \frac{\|ABv\|}{\|v\|} \leq \|A\| \cdot \|B\|.$$

Finally, we have

$$\begin{aligned}
\|A+B\| &= \max_{v\in C^K,\ v\neq 0} \frac{\|(A+B)v\|}{\|v\|} \\
&\leq \max_{v\in C^K,\ v\neq 0} \frac{\|Av\| + \|Bv\|}{\|v\|} \\
&\leq \max_{v\in C^K,\ v\neq 0} \frac{\|Av\|}{\|v\|} + \max_{v\in C^K,\ v\neq 0} \frac{\|Bv\|}{\|v\|} \\
&= \|A\| + \|B\|.
\end{aligned}$$

This completes the proof of the lemma.

The following lemma expresses $\|A\|_1$ in terms of the elements in A.

Lemma 2.3

$$\|A\|_1 = \max_{1\leq j\leq K} \sum_{i=1}^{K} |a_{i,j}|.$$

Proof. Clearly, the l_1 vector norm is convex. Indeed, for any two K-dimensional vectors u and v and a parameter $0 < \alpha < 1$,

$$\|\alpha u + (1-\alpha)v\|_1 \leq \|\alpha u\|_1 + \|(1-\alpha)v\|_1 = \alpha\|u\|_1 + (1-\alpha)\|v\|_1.$$

Thus, the vector v that satisfies $\|v\|_1 = 1$ and maximizes $\|Av\|_1$ is the standard unit vector $e^{(j)}$ for which the jth column in A has a maximum l_1 norm, which is also $\|A\|_1$. This completes the proof of the lemma.

The following lemma expresses $\|A\|_\infty$ in terms of the elements in A.

Lemma 2.4

$$\|A\|_\infty = \max_{1\leq i\leq K} \sum_{j=1}^{K} |a_{i,j}|.$$

Proof. Clearly, the vector v that satisfies $\|v\|_\infty = 1$ and maximizes $\|Av\|_\infty$ is the vector defined by

$$v_i = \bar{a}_{i,j}/|a_{i,j}|$$

for that i for which $\sum_j |a_{i,j}|$ is maximal (and therefore also equal to $\|A\|_\infty$). This completes the proof of the lemma.

Corollary 2.1

$$\|A^t\|_\infty = \|A\|_1.$$

Proof. The corollary follows from Lemmas 2.3 and 2.4.

Let D be an SPD matrix of order K. The inner product induced by D is defined by

$$(u,v)_D = (u, Dv)_2.$$

The vector norm induced by D is defined by

$$\|v\|_D = \sqrt{(v, v)_D}.$$

The norm of A with respect to $(\cdot, \cdot)_D$ is defined by

$$\|A\|_D = \max_{v \in C^K, \, v \neq 0} \frac{\|Av\|_D}{\|v\|_D} = \max_{v \in C^K, \, \|v\|_D = 1} \|Av\|_D.$$

The following lemma is analogous to Lemma 2.2.

Lemma 2.5 *Let D be an SPD matrix. Then*

$$\rho(A) \leq \|A\|_D.$$

Moreover, for any vector $v \in C^K$,

$$\|Av\|_D \leq \|A\|_D \|v\|_D.$$

Furthermore, for any matrix B of order K,

$$\|AB\|_D \leq \|A\|_D \|B\|_D$$

and

$$\|A + B\|_D \leq \|A\|_D + \|B\|_D.$$

Proof. The proof is similar to that of Lemma 2.2.

We say that the matrix A_D^t is the adjoint of A with respect to $(\cdot, \cdot)_D$ if

$$(u, A_D^{tv})_D = (Au, v)_D$$

for every two vectors $u, v \in C^K$. The following lemma states that the adjoint of the adjoint is the original matrix.

Lemma 2.6 *Let D be an SPD matrix. Then*

$$\left(A_D^t\right)_D^t = A.$$

Proof. The lemma follows from the fact that

$$(A_D^t v, u)_D = (v, Au)_D$$

for every two vectors $u, v \in C^K$.

We say that A is symmetric with respect to $(\cdot, \cdot)_D$ if

$$A_D^t = A.$$

Lemma 2.7 *Let D be an SPD matrix. Then $A_D^t A$ is symmetric with respect to $(\cdot, \cdot)_D$.*

Proof. The lemma follows from the fact that, for every two vectors $u, v \in C^K$,

$$(u, A_D^t A v)_D = (Au, Av)_D = (A_D^t Au, v)_D.$$

We say that A is orthogonal if

$$A^* A = I,$$

or, in other words, all the columns of A are orthonormal with respect to the usual inner product $(\cdot, \cdot)_2$.

Let A be a square or rectangular matrix, $A = (a_{i,j})_{1 \le i \le K_1, 1 \le j \le K_2}$. The transpose of A, A^t, is defined by

$$\left(A^t\right)_{j,i} = a_{i,j}, \quad 1 \le i \le K_1, \ 1 \le j \le K_2.$$

The following standard lemma gives an alternative definition to the notion of the transpose of a matrix.

Lemma 2.8 *Let $Z = (z_{i,j})_{1 \le i \le K_2, 1 \le j \le K_1}$ be a square or rectangular matrix. Then Z is the transpose of A if and only if*

$$(Au, v)_2 = (u, Zv)_2 \tag{2.3}$$

for every two vectors $u \in C^{K_2}$ and $v \in C^{K_1}$.

Proof. Let us first prove the "only if" part. Assume that $Z = A^t$. Then,

$$(Au, v)_2 = \sum_{i=1}^{K_1} \sum_{j=1}^{K_2} a_{i,j} u_j \bar{v}_i = \sum_{j=1}^{K_2} \sum_{i=1}^{K_1} u_j z_{j,i} \bar{v}_i = (u, Zv)_2.$$

Let us now prove the "if" part. Assume that (2.3) holds. Then, the assertion that $z_{j,i} = a_{i,j}$ follows by picking a vector u whose all components vanish except its jth component, u_j, which is equal to 1, and a vector v whose all components vanish except its ith component, v_i, which is equal to 1. This completes the proof of the lemma.

We say that the matrices A and Z are the transpose of each other with respect to $(\cdot, \cdot)_D$ if

$$(Au, v)_D = (u, Zv)_D$$

for every two vectors $u \in C^{K_2}$ and $v \in C^{K_1}$.

More standard linear algebra lemmas can be found in Section 2.3.

Define the absolute value of A by

$$|A|_{i,j} = |a_{i,j}|, \quad 1 \le i \le K_1, \ 1 \le j \le K_2. \tag{2.4}$$

Define the diagonal matrix of row sums of A by

$$rs(A) = diag\left(\sum_{j=1}^{K_2} a_{i,j}\right)_{1 \le i \le K_1}. \tag{2.5}$$

By "grid" we mean a finite set of points in R^2. Let g be a grid; for example, the grid in (3.6) below. Let s be a subset of g (a subgrid). Let $l_2(s)$ be the set of complex-valued functions defined on s. In particular, $l_2(g)$ is the set of complex-valued functions defined on g, also referred to as "grid functions."

Define the injection operator $J_s : l_2(g) \to l_2(s)$ by

$$(J_s v)(k) \equiv v(k), \quad v \in l_2(g), \ k \in s. \tag{2.6}$$

In particular, J_c is the injection onto the coarse grid c defined in (6.1) below.

2.2 Application in Pivoting

Here we present an application of the algorithm in Section 1.18 to the pivoting of tridiagonal matrix. Let

$$A = tridiag(b_i, 1, d_i)$$

be a tridiagonal matrix of order N. The pivoting process is defined by $p_1 = 1$ and, for $i = 2, 3, \ldots, N$,

$$p_i = 1 - \frac{b_i d_{i-1}}{p_{i-1}}. \tag{2.7}$$

The pivots p_i are useful in the factorization of A as the product of triangular matrices. Indeed, define the diagonal matrix

$$X = diag(p_1, p_2, \ldots, p_N).$$

Let L be the strictly lower triangular part of A, and U be the strictly upper triangular part of A. Then we have

$$A = (L + X)X^{-1}(X + U).$$

This factorization is useful in solving linear systems with A as a coefficient matrix.

Now, the pivoting process in (2.7) can be reformulated as a continued fraction. Indeed, define

$$q_i = p_i - 1.$$

The pivoting process in (2.7) is thus equivalent to the process defined by $q_1 = 0$ and, for $i = 2, 3, \ldots, N$,

$$q_i = \frac{-b_i d_{i-1}}{1 + q_{i-1}}. \tag{2.8}$$

It is well known (see, e.g., [93]) that the convergents in a continued fraction may be interpreted as a sequence of products of 2×2 matrices. In fact, the current convergent is obtained from the previous one by multiplying it on the right by some 2×2 matrix. Here, however, the q_is are the "mirror image" of a continued fraction, in which q_i corresponds to the 2×2 matrix obtained from multiplying the matrix corresponding to q_{i-1} by some 2×2 matrix on the left rather than on the right. Still, this structure is well suited for the parallel algorithm in Section 1.18. The p_is are then obtained from the q_is, and the pivoting is completed.

The solution of the tridiagonal linear system also requires forward elimination in $L + X$ and back substitution in $X + U$. These processes can by themselves be formulated as products of affine transformations, and thus are also suitable for the parallel algorithm in Section 1.18. (See Section 2.2 in [96] for the details.)

2.3 Standard Lemmas about Symmetric Matrices

This section contains standard lemmas that are useful in the sequel. Basically, these lemmas show that matrices that are symmetric with respect to an inner product of the form $(\cdot,\cdot)_D$ for some SPD matrix D enjoy the same properties as standard symmetric matrices. The first lemma shows that the inverse of a symmetric matrix is also symmetric.

Lemma 2.9 *Let D be an SPD matrix, and assume that A is symmetric with respect to $(\cdot,\cdot)_D$. Then A^{-1} is also symmetric with respect to $(\cdot,\cdot)_D$.*

Proof. For every two vectors x and y,

$$(x, A^{-1}y)_D = (AA^{-1}x, A^{-1}y)_D = (A^{-1}x, AA^{-1}y)_D = (A^{-1}x, y)_D.$$

This completes the proof of the lemma.

The next lemma shows that matrices that are symmetric with respect to an induced inner product have real spectra.

Lemma 2.10 *Let D be an SPD matrix, and assume that A is symmetric with respect to $(\cdot,\cdot)_D$. Then A has a real spectrum.*

Proof. If

$$Av = \lambda v$$

for a nonzero vector z and a scalar λ, then

$$\lambda(v, v)_D = (Av, v)_D = (v, Av)_D = \bar{\lambda}(v, v)_D.$$

This completes the proof of the lemma.

The next lemma says that a matrix is symmetric with respect to some inner product if and only if there exists a basis of orthonormal eigenvectors (with respect to this inner product) that correspond to real eigenvalues.

Lemma 2.11 *Let D be an SPD matrix. Then A is symmetric with respect to $(\cdot,\cdot)_D$ if and only if there exists a basis of eigenvectors of A that correspond to real eigenvalues and are orthonormal with respect to $(\cdot,\cdot)_D$ (i.e., they are orthogonal to each other with respect to $(\cdot,\cdot)_D$ and their D-induced norm is equal to 1).*

Proof. Let us first prove the "only if" part. Assume that A is symmetric with respect to $(\cdot,\cdot)_D$. Assume also that

$$Av = \lambda v \quad \text{and} \quad Au = \mu u$$

for some nonzero vectors v and u and scalars $\lambda \neq \mu$. Using Lemma 2.10, we then have

$$\lambda(v, u)_D = (Av, u)_D = (v, Au)_D = \bar{\mu}(v, u)_D = \mu(v, u)_D,$$

which implies

$$(v, u)_D = 0.$$

Consider now the Jordan block of A corresponding to the eigenvalue λ, and assume that the corresponding Jordan subspace is spanned by the vectors $w_1, w_2, \ldots,$ w_{n+1} in the Jordan basis. By induction, assume also that every invariant subspace of A of dimension at most n can be spanned by n eigenvectors of A that are orthonormal with respect to $(\cdot,\cdot)_D$. In particular, the span of $w_1, w_2, \ldots, w_n$ can be spanned by such orthonormal eigenvectors $v_1, v_2, \ldots, v_n$ corresponding to the eigenvalue λ. For some scalars $\alpha_1, \alpha_2, \ldots, \alpha_n$, we have

$$Aw_{n+1} = \lambda w_{n+1} + \sum_{i=1}^{n} \alpha_i v_i.$$

Therefore, for every $1 \le j \le n$, we have

$$\lambda(w_{n+1}, v_j)_D = (w_{n+1}, Av_j)_D = (Aw_{n+1}, v_j)_D = \lambda(w_{n+1}, v_j)_D + \alpha_j,$$

implying that

$$\alpha_j = 0.$$

Thus, w_{n+1} is also an eigenvector of A corresponding to the eigenvalue λ. By applying a Gramm–Schmidt process to w_{n+1} with respect to $(\cdot, \cdot)_D$, one obtains an eigenvector v_{n+1} that corresponds to the eigenvalue λ and is also orthogonal to $v_1, v_2, \ldots, v_n$ with respect to $(\cdot, \cdot)_D$ and satisfies $\|v_{n+1}\|_D = 1$. This completes the induction step and the proof of the "only if" part.

Let us now prove the "if" part. Assume that there exists a basis of eigenvectors of A that correspond to real eigenvalues and are orthonormal with respect to $(\cdot, \cdot)_D$. Let V be the matrix whose columns are these eigenvectors and Λ the diagonal matrix whose diagonal elements are the corresponding real eigenvalues. In other words,

$$AV = V\Lambda,$$

and, hence,

$$A = V\Lambda V^{-1}, \tag{2.9}$$

with

$$V^*DV = I. \tag{2.10}$$

From (2.10) we have

$$V^{-1} = V^*D. \tag{2.11}$$

By using (2.11) in (2.9), we get

$$A = V\Lambda V^*D. \tag{2.12}$$

As a result, for every two vectors x and y we have

$$\begin{aligned}
(Ax, y)_D &= (V\Lambda V^*Dx, y)_D \\
&= (DV\Lambda V^*Dx, y)_2 \\
&= (x, DV\Lambda V^*Dy)_2 \\
&= (Dx, V\Lambda V^*Dy)_2 \\
&= (x, V\Lambda V^*Dy)_D \\
&= (x, Ay)_D.
\end{aligned}$$

This completes the proof of the "if" part and the whole lemma.

The next lemma gives alternative definitions to the terms "positive definite" and "positive semidefinite" defined in Section 2.1 above. When the condition in the lemma is satisfied, we may say that A is SPD with respect to $(\cdot, \cdot)_D$.

Lemma 2.12 *Let D be an SPD matrix, and assume that A is symmetric with respect to $(\cdot, \cdot)_D$. Then A is positive definite (respectively, positive semidefinite) if and only if for every nonzero vector x $(Ax, x)_D$ is positive (respectively, positive or zero).*

Proof. Let us first prove the "only if" part. Let x be a nonzero vector. From Lemma 2.11, we have

$$x = \sum \alpha_i v_i,$$

where the α_is are some scalars and the v_is are the orthonormal eigenvectors of A with corresponding eigenvalues λ_i. Thus, we have

$$(Ax, x)_D = \sum \lambda_i |\alpha_i|^2.$$

Because x is nonzero, at least one of the α_is is also nonzero. Therefore, if A is positive definite (respectively, positive semidefinite), then all the λ_is are positive (respectively, positive or zero), implying that $(Ax, x)_D$ is positive (respectively, positive or zero). This completes the proof of the "only if" part.

Let us now prove the "if" part. Indeed, if $(Ax, x)_D$ is positive (respectively, positive or zero) for every nonzero vector x, then this is particularly true for an eigenvector of A, implying that the corresponding eigenvalue is positive (respectively, positive or zero). This completes the proof of the lemma. $\blacksquare$

The next lemma is a version of the "minimax" theorem.

Lemma 2.13 *Let D be an SPD matrix, and assume that A is symmetric with respect to $(\cdot, \cdot)_D$. Then*

$$\rho(A) = \max_{x \neq 0} \left| \frac{(Ax, x)_D}{(x, x)_D} \right|. \tag{2.13}$$

Proof. Let us solve a problem that is equivalent to the right-hand side in (2.13).

$$\text{maximize} \quad |(Ax, x)_D| \quad \text{subject to} \quad (x, x)_D = 1. \tag{2.14}$$

From Lemma 2.11, we can represent every vector x as

$$x = \sum \alpha_i v_i,$$

where the α_is are some scalars and the v_is are the orthonormal eigenvectors of A with corresponding eigenvalues λ_i. Furthermore, we have

$$(Ax, x)_D = \sum \lambda_i |\alpha_i|^2.$$

Thus, the problem in (2.14) can be reformulated as

$$\text{maximize} \quad \left| \sum \lambda_i |\alpha_i|^2 \right| \quad \text{subject to} \quad \sum |\alpha_i|^2 = 1, \tag{2.15}$$

where the λ_is are the eigenvalues of A and the α_is are the unknowns to be found. From Lagrange theory, we have that the extreme vectors $(\alpha_1, \alpha_2, \ldots)^t$ for the function to be maximized in (2.15) are the ones for which the gradient of that function is a scalar multiple of the gradient of the constraint; that is, $(\lambda_1 \alpha_1, \lambda_2 \alpha_2, \ldots)^t$ is a scalar multiple of $(\alpha_1, \alpha_2, \ldots)^t$. This could happen only when all the α_is vanish except those that correspond to one of the eigenvalues, say λ_j. At this extreme vector we have

$$\sum \lambda_i |\alpha_i|^2 = \lambda_j \sum |\alpha_i|^2 = \lambda_j.$$

This implies that the solution to (2.14) is the eigenvector of A with the eigenvalue of the largest possible magnitude, namely, $\rho(A)$. This completes the proof of the lemma.

Lemma 2.14 *Let D be an SPD matrix, and assume that A is symmetric with respect to $(\cdot, \cdot)_D$. Then*

$$\|A\|_D = \rho(A).$$

Proof. The proof is similar to the proof of Lemma 2.13, except that here one should solve the problems

$$\text{maximize} \quad (Ax, Ax)_D \quad \text{subject to} \quad (x, x)_D = 1 \qquad (2.16)$$

and

$$\text{maximize} \quad \sum \lambda_i^2 |\alpha_i|^2 \quad \text{subject to} \quad \sum |\alpha_i|^2 = 1 \qquad (2.17)$$

rather than (2.14) and (2.15), respectively. The solution of (2.16) and (2.17) leads to

$$\|A\|_D^2 = \rho(A)^2,$$

which completes the proof of the lemma.

The next lemma shows that the energy norm induced by a positive definite matrix is bounded in terms of the energy norm induced by a yet "more positive definite" matrix:

Lemma 2.15 *Let D be an SPD matrix, and assume that A, $\acute{A}$ and $A - \acute{A}$ are symmetric with respect to $(\cdot, \cdot)_D$ and positive semidefinite. Then for every vector x we have*

$$(x, \acute{A}x)_D \le (x, Ax)_D \qquad (2.18)$$

Furthermore,

$$\|\acute{A}\|_D \le \|A\|_D. \qquad (2.19)$$

Proof. Equation (2.18) follows from Lemma 2.12 and

$$(x, Ax)_D = (x, \acute{A}x)_D + (x, (A - \acute{A})x)_D \ge (x, \acute{A}x)_D.$$

Then, (2.19) follows from (2.18) and Lemmas 2.12 through 2.14. This completes the proof of the lemma.

The next lemma provides some properties for the square root of an SPD matrix.

Lemma 2.16 *Let D be an SPD matrix, and assume that A is symmetric with respect to $(\cdot, \cdot)_D$ and positive semidefinite. Then*

$$(A^{1/2})^2 = A, \qquad (2.20)$$

$A^{1/2}$ is also symmetric with respect to $(\cdot, \cdot)_D$, and

$$\|A^{1/2}\|_D^2 = \|A\|_D. \tag{2.21}$$

Proof. From the "only if" part in Lemma 2.11, we have that (2.9) holds (with the notation used there). From the definition of the square root of a matrix in Section 2.1, we have

$$A^{1/2} = V\Lambda^{1/2}V^{-1},$$

which implies (2.20). From the "if" part in Lemma 2.11, we have that $A^{1/2}$ is also symmetric with respect to $(\cdot, \cdot)_D$. Finally, from Lemma 2.14, we have

$$\|A^{1/2}\|_D^2 = \rho(A^{1/2})^2 = \rho(A) = \|A\|_D,$$

which completes the proof of the lemma.

The following lemma gives an alternative definition to the norm of a matrix.

Lemma 2.17 *Let D be an SPD matrix. Then*

$$\|A\|_D = \|D^{1/2}AD^{-1/2}\|_2.$$

Proof. Using Lemma 2.16 (with D there being the identity matrix I and A there being the SPD matrix D from this lemma), we have

$$\begin{aligned}
\|A\|_D &= \max_{v \neq 0} \frac{\|Av\|_D}{\|v\|_D} \\
&= \max_{D^{1/2}v \neq 0} \frac{\|D^{1/2}AD^{-1/2}(D^{1/2}v)\|_2}{\|D^{1/2}v\|_2} \\
&= \|D^{1/2}AD^{-1/2}\|_2.
\end{aligned}$$

This completes the proof of the lemma.

We conclude with two lemmas regarding more general matrices, which are not necessarily symmetric with respect any inner product. The following lemma states that the norm of the adjoint is the same as the norm of the original matrix:

Lemma 2.18 *Let D be an SPD matrix, and let A_D^t be the adjoint of A with respect to $(\cdot, \cdot)_D$. Then*

$$\|A_D^t\|_D = \|A\|_D.$$

Proof. On one hand, we have from Lemmas 2.5, 2.7, and 2.13 that

$$\begin{aligned}
\|A\|_D^2 &= \max_{v \in C^K, \, v \neq 0} \frac{\|Av\|_D^2}{\|v\|_D^2} \\
&= \max_{v \in C^K, \, v \neq 0} \frac{(Av, Av)_D}{(v, v)_D} \\
&= \max_{v \in C^K, \, v \neq 0} \frac{(A_D^t Av, v)_D}{(v, v)_D} \\
&= \rho(A_D^t A) \\
&\leq \|A_D^t A\|_D \\
&\leq \|A_D^t\|_D \|A\|_D,
\end{aligned}$$

which implies

$$\|A\|_D \leq \|A_D^t\|_D.$$

On the other hand, using Lemma 2.6, we also have

$$
\begin{aligned}
\|A_D^t\|_D^2 &= \max_{v \in C^K,\ v \neq 0} \frac{\|A_D^{tv}\|_D^2}{\|v\|_D^2} \\
&= \max_{v \in C^K,\ v \neq 0} \frac{(A_D^{tv}, A_D^{tv})_D}{(v, v)_D} \\
&= \max_{v \in C^K,\ v \neq 0} \frac{(AA_D^{tv}, v)_D}{(v, v)_D} \\
&= \rho(AA_D^t) \\
&\leq \|AA_D^t\|_D \\
&\leq \|A\|_D \|A_D^t\|_D,
\end{aligned}
$$

which implies

$$\|A_D^t\|_D \leq \|A\|_D.$$

This completes the proof of the lemma.

Finally, the following lemma bounds the norm induced by an SPD matrix D.

Lemma 2.19 *Let D be an SPD matrix, and let A_D^t be the adjoint of A with respect to $(\cdot, \cdot)_D$. Then*

$$\|A\|_D \leq \sqrt{\|A_D^t\|_\infty \|A\|_\infty}.$$

Proof. From Lemmas 2.2, 2.7, and 2.13, we have that

$$
\begin{aligned}
\|A\|_D^2 &= \max_{v \in C^K,\ v \neq 0} \frac{\|Av\|_D^2}{\|v\|_D^2} \\
&= \max_{v \in C^K,\ v \neq 0} \frac{(Av, Av)_D}{(v, v)_D} \\
&= \max_{v \in C^K,\ v \neq 0} \frac{(A_D^t Av, v)_D}{(v, v)_D} \\
&= \rho(A_D^t A) \\
&\leq \|A_D^t A\|_\infty \\
&\leq \|A_D^t\|_\infty \|A\|_\infty.
\end{aligned}
$$

This completes the proof of the lemma.

Corollary 2.2 *Let D be an SPD matrix, and assume that A is symmetric with respect to $(\cdot, \cdot)_D$. Then*

$$\|A\|_D \leq \|A\|_\infty.$$

Proof. The corollary follows from Lemma 2.19 by setting $A_D^t = A$.

Corollary 2.3 *Assume that $A = (a_{i,j})$ is symmetric and diagonally dominant with positive main-diagonal elements. Define $D = diag(A)$. Then*

$$\|D^{-1/2}AD^{-1/2}\|_2 = \|D^{-1}A\|_D \leq 2.$$

Proof. Because A is symmetric, $D^{-1}A$ is symmetric with respect to $(\cdot, \cdot)_D$. The corollary follows from Lemma 2.17, Corollary 2.2, and Lemma 2.4, when applied to $D^{-1}A$.

2.4 The Fourier Transform

In many problems in pure and applied science, the notion of "scales" or "levels" is helpful not only in the solution process but also in a deep understanding of the problem and its complexity. Most often, each scale has its own contribution to and influence on the mathematical or physical phenomenon. A most useful tool in distinguishing between different scales is the Fourier transform. This transform interprets a given function in terms of frequencies or waves rather than numerical nodal values. Low frequencies or long waves describe the variation of the function in coarse scales, whereas high frequencies or short waves correspond to fine-scale, delicate changes in the function. The Fourier transform has many applications in pure and applied mathematics, from classical, analytic fields such as functional analysis to modern, practical fields such as digital signal processing.

The hierarchy of Fourier functions can be introduced as eigenfunctions of a boundary-value problem. Consider the ordinary differential equation

$$-u''(x) = \mathcal{F}(x), \qquad 0 < x < 1, \tag{2.22}$$

where $\mathcal{F}$ is a given function on $(0, 1)$ and u is the unknown function on $[0, 1]$. Assume also that Dirichlet boundary conditions that specify the values of u at the endpoints are given:

$$u(0) = u(1) = 0. \tag{2.23}$$

The boundary-value problem (2.22) and (2.23) is called a Sturm–Liouville problem. This problem has the set of eigenfunctions

$$\{\sin(\pi kx)\}_{k=1}^{\infty} \tag{2.24}$$

with the corresponding eigenvalues $\pi^2 k^2$. In other words, the functions in (2.24) satisfy the boundary conditions in (2.23) and are also eigenfunctions of the second-derivative operator in (2.22):

$$-\sin''(\pi kx) = \pi^2 k^2 \sin(\pi kx).$$

It is well known that the eigenfunctions of a Sturm–Liouville problem are orthogonal to each other with respect to integration. Indeed, for every distinct $k \geq 1$ and $l \geq 1$,

$$\int_0^1 \sin(\pi kx)\sin(\pi lx)dx = \frac{1}{2}\int_0^1 (\cos(\pi(k-l)x) - \cos(\pi(k+l)x))dx$$

$$= \frac{1}{2\pi(k-l)}[\sin(\pi(k-l)x)]_0^1 - \frac{1}{2\pi(k+l)}[\sin(\pi(k+l)x)]_0^1$$

$$= 0 - 0 = 0.$$

Let us now describe the discrete Fourier (sine) transform in an N-dimensional vector space. Let N be a positive integer, and define $h = 1/(N+1)$. The kth Fourier vector, or mode, is obtained by sampling the kth eigenfunction in (2.24) at the points

$$h, 2h, 3h, \ldots, Nh$$

In other words, for every $1 \le k \le N$, the Fourier mode (discrete wave) $v^{(k)}$ is defined by

$$v^{(k)} \equiv (2h)^{1/2}(\sin(\pi kh), \sin(2\pi kh), \ldots, \sin(N\pi kh))^t. \tag{2.25}$$

Define the symmetric, tridiagonal matrix of order N

$$A = tridiag(-1, 2, -1). \tag{2.26}$$

In fact, the matrix A is obtained from the finite-difference discretization method applied to the boundary-value problem (2.22) and (2.23) [see Section 3.5 below]. The Fourier modes $v^{(k)}$ are eigenvectors of A:

$$Av^{(k)} = 4\sin^2(\pi kh/2)v^{(k)}. \tag{2.27}$$

From the symmetry of A, we have that the $v^{(k)}$s form an orthogonal basis in C^N (Lemma 2.11).

Let $\mathcal{V}$ be the matrix whose columns are the $v^{(k)}$s. That is,

$$\mathcal{V} = \left(v^{(1)} \mid v^{(2)} \mid \cdots \mid v^{(N)}\right). \tag{2.28}$$

The matrix $\mathcal{V}$ represents an operator in C^N defined by $\mathcal{V}: C^N \to C^N$ defined by

$$u \to \mathcal{V}u, \quad u \in C^N. \tag{2.29}$$

This operator is referred to as the sine transform in one dimension. The following standard lemma shows that $\mathcal{V}$ is symmetric and orthogonal, thus equal to its inverse.

Lemma 2.20 $\mathcal{V}$ *is an orthogonal symmetric matrix. That is,*

$$\mathcal{V}^{-1} = \mathcal{V}^t = \mathcal{V}.$$

Proof. The symmetry of $\mathcal{V}$ follows from

$$\mathcal{V}_{j,k} = (2h)^{1/2}\sin(\pi jkh) = (2h)^{1/2}\sin(\pi kjh) = \mathcal{V}_{k,j}.$$

To show that $\mathcal{V}$ is also orthogonal, one needs to show that the $v^{(k)}$s are orthonormal, that is,

$$(v^{(k)}, v^{(l)})_2 = \begin{cases} 0 & \text{if } k \ne l \\ 1 & \text{if } k = l. \end{cases}$$

Now, the orthogonality of the $v^{(k)}$s follows from the discussion that follows (2.26) and (2.27) above. It is only left to show that these vectors are also normal, that is,

$$(v^{(k)}, v^{(k)})_2 = 1$$

for every $1 \le k \le N$. Indeed,

$$(v^{(k)}, v^{(k)})_2 = 2h \sum_{j=1}^{N} \sin^2(\pi kjh)$$

$$= h \sum_{j=1}^{N} \left(1 - \Re\left(\exp(2\sqrt{-1}\pi kjh)\right)\right)$$

$$= hN + h - h \sum_{j=0}^{N} \Re\left(\exp(2\sqrt{-1}\pi kjh)\right)$$

$$= 1 - h\Re\left(\sum_{j=0}^{N} \exp(2\sqrt{-1}\pi kjh)\right)$$

$$= 1 - h\Re\left(\frac{1 - \exp(2\sqrt{-1}\pi k)}{1 - \exp(2\sqrt{-1}kh)}\right)$$

$$= 1 - 0 = 1.$$

This completes the proof of the lemma.

Lemma 2.20 shows that the sine transform in (2.29) may be thought of as a change of coordinates or basis. Instead of specifying the nodal values in the original vector u in (2.29), one obtains the various frequencies contained in u in the vector $\mathcal{V}u$. Each coordinate in $\mathcal{V}u$ provides the amplitude of a certain wave contained in u.

The Fourier transform defined above can also be slightly modified to handle boundary conditions different from those in (2.23). Consider, for example, the Dirichlet–Neumann boundary conditions

$$u(0) = u'(1) = 0 \tag{2.30}$$

that specify the value of u at one end point and the value of its derivative at the other end point. The eigenfunctions of the corresponding Sturm–Liouville problem (2.22), (2.30) are

$$\{\sin(\pi(k - 1/2)x)\}_{k=1}^{\infty}. \tag{2.31}$$

If h is redefined as $h = 1/(N + 1/2)$, then the Fourier modes $v^{(k)}$ are obtained from sampling the kth function in (2.31) at the points

$$h, 2h, 3h, \ldots, Nh.$$

These modes are the eigenvectors of the matrix whose elements are the same as in A, except of its lower-right element, which is equal to 1, rather than $A_{N,N} = 2$. Because this matrix is still symmetric, its eigenvectors are still orthogonal to each other as before.

The sine transform provides the representation of a vector in C^N in terms of its oscillations rather than its components or nodal values. In many cases, however, one is interested in oscillations in more than one spatial direction. Consider for example, a rectangular, uniform $N \times N$ grid and vectors in C^{N^2} that are defined on this grid, hence also referred to as grid functions. For $1 \le k, l \le N$, define the (k, l)-wave or mode to be the grid function that is the tensor product of $v^{(k)}$ and $v^{(l)}$:

$$v_{j,m}^{(k,l)} = v_j^{(k)} v_m^{(l)}, \quad 1 \le j, m \le N.$$

Let $\mathcal{V}$ denote now the matrix of order N^2 whose columns are the $v^{(k,l)}$s, $1 \le k, l \le N$. The transform

$$u \to \mathcal{V}u, \quad u \in C^{N^2} \tag{2.32}$$

is the 2-D (two-dimensional) sine transform. The following standard lemma establishes that the 2-D sine-transform matrix $\mathcal{V}$ is also orthogonal, so the 2-D sine transform in (2.32) may also be thought of as a change of basis in C^{N^2} from the usual nodal basis to the Fourier basis of waves or modes.

Lemma 2.21 *The 2-D sine-transform matrix $\mathcal{V}$ is an orthogonal symmetric matrix. That is,*

$$\mathcal{V}^{-1} = \mathcal{V}^t = \mathcal{V}.$$

Proof. The symmetry of $\mathcal{V}$ follows from the symmetry of $\mathcal{V}$ established in Lemma 2.20.

$$v_{m,n}^{(k,l)} = v_m^{(k)} v_n^{(l)} = v_k^{(m)} v_l^{(n)} = v_{k,l}^{(m,n)}.$$

The orthonormality of the $v^{(k,l)}$s also stems from Lemma 2.20. Indeed, consider the (k, l)- and (m, n)-waves:

$$
\begin{aligned}
\left(v^{(k,l)}, v^{(m,n)} \right)_2 &= \sum_{1 \le i,j \le N} v_{i,j}^{(k,l)} v_{i,j}^{(m,n)} \\
&= \sum_{i=1}^N v_i^{(k)} v_i^{(m)} \sum_{j=1}^N v_j^{(l)} v_j^{(n)} \\
&= \left(v^{(k)}, v^{(m)} \right)_2 \left(v^{(l)}, v^{(n)} \right)_2 \\
&= \begin{cases} 1 & \text{if } k = m \text{ and } l = n \\ 0 & \text{otherwise.} \end{cases}
\end{aligned}
$$

This completes the proof of the lemma.

2.5 Exercises

1. Repeat the exercises at the end of Chapter 1, only this time replace the abstract problem in Section 1.18 by the more concrete pivoting problem in Section 2.2. (If your code from these exercises is written as a template function in C++, then all you have to do is to use it in conjunction with the "matrix2" class in Section 2.20 in [103].)
2. Define the Fourier vectors $w^{(k)}$ ($0 \le k < N$) by

$$w_j^{(k)} = N^{-1/2} \exp(2\pi\sqrt{-1}jkh), \quad 1 \le j \le N,$$

where $h = 1/N$. Show that the $w^{(k)}$s are the eigenvectors of the symmetric $N \times N$ matrix

$$A = \begin{pmatrix} 2 & -1 & & & & & & -1 \\ -1 & 2 & -1 & & & & & \\ & -1 & 2 & \ddots & & & & \\ & & \ddots & \ddots & -1 & & & \\ & & & -1 & 2 & -1 & & \\ & & & & -1 & 2 & -1 & \\ -1 & & & & & -1 & 2 \end{pmatrix}$$

with distinct positive (or zero) eigenvalues. Conclude that A is positive semidefinite, and that the $w^{(k)}$s form an orthonormal basis of C^N.

3. Define the Fourier (cosine) vectors $w^{(k)}$ $(0 \le k < N)$ by

$$w_j^{(k)} = \sin(\pi/2 + \pi(j - 1/2)kh), \qquad 1 \le j \le N,$$

where $h = 1/N$. Show that the $w^{(k)}$s are the eigenvectors of the symmetric tridiagonal $N \times N$ matrix

$$A = \begin{pmatrix} 1 & -1 & & & & & \\ -1 & 2 & -1 & & & & \\ & -1 & 2 & \ddots & & & \\ & & \ddots & \ddots & -1 & & \\ & & & -1 & 2 & -1 & \\ & & & & -1 & 2 & -1 \\ & & & & & -1 & 1 \end{pmatrix}$$

with distinct positive (or zero) eigenvalues. Conclude that A is positive semidefinite, and that the $w^{(k)}$s form an orthogonal basis of R^N. Normalize the $w^{(k)}$s so that they are also orthonormal.

4. Define the 2-D Fourier vectors $w^{(k,l)}$ $(0 \le k, l < N)$ by

$$w_{j,m}^{(k,l)} = w_j^{(k)} w_m^{(l)}, \qquad 1 \le j, m \le N.$$

Define the operators $X : R^{N^2} \to R^{N^2}$ and $Y : R^{N^2} \to R^{N^2}$ by

$$\begin{aligned} (Xv)_{i,j} &= v_{i,j} - v_{i,j+1} & \text{if } j = 0 \\ (Xv)_{i,j} &= 2v_{i,j} - v_{i,j-1} - v_{i,j+1} & \text{if } 0 < j < N \\ (Xv)_{i,j} &= v_{i,j} - v_{i,j-1} & \text{if } j = N \\ (Yv)_{i,j} &= v_{i,j} - v_{i+1,j} & \text{if } i = 0 \\ (Yv)_{i,j} &= 2v_{i,j} - v_{i-1,j} - v_{i+1,j} & \text{if } 0 < i < N \\ (Yv)_{i,j} &= v_{i,j} - v_{i-1,j} & \text{if } i = N, \end{aligned}$$

for every 2-D vector $v \in R^{N^2}$. Show that the $w^{(k,l)}$s are the eigenvectors of the symmetric operator $X + Y$ with distinct positive (or zero) eigenvalues. Conclude that $X + Y$ is positive semidefinite, and that the $w^{(k,l)}$s form an orthonormal basis of R^{N^2} (provided that the $w^{(k)}$s have been normalized in the previous exercise).

5. Define the Fourier (sine) vectors $w^{(k)}$ $(0 \leq k < N)$ by

$$w_j^{(k)} = \sin(\pi(k+1/2)jh), \qquad 1 \leq j \leq N,$$

where $h = 1/(N+1/2)$. Show that the $w^{(k)}$s are the eigenvectors of the symmetric tridiagonal $N \times N$ matrix

$$A = \begin{pmatrix} 2 & -1 & & & & & \\ -1 & 2 & -1 & & & & \\ & -1 & 2 & \ddots & & & \\ & & \ddots & \ddots & -1 & & \\ & & & -1 & 2 & -1 & \\ & & & & -1 & 2 & -1 \\ & & & & & -1 & 1 \end{pmatrix}$$

with distinct positive eigenvalues. Conclude that A is SPD, and that the $w^{(k)}$s form an orthogonal basis of R^N. Normalize the $w^{(k)}$s so that they are also orthonormal.

6. Define the 2-D Fourier vectors $w^{(k,l)}$ $(0 \leq k, l < N)$ by

$$w_{j,m}^{(k,l)} = w_j^{(k)} w_m^{(l)}, \qquad 1 \leq j, m \leq N.$$

Define the operators $X : R^{N^2} \to R^{N^2}$ and $Y : R^{N^2} \to R^{N^2}$ by

$$
\begin{aligned}
(Xv)_{i,j} &= 2v_{i,j} - v_{i,j+1} & \text{if } j = 0 \\
(Xv)_{i,j} &= 2v_{i,j} - v_{i,j-1} - v_{i,j+1} & \text{if } 0 < j < N \\
(Xv)_{i,j} &= v_{i,j} - v_{i,j-1} & \text{if } j = N \\
(Yv)_{i,j} &= 2v_{i,j} - v_{i+1,j} & \text{if } i = 0 \\
(Yv)_{i,j} &= 2v_{i,j} - v_{i-1,j} - v_{i+1,j} & \text{if } 0 < i < N \\
(Yv)_{i,j} &= v_{i,j} - v_{i-1,j} & \text{if } i = N,
\end{aligned}
$$

for every 2-D vector $v \in R^{N^2}$. Show that the $w^{(k,l)}$s are the eigenvectors of the symmetric operator $X + Y$ with distinct positive eigenvalues. Conclude that $X + Y$ is SPD, and that the $w^{(k,l)}$s form an orthonormal basis of R^{N^2} (provided that the $w^{(k)}$s have been normalized in the previous exercise).

Part II

Partial Differential Equations and Their Discretization

So far, we've discussed the multilevel (or multiscale) approach, and illustrated its usefulness in several branches of mathematics. In the rest of this book, we'll focus on the main purpose of the multilevel concept: the numerical solution of partial differential equations (PDEs).

In this part, we focus on scalar elliptic PDEs such as the Poisson equation and the diffusion equation in two spatial dimensions. We consider several discretization methods for approximating the original PDE on a discrete grid. This discretization produces a large system of algebraic equations that can be solved numerically. The solution to this system is then accepted as the numerical solution to the original PDE at the gridpoints.

Thus, the original physical phenomenon that has been reduced to a mathematical model in the form of a PDE is now reduced further to a system of linear algebraic equations. This algebraic system can then be solved by iterative methods such as multigrid later on in this book.

In this part, we describe three discretization methods. In the first chapter in it (Chapter 3), we describe the finite-difference and finite-volume discretization methods, and discuss their accuracy and adequacy. In the second one (Chapter 4), we describe the finite-element discretization method for elliptic PDEs, including local and adaptive mesh refinement.

3

Finite Differences and Volumes

In this chapter, we describe the finite-difference and finite-volume discretization methods for scalar second-order elliptic PDEs such as the Poisson equation, highly anisotropic diffusion equations, the indefinite Helmholtz equation, and the convection diffusion equation in two spatial dimensions. We study not only the accuracy of the discretization method (the rate by which the discretization error approaches zero together with the meshsize) but also its adequacy (the rate by which the discretization error approaches zero together with the meshsize and the parameter used in the PDE). Finally, we present the finite-volume discretization method for diffusion problems with variable or even discontinuous coefficients.

3.1 Elliptic PDEs

Many important problems in applied science and engineering are modeled in the form of partial differential equations (PDEs). Realistic phenomena are often modeled by systems of (time-dependent, nonlinear) PDEs. Here, however, we restrict the discussion to scalar, second-order, linear PDEs of the form (3.1) below.

The discussion in this chapter uses some of the standard lemmas in Section 2.3 (with D there being just the identity matrix I). The Nabla operator "∇" used below stands for either the divergence operator or the gradient operator, as appropriate. More concretely, if $s(x, y)$ is a scalar function and $v(x, y) = (v_1(x, y), v_2(x, y))$ is a vector function, then we have

$$\nabla s = (s_x(x, y), s_y(x, y))^t$$
$$\nabla \cdot v = v_{1\ x}(x, y) + v_{2\ y}(x, y)$$
$$\nabla v = (\nabla v_1 \quad \nabla v_2).$$

In particular, the first ∇-symbol in (3.1) stands for the divergence operator, whereas the second and third ones stand for the gradient operator. Let $\Omega \subset R^2$ be a domain in the 2-D (two-dimensional) Cartesian plane. The PDE is defined in Ω by

$$-\nabla \cdot \mathcal{D}\nabla u + (a_1, a_2) \cdot \nabla u + \beta u = \mathcal{F}. \tag{3.1}$$

Here $\mathcal{D}$ is a symmetric 2×2 matrix whose elements are scalar real functions defined in Ω; a_1, a_2, β, and $\mathcal{F}$ are also real scalar functions defined in Ω; and u is the real, unknown scalar function in Ω that solves (3.1).

Recall that the determinant of a matrix is equal to the product of its eigenvalues. Because $\mathcal{D}$ is symmetric, it has two real eigenvalues (Lemma 2.10). We say that the equation in (3.1) is elliptic, hyperbolic, or parabolic depending on whether the determinant of $\mathcal{D}$ is positive, negative, or zero, respectively. In other words, (3.1) is elliptic if the eigenvalues of $\mathcal{D}$ are both positive or both negative, hyperbolic if one of them is positive and the other is negative, and parabolic if one of them is zero. Because $\mathcal{D}$ depends on the independent variables x and y, the above conditions must hold uniformly throughout Ω. For example, for (3.1) to be elliptic, the determinant of $\mathcal{D}$ must be positive and bounded away from zero throughout Ω:

$$\det(\mathcal{D}) \geq \eta,$$

where η is some positive constant.

In the sequel, we focus on elliptic boundary-value problems. Such problems are obtained when (3.1) is accompanied with boundary conditions, which must be satisfied by the solution $u(x, y)$ at boundary points $(x, y) \in \partial\Omega$.

3.2 The Diffusion Equation

Recall that, because $\mathcal{D}$ is symmetric, it has two orthonormal eigenvectors (Lemma 2.11). Therefore, one could apply an orthogonal transformation to Ω, with which the 2×2 matrix in the first term in (3.1) is diagonal. Furthermore, we are particularly interested in symmetric PDEs in the sense in Section 4.1 below. Therefore, we consider particularly the symmetric elliptic diffusion equation that is obtained from (3.1) by setting $a_1 \equiv a_2 \equiv \beta \equiv 0$ and assuming that $\mathcal{D}$ is diagonal. This equation arises in many applications, and is thus especially important:

$$-\nabla\mathcal{D}\nabla u \equiv -(D_1 u_x)_x - (D_2 u_y)_y = \mathcal{F} \qquad (3.2)$$

in Ω, where $\mathcal{D} = diag(D_1, D_2)$ is a diagonal matrix of order 2, and D_1 and D_2 are uniformly positive functions in Ω.

When $D_1 \equiv D_2$, we say that the problem is isotropic. When, on the other hand, the diffusion coefficients D_1 and D_2 in (3.2) are different from each other, we say that the problem is anisotropic. In particular, when $D_1 \gg D_2$ or $D_1 \ll D_2$, we say that the problem is highly anisotropic.

In order to complete the PDE into a boundary-value problem, we must also specify conditions for the behavior of the solution $u(x, y)$ at boundary points $(x, y) \in \partial\Omega$. Let $\Gamma_D \subset \partial\Omega$ be a subset of the boundary of Ω, and define

$$\Gamma = \partial\Omega \setminus \Gamma_D.$$

The boundary conditions are as follows.

$$u = \mathcal{F}_1 \qquad (3.3)$$

on Γ_D, and

$$(\mathcal{D}\nabla u) \cdot \mathbf{n} + \mathcal{G}_1 u = \mathcal{G}_2 \qquad (3.4)$$

on Γ, where $\mathcal{F}_1$ is a real function in Γ_D, $\mathcal{G}_1$ and $\mathcal{G}_2$ are real functions in Γ (with $\mathcal{G}_1$ being also nonnegative), and $\mathbf{n}$ is the unit vector that is the outer normal vector to Ω in Γ.

The boundary conditions in Γ_D, where u is specified, are called Dirichlet boundary conditions or boundary conditions of the first kind. The boundary conditions in Γ can be of two possible kinds: where $\mathcal{G}_1$ vanishes, the boundary conditions are called Neumann boundary conditions or boundary conditions of the second kind, and where $\mathcal{G}_1$ does not vanish the boundary conditions are called mixed boundary conditions or boundary conditions of the third kind.

In the next section we go ahead and discretize the boundary-value problem, so that it can be solved numerically on the computer.

3.3 The Finite-Difference Discretization Method

A discretization method approximates the boundary-value problem as in (3.2)–(3.4) by a discrete system of algebraic equations of the form

$$Ax = b, \tag{3.5}$$

where $A = (a_{i,j})$ is the coefficient matrix, b is the known right-hand side, and x is the vector of unknowns.

In the sequel, we derive the algebraic system in (3.5) from the original boundary-value problem. This process is called *discretization*, because it approximates the original problem, defined in the continuous domain Ω, on a finite discrete grid, with a finite number of degrees of freedom or unknowns.

In this chapter, we discuss only the construction of the algebraic system (3.5). The actual numerical solution of (3.5) is discussed later on in this book. This task is also called "inversion" of A. Of course, A is never inverted explicitly, because this is prohibitively expensive. The term "inversion" only means finding the algebraic solution x in (3.5).

In the sequel, we show how (3.5) is obtained from (3.2)–(3.4) by the finite-difference discretization method. Assume, for simplicity, that Ω is the unit square $[0,1] \times [0,1]$. Let g be the uniform $n \times n$ grid that approximates Ω:

$$g = \{(i,j) \mid 1 \le i, j \le n\} \tag{3.6}$$

(Figure 3.1).

The gridpoints in g are ordered row by row from bottom to top and from left to right. The algebraic equation that corresponds to the (i,j)th gridpoint in g is contained in the $((i-1)n+j)$th row in A. Similarly, the value of the $((i-1)n+j)$th component in x is the numerical approximation to $u(jh, ih)$, and the $((i-1)n+j)$th component in b is $\mathcal{F}(jh, ih)$. Here, the coefficient matrix A is of order n^2, and the given right-hand side vector b and the unknown vector x are of dimension n^2.

The discretization method should preserve symmetry: when the PDE is symmetric in the sense in Section 4.1 below, the coefficient matrix A should be symmetric as well. This property indeed holds for the finite-difference discretization method [118]. In fact, in this method, a derivative in (3.2) is approximated by a symmetric finite difference in the relevant spatial direction, divided by the mesh size $h = 1/(n+1)$.

Fig. 3.1. The uniform $n \times n$ grid for the finite-difference discretization method.

Let us construct the kth equation in the linear system (3.5), or the kth row in A. Assume that $k = (i-1)n+j$, where (i, j) is an interior point in the grid g. In symmetric finite differencing, the divided difference between the values of two adjacent gridpoints is used to approximate the derivative at the midpoint in between them. This divided difference is then multiplied by the corresponding diffusion coefficient D_1 or D_2 at that midpoint. Finally, the divided difference between two such midpoints is used to approximate $(D_1 u_x)_x$ [or $(D_2 u_y)_y$] at the gridpoint in between them. For example, $-(D_1 u_x)_x$ is approximated at an interior gridpoint $(i, j) \in g$ as follows.

$$(D_1 u_x)((j + 1/2)h, ih) \doteq h^{-1} D_1((j + 1/2)h, ih)(x_{k+1} - x_k)$$
$$(D_1 u_x)((j - 1/2)h, ih) \doteq h^{-1} D_1((j - 1/2)h, ih)(x_k - x_{k-1})$$
$$((D_1 u_x)_x)(jh, ih) \doteq h^{-1}((D_1 u_x)((j + 1/2)h, ih) - (D_1 u_x)((j - 1/2)h, ih))$$
$$\doteq h^{-2}(D_1((j + 1/2)h, ih)x_{k+1} - (D_1((j + 1/2)h, ih)$$
$$+ D_1((j - 1/2)h, ih))x_k + D_1((j - 1/2)h, ih)x_{k-1}).$$

The kth equation in (3.5) (or the kth row in A) is, thus,

$$Nx_{k+n} + Ex_{k+1} + Cx_k + Wx_{k-1} + Sx_{k-n} = b_k, \tag{3.7}$$

where N (North), W (West), C (Center), E (East), S (South), and b_k are scalars depending on i and j and are defined as follows:

$$N = -h^{-2}D_2(jh, (i + 1/2)h)$$
$$S = -h^{-2}D_2(jh, (i - 1/2)h)$$
$$E = -h^{-2}D_1((j + 1/2)h, ih)$$
$$W = -h^{-2}D_1((j - 1/2)h, ih)$$
$$C = -(N + S + E + W)$$
$$b_k = \mathcal{F}(jh, ih).$$

(Don't confuse the scalar N that stands for the Northern coefficient with the integer N that is sometimes used to denote the order of A.)

When (i, j) is a boundary point of g, the above equation must be modified using discrete boundary conditions. For example, when mixed or Neumann boundary conditions are imposed on the left edge in Figure 3.1, then for gridpoints of the form $(i, 1)$ we have

$$N = -h^{-2}D_2(h, (i + 1/2)h)$$
$$S = -h^{-2}D_2(h, (i - 1/2)h)$$
$$E = -h^{-2}D_1((3/2)h, ih)$$
$$W = 0$$
$$C = -(N + S + E) + h^{-1}\mathcal{G}_1(h/2, ih)$$
$$b_k = \mathcal{F}(h, ih) + h^{-1}\mathcal{G}_2(h/2, ih).$$

Clearly, A is a symmetric, diagonally dominant L-matrix. Furthermore, it is irreducible in the sense in [118], hence also nonsingular and M-matrix. From Lemmas 2.1 and 2.10, it follows that it is also SPD.

The M-matrix property of A means that A^{-1} is nonnegative in the sense that all its elements are positive or zero. This property indicates that A is indeed a good approximation to the differential operator in the original boundary-value problem (3.2)–(3.4). Indeed, the Green function (the inverse of the differential operator) is nonnegative as well.

The structure of coupling between gridpoints in the finite-difference discretization method is that of immediate-neighbor coupling only. We refer to this structure as the 5-coefficient stencil (or 5-point stencil) and denote it by the bracketed 3×3 matrix

$$\begin{bmatrix} & N & \\ W & C & E \\ & S & \end{bmatrix}. \tag{3.8}$$

3.4 Error Estimate

Here we estimate the discretization error, that is, the difference between the numerical solution x and the solution u of the original boundary-value problem in the grid.

Let $\mathbf{u}$ be the vector obtained from the restriction of u to the grid g. In other words, $\mathbf{u}$ is the vector (or grid function) that agrees with u on the grid:

$$\mathbf{u}_k = u(jh, ih), \quad k = (i - 1)n + j.$$

The truncation error vector $\mathbf{t}$ is the residual of the discrete system (3.5) at $\mathbf{u}$:

$$\mathbf{t} \equiv b - A\mathbf{u}. \tag{3.9}$$

The discretization error is now defined by

$$x - \mathbf{u},$$

where x is the solution in (3.5). (We also refer to the discretization error as simply the error.) The discretization error can be estimated by the truncation error as follows.

$$\|x - \mathbf{u}\|_2 = \|A^{-1}(b - A\mathbf{u})\|_2 = \|A^{-1}\mathbf{t}\|_2 \leq \|A^{-1}\|_2\|\mathbf{t}\|_2. \tag{3.10}$$

It is well known that, when u is differentiable to order 4 and $h \to 0$, the components of $\mathbf{t}$ are as small as $O(h^2)$ (or h^2 times a constant independent of h). More precisely, the truncation error is of order

$$h^2(u_{xxxx} + u_{yyyy}),$$

where these fourth derivatives are taken at intermediate points in each cell. The differentiability of u guarantees that these derivatives are indeed bounded, so the components of $\mathbf{t}$ are indeed of order h^2 as $h \to 0$. As a consequence, we have

$$\|x - \mathbf{u}\|_2 \le \|A^{-1}\|_2 \|\mathbf{t}\|_2 = O(h\|A^{-1}\|_2). \tag{3.11}$$

In the next section, we show that, for the Poisson equation, $\|A^{-1}\|_2 = O(1)$ as $h \to 0$. This will guarantee that the discretization error is of order h^2, which means that the finite-difference scheme is of second-order accuracy.

3.5 Finite Differences for the Poisson Equation

A particularly important PDE that arises often in applications is the Poisson equation:

$$-\Delta u \equiv -\nabla \cdot \nabla u = -u_{xx} - u_{yy} = \mathcal{F}, \quad (x,y) \in \Omega. \tag{3.12}$$

Assume that Ω is the unit square, and the finite-difference discretization method uses the uniform grid in (3.6). The resulting 5-point stencil is:

$$h^{-2} \begin{bmatrix} & -1 & \\ -1 & 4 & -1 \\ & -1 & \end{bmatrix}. \tag{3.13}$$

Assume also that Dirichlet boundary conditions are imposed. In this case, the above stencil is valid at the interior gridpoints only, whereas at the boundary gridpoints it must be modified in such a way that the coefficient that corresponds to coupling with a point that lies outside the grid is set to zero.

The Poisson equation is actually a special case of the diffusion equation (3.2), and the resulting linear system (3.5) with the stencil in (3.13) is also a special model problem, in which the stencil is constant and independent of the particular gridpoint under consideration. Still, this problem is particularly important from both practical and theoretical points of view. Indeed, it is of second-order accuracy, which indicates that the finite-difference scheme in the more general case (3.7) should be accurate as well.

The eigenvectors of the matrix A that corresponds to the stencil in (3.13) are the 2-D sine modes $v^{(k,l)}$ in Section 2.4:

$$Av^{(k,l)} = 2h^{-2}(1 - \cos(\pi kh) + 1 - \cos(\pi lh))v^{(k,l)}$$
$$= 4h^{-2}(\sin^2(\pi kh/2) + \sin^2(\pi lh/2))v^{(k,l)}.$$

Note that A is actually diagonalized by the sine transform, which implies that the sine transform can actually be used as a direct algebraic solver for (3.5). Unfortunately, this solver is limited to the present model problem only: using a slightly different domain, boundary conditions, or diffusion coefficient, would make it inapplicable. This is why the Fourier transform is used here for theoretical purposes only.

The smallest eigenvalue of A is obtained for the eigenvector $v^{(1,1)}$. This is why we call $v^{(1,1)}$ the nearly-singular eigenvector of A. This vector is most important not only in the present error estimates but also in the construction of multigrid linear-system solvers in Chapter 6 below. Indeed, a good coarse-grid approximation must also have a similar nearly singular eigenvector (with practically the same eigenvalue) to be able to supply a good correction term to the original linear system (3.5).

As $h \to 0$, the smallest eigenvalue of A approaches $2\pi^2$. Because A is symmetric, A^{-1} is symmetric as well (Lemma 2.9). Using also Lemma 2.14, we have

$$\|A^{-1}\|_2 = \rho(A^{-1}) \sim 1/(2\pi^2)$$

as $h \to 0$. Using this result in (3.11), we have that the finite-difference discretization method is second-order accurate for the Poisson equation, provided that u is differentiable to order 4 in Ω.

3.6 Error Estimate for Diffusion Problems

The finite-difference discretization method is also accurate for more general diffusion problems as in (3.2), provided that D_1 and D_2 are differentiable to order 3. Indeed, assume that Dirichlet boundary conditions are imposed, A is the corresponding coefficient matrix, and $\acute{A}$ is the coefficient matrix for the Poisson equation (3.13). Because D_1 and D_2 are uniformly positive in Ω, there is a constant $\eta > 0$ such that $D_1 \geq \eta$ and $D_2 \geq \eta$ throughout Ω. Because A, $\eta\acute{A}$, and $A - \eta\acute{A}$ are symmetric and diagonally dominant, they are also positive semidefinite (Lemmas 2.10 and 2.1). Using also Lemma 2.15 and the proof of Lemma 2.13, we have that the minimal eigenvalue of A is equal to

$$\min_{\|v\|_2=1} (Av,v)_2 \geq \min_{\|v\|_2=1} (\eta\acute{A}v,v)_2 = 8\eta h^{-2} \sin^2(\pi h/2).$$

As a consequence, we have that, as $h \to 0$, the minimal eigenvalue of A is asymptotically larger than or equal to $2\eta\pi^2$, and, hence, $\|A^{-1}\|_2$ is asymptotically at most $1/(2\eta\pi^2)$. Using this result in (3.11), we have that the finite-difference scheme is of second-order accuracy, provided that u is differentiable to order 4 and D_1 and D_2 are differentiable to order 3 in Ω.

Unfortunately, these differentiability assumptions do not always hold. Therefore, the finite-difference discretization method may no longer be accurate for diffusion problems with nondifferentiable, let alone discontinuous, coefficients. For such difficult cases, one should use discretization methods based on integration (Section 3.12 and Chapter 4 below).

3.7 The Indefinite Helmholtz Equation

Here consider the Helmholtz equation

$$-u_{xx} - u_{yy} + \beta u = \mathcal{F} \quad (x,y) \in \Omega, \tag{3.14}$$

with suitable boundary conditions. We are particularly interested in the case $\beta < 0$, to which we refer as the indefinite Helmholtz equation.

When Ω is the unit square and Dirichlet boundary conditions are imposed, the eigenfunctions of the boundary-value problem are

$$\{\sin(\pi kx)\sin(\pi ly)\}_{1\leq k,l<\infty}. \tag{3.15}$$

The eigenvalue of an eigenfunction in (3.15) with respect to the differential operator in (3.14) is

$$\pi^2(k^2 + l^2) + \beta. \tag{3.16}$$

We assume that the boundary-value problem in (3.14) is well-posed in the sense that it has a unique solution. In other words, the differential operator has a bounded inverse. Since the differential operator in (3.14) is symmetric in the sense in Section 4.1, the above assumption is equivalent to assuming that all its eigenvalues are bounded away from zero (in magnitude):

$$|\pi^2(k^2 + l^2) + \beta| \geq m_0, \quad k \in Z, \ l \in Z, \ k > 0, \ l > 0, \tag{3.17}$$

where m_0 is a positive constant. In other words, the circle denoted by the bullets in Figure 3.2 is bounded away from $(Z^+)^2$, independently of β. This assumption holds for all nonnegative β's and most negative ones.

When (3.14) is discretized by the finite-difference discretization method in Section 3.3 (or the finite-volume discretization method in Section 3.12 below) on a uniform $n \times n$ grid as in (3.6), one obtains the stencil

$$h^{-2}\begin{bmatrix} 0 & -1 & 0 \\ -1 & 4+\beta h^2 & -1 \\ 0 & -1 & 0 \end{bmatrix}. \tag{3.18}$$

The eigenvectors of the coefficient matrix A that uses this stencil are the discrete counter parts of the functions in (3.15), namely, the two-dimensional Fourier sine modes $v^{(k,l)}$ defined in Section 2.4. In fact, we have

$$Av^{(k,l)} = (4h^{-2}(\sin^2(\pi kh/2) + \sin^2(\pi lh/2)) + \beta)v^{(k,l)}. \tag{3.19}$$

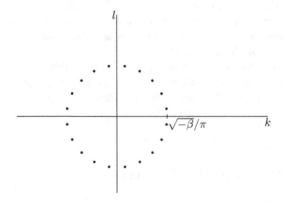

Fig. 3.2. Frequencies (k,l) for which $\sin(\pi kx)\sin(\pi ly)$ is a nearly singular eigenfunction of the indefinite Helmholtz equation in the unit square with Dirichlet boundary conditions.

As mentioned above, we are particularly interested in the more difficult case $\beta < 0$. In this case, A can be indefinite: it may have both positive and negative eigenvalues. In the sequel, we study not only the accuracy but also the adequacy of the finite-difference discretization in (3.18) for this kind of problem.

3.8 Adequate Discretization of the Helmholtz Equation

The "accuracy" of a discretization method is evaluated in terms of the asymptotic behavior of the discretization error as $h \to 0$. In order to obtain an estimate of the average error at the points in the grid, we use the average norm, which is just the $\|\cdot\|_2$ norm divided by the square root of the number of points in the grid. The average norm of the discretization error (also referred to as the average discretization error) can be estimated by multiplying (3.10) throughout by h, which gives the estimate

$$h\|x - \mathbf{u}\|_2 \leq \|A^{-1}\|_2 h \|\mathbf{t}\|_2, \tag{3.20}$$

where $\mathbf{u}$ is the solution of the original boundary-value problem confined to the grid, x is the solution to the discrete system (3.5), and $\mathbf{t}$ is the truncation error vector defined in (3.9). For the Poisson equation and the definite Helmholtz equation [$\beta \geq 0$ in (3.14)], the right-hand side in (3.20) is indeed of order h^2 as $h \to 0$, implying that the discretization error on the left-hand side in (3.20) approaches zero rapidly (see Section 3.5). Although this asymptotic result is true also in the indefinite case $\beta < 0$, it may be of little practical significance when very large βs are considered.

In many practical problems, the PDE may use a very small parameter, which may well take values that are far smaller than the meshsize h. In this case, the standard limit process $h \to 0$ no longer tells the whole story, and may well be misleading. In fact, discretization methods can then exhibit large errors in practice even when they are accurate in theory [27]. A much more informative limit process is the process in which both the parameter in the PDE and the meshsize h approach zero at the same time, and x, $\mathbf{u}$, A, and $\mathbf{t}$ in (3.20) depend not only on h but also on that parameter [102]. When the average discretization error on the left-hand side of (3.20) approaches zero for a particular limit case of the above kind, we say that the discretization is adequate in this particular limit case.

In the indefinite Helmholtz equation, a most interesting and relevant limit case is the one in which $h \to 0$ and $\beta \to -\infty$ at the same time. In this limit process, however, β can't change continuously; it must take discrete values in such a way that m_0 in (3.17) is independent of β, so (3.17) holds then uniformly in the entire limit process. The limit process $\beta \to -\infty$ is thus actually a sequence of βs for which (3.17) holds uniformly for some predetermined m_0.

The asymptotic notations "$\sim$", "$\ll$", "$O()$", and "$o()$" used below are thus interpreted in the limit case

$$(\beta, h) \to (-\infty, 0), \tag{3.21}$$

with discrete βs that satisfy (3.17) for some predetermined m_0 independent of the limit process.

A particularly relevant sequence of βs is

$$\{\beta_j = -\pi^2 j^2\}_{j \in Z,\ j \geq 1}.$$

In this case, the pair of integers that minimizes the left-hand side in (3.17) is clearly $(k, l) = (j, 1)$ $(k, l) = (j, 1)$ or $(k, l) = (1, j)$, so m_0 can be as large as

$$m_0 = \pi^2.$$

In order to obtain a meaningful estimate in (3.20), we first estimate the truncation-error vector $\mathbf{t}$. To this end, assume that the solution u in (3.14) is a nearly-singular eigenfunction, namely, an eigenfunction as in (3.15) with

$$|\pi^2(k^2 + l^2) + \beta| \ll |\beta|. \tag{3.22}$$

This is indeed a fair assumption: because the right-hand side $\mathcal{F}$ in (3.14) is independent of β, so u must indeed contain such eigenfunctions in its Fourier expansion. Assume further that the limit process in (3.21) is done in such a way that

$$|\beta|h^2 = O(1). \tag{3.23}$$

With these assumptions, the Taylor expansion of u in a circle of (at least) radius h around each gridpoint converges. Furthermore, when this expansion is used in the stencil in (3.18), the resulting truncation error is

$$|\mathbf{t}| = \left| \frac{u_{xxxx} + u_{yyyy}}{12} \right| h^2 \le \frac{\beta^2 h^2}{12}, \tag{3.24}$$

pointwise in the grid.

Let us now estimate the factor $\|A^{-1}\|_2$ in the right-hand side in (3.20). Since A is symmetric, it is sufficient to bound the magnitude of its eigenvalues from below by a positive constant.

Let us first show that the eigenvalues of A are asymptotically the same as their continuous counterparts in (3.16). For this purpose, let us first assume that the limit process in (3.21) is done in such a way that

$$|\beta|h^2 \ll 1. \tag{3.25}$$

Consider a nearly singular eigenvector of A, namely, an eigenvector of A with an eigenvalue of order $o(|\beta|)$. From the assumption in (3.25), we must have in the right-hand side in (3.19) that $\sin^2(\pi k h/2) \ll 1$ and $\sin^2(\pi l h/2) \ll 1$. Therefore, we must also have $\pi k h/2 \ll 1$ and $\pi l h/2 \ll 1$, which imply that the eigenvalue of the nearly singular eigenvector of A in the right-hand side in (3.19) is asymptotically the same as its continuous counter part in (3.16).

Using also (3.17), the minimal magnitude of an eigenvalue of A is (asymptotically) at least m_0. From this and (3.24), it follows that (3.20) takes the form

$$h\|x - \mathbf{u}\|_2 \le \frac{\beta^2 h^2}{12 m_0}. \tag{3.26}$$

Thus, if an average discretization error of 10^{-1} is required, then $|\beta| \le m_0^{1/2} h^{-1}$ should be used.

In the tests in Chapter 10 below, we use $|\beta|$ as large as 790. (This value is chosen intentionally to produce a nearly singular coefficient matrix A, which is most difficult and challenging.) Although the meshsize used in these tests is too large in terms of (3.26), it is also shown there that using a smaller meshsize doesn't affect the convergence of the multigrid linear system solver tested there. Further numerical tests with variable β in two and three spatial dimensions can be found in [92] and [94].

In Chapter 22 in [103] it is shown that the bilinear finite element discretization method is as adequate as the finite-difference discretization method used here. A more adequate discretization method is proposed in [67]. However, it is tested only for $|\beta|$ as small as $|\beta| \leq 64$, and with boundary conditions of the third kind only.

3.9 Adequate Discretization of Highly Anisotropic Equations

In this section, we study the adequacy of the finite-difference scheme for highly anisotropic diffusion equations. Such equations arise often in diffusion problems in severely stretched domains and grids. Our model equation is of the form

$$-\varepsilon u_{xx} - u_{yy} = \mathcal{F} \tag{3.27}$$

in the unit square (with, say, Dirichlet boundary conditions), where ε is a small positive parameter. Here, the x spatial direction is the direction of weak diffusion, whereas the y spatial direction is the direction of strong diffusion.

Let h_x be the meshsize in the x spatial direction, and h_y the meshsize in the y spatial direction. The finite-difference discretization method for this equation gives the 5-point stencil

$$\begin{bmatrix} 0 & -h_y^{-2} & 0 \\ -\varepsilon h_x^{-2} & 2h_y^{-2} + 2\varepsilon h_x^{-2} & -\varepsilon h_x^{-2} \\ 0 & -h_y^{-2} & 0 \end{bmatrix}. \tag{3.28}$$

It is easy to show that the finite-difference discretization method is accurate in the sense that the discretization error approaches zero when the meshsizes h_x and h_y do (see Section 3.5). Unfortunately, the limit process $(h_x, h_y) \to (0,0)$ used in this analysis assumes that ε is kept fixed; this assumption may be unrealistic in highly anisotropic equations, where the parameter ε may easily be as small as h_x and h_y. A more relevant limit process is, therefore, the process introduced in [102], in which ε, h_x, and h_y approach zero at the same time:

$$(\varepsilon, h_x, h_y) \to (0,0,0). \tag{3.29}$$

Our aim is thus to give conditions on ε, h_x, and h_y to guarantee an asymptotically small discretization error in the limit process in (3.29).

In the sequel, the notations "$\sim$", "$\ll$", "$o()$", and "$O()$" are with respect to the limit process in (3.29). In order to estimate the average norm of the discretization error, we use the estimate [also used in (3.20) above]

$$h\|x - \mathbf{u}\|_2 \leq \|A^{-1}\|_2 h\|\mathbf{t}\|_2, \tag{3.30}$$

where x, $\mathbf{u}$, A, and $\mathbf{t}$ depend also on ε. From (3.30), it is clear that we actually need to estimate $\|A^{-1}\|_2$ and the truncation error only. This is done below.

As in Sections 3.5 and 3.6, it is easy to see that

$$\|A^{-1}\|_2 \sim \pi^{-2}.$$

In order to estimate the truncation error vector $\mathbf{t}$, let us assume that the solution u in (3.27) is a nearly singular eigenfunction of the differential operator in (3.27). This is indeed a fair assumption: because the right-hand side $\mathcal{F}$ in (3.27) is independent of ε, h_x, or h_y, u must indeed contain such eigenfunctions in its Fourier expansion. Now, such an eigenfunction must have the form $\sin \pi k x \sin(\pi l y)$, with $l = O(1)$ and $k = O(\varepsilon^{-1/2})$. Assuming further that

$$\varepsilon^{-1/2} h_x = O(1),$$

we have that the Taylor expansion of u in a circle of (at least) radius h around each gridpoint converges. Using this expansion in the stencil in (3.28), we can estimate the truncation error by

$$|\mathbf{t}| = \left| \frac{\varepsilon u_{xxxx} h_x^2 + u_{yyyy} h_y^2}{12} \right| = O(\varepsilon^{-1} h_x^2 + h_y^2)$$

pointwise in the grid. Using this estimate in (3.30), we have the estimate

$$h\|x - \mathbf{u}\|_2 = O(\varepsilon^{-1} h_x^2 + h_y^2), \tag{3.31}$$

so the discretization in (3.28) is adequate as long as

$$\varepsilon \gg h_x^2. \tag{3.32}$$

It can also be seen from (3.31) that h_y can be much larger than h_x, resulting in a more efficient discretization that uses a smaller grid. In fact, h_y can be as large as

$$h_y = O(h_x \varepsilon^{-1/2}),$$

with the estimate in (3.31) still remaining the same. This observation is further used in Section 4.7 in problems with anisotropic and discontinuous coefficients.

3.10 Oblique Anisotropy

The above analysis uses the fact that the strong and weak diffusion direction in (3.27) are just the standard x and y spatial directions, These directions are also used to form the uniform grid; in other words, the strong and weak diffusion directions align with the grid.

Unfortunately, this is not always the case. Consider, for example, the equation

$$-\varepsilon u_{\xi\xi} - u_{\eta\eta} = \mathcal{F} \tag{3.33}$$

in the unit square $0 < x, y < 1$ (with, say, Dirichlet boundary conditions), where ξ and η are some orthonormal directions in the Cartesian plane that do not coincide with the standard x and y directions. Here, because $\mathcal{F}$ is independent of ε, the solution u of (3.33) must be smooth in the η spatial direction, but could oscillate rapidly (with frequency up to $\varepsilon^{-1/2}$) in the ξ spatial direction. This rapid oscillation is noticeable in both the x and y spatial direction, which changes the adequacy analysis considerably.

In order to be discretized on a uniform grid that aligns with the x-y coordinates, the equation (3.33) is reformulated as

$$-\nabla \cdot \mathcal{D}\nabla u = \mathcal{F}, \tag{3.34}$$

where the 2×2 diffusion matrix $\mathcal{D}$ is given by

$$\mathcal{D} = \mathcal{O}^t \begin{pmatrix} \varepsilon & 0 \\ 0 & 1 \end{pmatrix} \mathcal{O},$$

where $\mathcal{O}$ is the 2×2 orthogonal matrix that transforms the standard x-y coordinates into the ξ-η coordinates. Now, the left-hand side in (3.34) is just a linear combination of u_{xx}, u_{yy}, and u_{xy}. In the finite-difference scheme, u_{xx} and u_{yy} are discretized as before, whereas the mixed derivative u_{xy} is discretized by the stencil

$$(4h_x h_y)^{-1} \begin{bmatrix} -1 & 0 & 1 \\ 0 & 0 & 0 \\ 1 & 0 & -1 \end{bmatrix}.$$

Thus, the truncation error is a homogeneous polynomial in h_x and h_y, with coefficients that contain derivatives of order 4 of u. Because u may oscillate rapidly [with frequency of $O(\varepsilon^{-1/2})$] in the ξ spatial direction, these derivatives can be as large as ε^{-2}. In summary, because the truncation error is of order $O(\varepsilon^{-2}(h_x + h_y)^2)$, the adequacy condition is

$$\varepsilon \gg h_x + h_y. \tag{3.35}$$

This condition is much more restrictive than the one in (3.32). In fact, it requires very small meshsizes h_x and h_y, to help capture the subtle variation of u in the oblique direction ξ.

3.11 Finite Differences for the Convection-Diffusion Equation

The convection-diffusion equation is obtained from the elliptic PDE (3.1) by setting $\beta \equiv 0$ and choosing the 2×2 matrix $\mathcal{D}$ to be the identity matrix multiplied by a small positive parameter ε:

$$-\varepsilon(u_{xx} + u_{yy}) + a_1 u_x + a_2 u_y = \mathcal{F} \tag{3.36}$$

in $\Omega \subset R^2$, with the boundary conditions (3.3) and (3.4). The two-dimensional field (a_1, a_2) is referred to as the field of the characteristic directions.

Assume that Ω is the unit square $0 < x, y < 1$. The convection-diffusion equation can then be discretized on the uniform $n \times n$ grid g in (3.6) by finite differences. Clearly, the diffusion term [the first term in (3.36)] is discretized by ε times the stencil in (3.13). The only remaining task is to discretize the convection term $a_1 u_x + a_2 u_y$. Consider, for example, an interior point $(i, j) \in g$, and let us construct the kth equation in (3.5), where $k = (i-1)n + j$. In other words, let us construct the kth row in A. Let us first consider a naive, symmetric discretization for the first convection term:

$$(a_1 u_x)(jh, ih) \doteq (2h)^{-1} a_1(jh, ih)(x_{k+1} - x_{k-1}). \tag{3.37}$$

Unfortunately, this approach fails for small ε, because it introduces high-frequency nonphysical oscillations in the numerical solution. Indeed, consider the vector with the values $(1, -1, 1, -1, \ldots)$ along each x-line in the grid. This vector is completely unnoticed by the discrete system representing the discrete convection term in (3.37), and may lead to $O(1)$ discretization error.

In order to cure this problem, it is necessary to add to (3.37) also an artificial diffusion term in the amount of

$$(2h)^{-1} |a_1(jh, ih)| (2x_k - x_{k+1} - x_{k-1})$$

to help stabilize the coefficient matrix and make it a diagonally dominant M-matrix. A similar cure is used in the discretization of the convection term in the y spatial direction. The resulting 5-point stencil (3.8) contains the coefficients

$$N = -h^{-2}\varepsilon + (2h)^{-1}(a_2(jh, ih) - |a_2(jh, ih)|)$$
$$S = -h^{-2}\varepsilon + (2h)^{-1}(-a_2(jh, ih) - |a_2(jh, ih)|)$$
$$E = -h^{-2}\varepsilon + (2h)^{-1}(a_1(jh, ih) - |a_1(jh, ih)|)$$
$$W = -h^{-2}\varepsilon + (2h)^{-1}(-a_1(jh, ih) - |a_1(jh, ih)|)$$
$$C = -(N + S + E + W).$$

The artificial diffusion term reduces the accuracy of the discretization to $O(h)$, meaning that the truncation error is of $O(h)$ pointwise in the grid, provided that the solution u is differentiable to order 2. (See [102] and the references therein for error analysis.) This is why the above discretization is known as the *first-order upwind scheme*. We also refer to it simply as the *upwind scheme*.

In singularly perturbed problems, the diffusion coefficient ε is very small. In fact, on most practical grids, it can be as small as the meshsize h. In this kind of problem, adequacy is a more relevant property than accuracy. This means that the discretization error should be estimated in the limit case in which both h and ε approach zero at the same time (see Section 3.8). Fortunately, the upwind scheme for (3.36) is indeed adequate, provided that no closed characteristics are present [102]. This is because, unlike the problem in Section 3.9 whose solution may oscillate frequently in the weak diffusion direction, here the solution $u \equiv u(\varepsilon)$ in (3.36) must be smooth not only in the characteristic direction but also in the cross-characteristic direction (thanks to the piecewise smooth boundary conditions that are carried to the entire domain through the characteristic curves). In fact, the solution may vary sharply only in boundary layers at the far end of the characteristic curves. Indeed, the right-hand side and boundary conditions in (3.36) must be independent of ε to allow a meaningful solution also in the incompressible limit $\varepsilon \to 0$. The analysis in [102] shows that the upwind scheme is indeed adequate for the convection-diffusion equation as long as $\varepsilon^{-1}h^2 \ll 1$ as both ε and h approach zero at the same time.

3.12 The Finite-Volume Discretization Method

As is evident from the above definition, the finite-difference discretization method uses pointwise values of the functions in the PDE. In particular, in the diffusion problem in Section 3.3, it uses values of the diffusion coefficients D_1 and D_2 at midpoints of the form $((j+1/2)h, ih)$, $((j-1/2)h, ih)$, $(jh, (i+1/2)h)$, and so on. and $(jh, (i-1/2)h)$. This definition makes sense as long as D_1 and D_2 are rather smooth. In fact, it leads to second-order accuracy as long as D_1 and D_2 are differentiable to order 3 (Section 3.6). Unfortunately, when D_1 or D_2 is not sufficiently differentiable (and, in particular, when it is discontinuous), the finite-difference discretization method is no longer accurate. In such cases, one should better turn to discretization methods based on integration rather than differentiation, such as the finite-volume and finite-element discretization methods.

In the following, we describe the finite-volume discretization method (see, e.g., [2] and the references therein). In this discretization method, the grid is as in Figure 3.3: the grid is uniform, and the gridpoints may also lie on boundary segments on which Neumann or mixed boundary conditions are imposed.

Let us first discretize the PDE at gridpoints that don't lie on the boundary. For this purpose, let (i, j) denote such a gridpoint, and let e denote the small $h \times h$ square (or volume) around it (Figure 3.4). The boundary of e, ∂e, may be written as the union of the north, south, east, and west edges of e:

$$\partial e = (\partial e)_N \cup (\partial e)_S \cup (\partial e)_E \cup (\partial e)_W.$$

By integrating the diffusion equation (3.2) over e and using Green's formula, we have

$$-\left(\int_{(\partial e)_N} D_2 u_y dx - \int_{(\partial e)_S} D_2 u_y dx + \int_{(\partial e)_E} D_1 u_x dy - \int_{(\partial e)_W} D_1 u_x dy \right)$$

$$= -\int_{\partial e} (D_1 u_x, D_2 u_y) \cdot \mathbf{n} ds$$

$$= -\int_e ((D_1 u_x)_x + (D_2 u_y)_y) \, dx dy = \int_e \mathcal{F} dx dy,$$

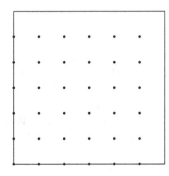

Fig. 3.3. The uniform $n \times n$ grid for the finite-volume discretization method. It is assumed that Dirichlet boundary conditions are imposed on the right and upper edges, and Neumann or mixed boundary conditions are imposed on the left and lower edges.

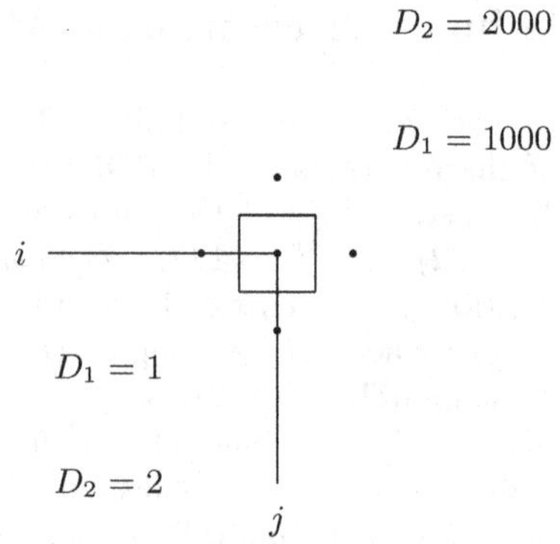

$$D_2 = 2000$$

$$D_1 = 1000$$

Fig. 3.4. The finite volume e surrounding the interior gridpoint (i, j) that may lie on a discontinuity line in the diffusion coefficients D_1 or $D_@$.

where $\mathbf{n}$ is the outer normal vector and ds is the length element in ∂e. By replacing the derivatives by corresponding finite differences, we obtain a 5-point stencil as in (3.7) and (3.8), with the following coefficients,

$$N = -h^{-1} \int_{(\partial e)_N} D_2 dx$$

$$S = -h^{-1} \int_{(\partial e)_S} D_2 dx$$

$$E = -h^{-1} \int_{(\partial e)_E} D_1 dy$$

$$W = -h^{-1} \int_{(\partial e)_W} D_1 dy$$

$$C = -(N + S + E + W) = h^{-1} \int_{\partial e} (D_1, D_2) \cdot \mathbf{n} ds$$

$$b_k = \int_e \mathcal{F} dx dy,$$

where $k = (i-1)n + j$. Note that when D_1 and D_2 are constant in each edge of e, this stencil coincides with the stencil used in the finite-difference discretization method in Section 3.3. This happens, for example, when D_1 and D_2 are piecewise constant, with discontinuity lines that align with the grid. Later on in this book we show several examples of this kind.

Of course, if (i, j) lies next to a Dirichlet boundary, then no coupling to this boundary is needed. For example, if $j = 1$, then W is dropped from the 5-point stencil, and the amount $-W\mathcal{F}_1(0, ih)$ is added to b_k.

A more complicated case arises when a gridpoint in Figure 3.3 lies on a boundary segment on which Neumann or mixed boundary conditions are imposed (e.g., the left edge in Figure 3.3, where $j = 0$). In this case, a "half volume" of size $(h/2) \times h$ is used in the above integration rather than the usual $h \times h$ volume (see Figure 3.5). Thus, the north and south edges of e are of length $h/2$ rather than h. The resulting

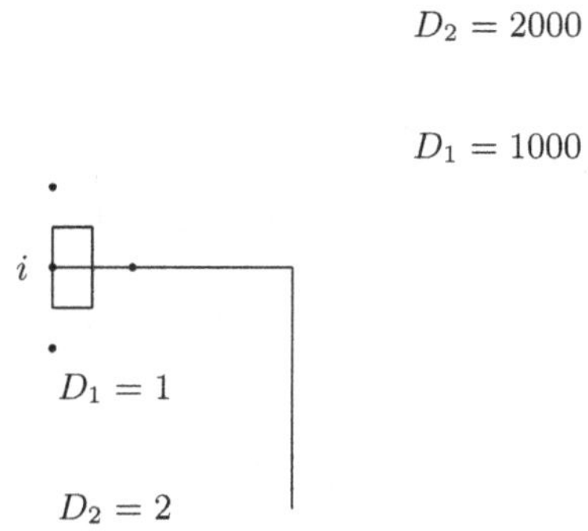
$D_2 = 2000$
$D_1 = 1000$
i
$D_1 = 1$
$D_2 = 2$

Fig. 3.5. The half finite volume e at the boundary point $(i, 1)$ that lies on the boundary segment on which Neumann or mixed boundary conditions are imposed.

5-point stencil at the boundary point $(i, 1)$ is

$$
\begin{aligned}
N &= -h^{-1} \int_{(\partial e)_N} D_2 dx \\
S &= -h^{-1} \int_{(\partial e)_S} D_2 dx \\
E &= -h^{-1} \int_{(\partial e)_E} D_1 dy \\
W &= 0 \\
C &= -(N + S + E) + \int_{(\partial e)_W} \mathcal{G}_1 dy \\
b_k &= \int_e \mathcal{F} dx dy + \int_{(\partial e)_W} \mathcal{G}_2 dy.
\end{aligned}
$$

As with finite differences, here also the coefficient matrix A is a symmetric, diagonally dominant L-matrix, hence also an SPD M-matrix. These desirable properties show that the finite-volume discretization method indeed makes sense. This discretization method is used later in this book to form challenging linear systems to test the present multigrid linear system solvers.

3.13 Exercises

1. Show that the matrix A resulting from the finite-difference or finite-volume discretization method is a diagonally dominant L-matrix.
2. We say that the $N \times N$ matrix A is reducible if there is a nontrivial proper subset of unknowns

$$s \subset \{1, 2, \ldots, N\}$$

that are decoupled from the rest of the unknowns in the sense that

$$a_{i,j} = 0 \text{ for every } i \in s \text{ and } j \notin s.$$

Show that the matrix A resulting from the finite-difference or finite-volume discretization method is irreducible (not reducible).

3. We say that the $N \times N$ matrix A is irreducibly diagonally dominant if it is irreducible, diagonally dominant, and at least in one row in A the diagonal dominance is strict; that is, there exists an index $1 \le k \le N$ for which

$$a_{k,k} > \sum_{1 \le j \le N, \; j \ne k} |a_{k,j}|.$$

Show that the matrix A resulting from the finite-difference or finite-volume discretization of a well-posed boundary-value problem is indeed irreducibly diagonally dominant.

4. It follows from the theory in [118] that an irreducibly diagonally dominant matrix is nonsingular. Conclude that A in the previous exercise is indeed nonsingular.

5. It follows from the theory in [118] that an irreducibly diagonally dominant L-matrix is also an M-matrix. Conclude that the matrix A obtained from the finite difference or finite volume discretization of a well-posed boundary-value problem is indeed an M-matrix.

6. Conclude from the theory in [118] that the matrix A in the previous exercise, being an M-matrix, has a unique eigenvector with positive components and positive eigenvalue, which is the minimal eigenvalue of A in terms of magnitude. This eigenvector is called the nearly singular eigenvector of A.

7. Assume that A is also symmetric. Conclude from Lemma 2.1 that it is also SPD. Conclude that the eigenvalue of A with respect to its nearly singular eigenvector is its minimal eigenvalue.

8. Show that the matrix A obtained from finite-difference or finite-volume discretization of (3.2) (with any kind of boundary conditions that make the boundary-value problem well-posed) is indeed symmetric. Conclude from the previous exercise that it is also SPD, and that the eigenvalue corresponding to its nearly singular eigenvector is its minimal eigenvalue.

9. Show that the nearly singular eigenvector of the matrix A resulting from the finite-difference discretization of the Poisson equation with Dirichlet boundary conditions [as in (3.13)] is $v^{(1,1)}$ in Section 2.4.

10. Repeat the above exercise, only this time use an IMSL[1] routine to compute the nearly singular eigenvector and the corresponding eigenvalue numerically. Verify that the results are indeed the same as in the previous exercise.

11. Use the above IMSL routine to compute the nearly singular eigenvector and its corresponding eigenvalue for the matrix A obtained from the finite-volume discretization of a PDE as in (3.2) with discontinuous diffusion coefficients and various kinds of boundary conditions. Verify that the eigenvalue and the components of the nearly singular eigenvector are indeed positive.

---

[1] http://www.vni.com/search/index.html.

4
Finite Elements

In this chapter, we present the weak formulation of the boundary-value problem and the finite-element discretization method derived from it. We start from structured bilinear finite-element meshes, and proceed to highly unstructured linear finite-element triangulations, which use local and adaptive refinement to approximate complicated domains and irregular solutions efficiently.

4.1 The Finite Element Discretization Method

As we've seen above, the finite-difference discretization method is accurate only when the diffusion coefficients (and the solution) are sufficiently smooth. Unfortunately, this is not always the case. Many relevant models in applied science and engineering are formulated as PDEs with not only nondifferentiable but also discontinuous coefficients. For these problems, the finite-difference discretization method may be inaccurate or even inapplicable.

The finite-volume discretization method is well defined also for problems with nondifferentiable or even discontinuous diffusion coefficients. Still, it uses uniform grids, which are suitable for rectangular domains but not for more complicated domains arising frequently in practical applications. Furthermore, these grids don't support local refinement, which is an essential tool in the efficient approximation of irregular domains and solutions.

Fortunately, the finite-element discretization method [112] [31] can handle not only discontinuous coefficients but also complicated domains and irregular solutions.

4.2 The Weak Formulation

The main problem with the finite-difference and finite-volume discretization methods is that they are based on the so-called "strong" formulation of the boundary-value problem in (3.2) through (3.4). This formulation requires that the flux vector

$$(D_1 u_x, D_2 u_y)$$

be not only continuous but also differentiable everywhere in Ω. This requirement is so strong that the problem in its strong formulation may have no solution at all.

The finite-element discretization method, on the other hand, is based on the so-called "weak" formulation rather than the strong one. In this formulation, Green's formula is used to integrate (3.2) over Ω and obtain a new problem, which accepts also a nondifferentiable flux, provided it is square-integrable (namely, it is in $L_2(\Omega)$).

The weak formulation is as follows. Let H be a Hilbert space, with the corresponding inner product $(\cdot, \cdot)$. Let $a(u, v)$ be a bilinear form in $H \times H$. Assume that $a(\cdot, \cdot)$ is bounded; that is, there exists a positive constant η_a such that

$$|a(v, w)| \leq \eta_a \sqrt{(v, v)(w, w)}$$

for every $v, w \in H$. Assume also that the quadratic form $a(\cdot, \cdot)$ is coercive in the sense that there exists a positive constant η_a such that

$$a(v, v) \geq \eta_a(v, v) \tag{4.1}$$

for every $v \in H$. Let $f(\cdot)$ be a bounded functional in H, i.e., there exists a positive constant η_f such that

$$|f(v)| \leq \eta_f \sqrt{(v, v)}$$

for every $v \in H$.

The weak formulation is the following problem. Let $H_0 \subset H$ be a subspace, and let $H_1 \subset H$ be a shift of H_0 (namely, $H_1 + H_0 = H_1$);

$$\text{find } u \in H_1 \text{ such that } a(u, v) = f(v) \quad \text{for every } v \in H_0. \tag{4.2}$$

Let us make this problem a little more concrete. Let $\Omega \subset R^2$ be a bounded domain. Let $\mathcal{D} \equiv \mathcal{D}(x, y)$ be a 2×2 matrix [with elements in $L_\infty(\Omega)$] that is symmetric and uniformly positive definite in Ω. Let $\Gamma_D \subset \partial\Omega$ be a subset of the boundary of Ω, and define

$$\Gamma = \partial\Omega \setminus \Gamma_D.$$

Let $\mathcal{F}, \beta \in L_2(\Omega)$, $\mathcal{F}_1$ in the Sobolev space of order $1/2$ on Γ_D (see [18]), $\mathcal{G}_1 \in L_\infty(\Gamma)$, and $\mathcal{G}_2 \in L_2(\Gamma)$ (β and $\mathcal{G}_1$ are also nonnegative). Let H be the Sobolev space of order 1 in Ω [the space of functions with derivatives in $L_2(\Omega)$]. Let H_0 be the space of functions in H that vanish in Γ_D. Let H_1 be the set of functions in H that agree with $\mathcal{F}_1$ in Γ_D.

Let us now show the connection between the above weak formulation and the strong formulation

$$-\nabla(\mathcal{D}\nabla u) + \beta u = \mathcal{F} \tag{4.3}$$

in Ω,

$$u = \mathcal{F}_1 \tag{4.4}$$

in Γ_D, and

$$(\mathcal{D}\nabla u) \cdot \mathbf{n} + \mathcal{G}_1 u = \mathcal{G}_2 \tag{4.5}$$

in Γ (where $\mathbf{n}$ is the outer normal vector). In fact, using Green's formula, a solution u of (4.3) through (4.5) also solves (4.2), with the definitions

$$a(u, v) = \int_\Omega (\mathcal{D}\nabla u) \cdot \nabla v \, d\Omega + \int_\Omega \beta u v \, d\Omega + \int_\Gamma \mathcal{G}_1 u v \, d\Gamma \tag{4.6}$$

and

$$f(v) = \int_\Omega \mathcal{F}v \, d\Omega + \int_\Gamma \mathcal{G}_2 v \, d\Gamma. \tag{4.7}$$

Let us now make sure that the quadratic form $a(\cdot, \cdot)$ is indeed coercive, that is, that (4.1) indeed holds. For this purpose, we also assume that

- either the support of β has a positive measure in Ω:

$$\int_\Omega \beta \, d\Omega > 0,$$

- or the support of $\mathcal{G}_1$ has a positive measure in Γ:

$$\int_\Gamma \mathcal{G}_1 \, d\Gamma > 0,$$

- or Γ_D has a positive measure in $\partial\Omega$:

$$\int_{\Gamma_D} d\Gamma_D > 0.$$

Under this assumption, the quadratic form $a(\cdot, \cdot)$ is indeed coercive, and, therefore, the weak formulation is indeed well-posed in the sense that it has a unique solution. (See, e.g., Chapter 11 in [103] for the detailed proof.) As a conclusion, the weak formulation is indeed better suited to model the original physical phenomenon, which must also have a unique solution. The strong formulation, on the other hand, is not as suitable, because it may have no solution at all. If, however, it does have a solution, then it also solves the weak formulation, and hence must coincide with the unique solution of the weak formulation.

In the following, we describe a discrete version of the weak formulation, which leads to the finite-element discretization.

4.3 The Discrete Weak Formulation

The finite-element discretization method is based on the following discrete approximation of (4.2).

$$\text{find } \tilde{u} \in V_1 \text{ such that } a(\tilde{u}, v) = f(v) \quad \text{for every } v \in V_0, \tag{4.8}$$

where V_0 is a finite-dimensional subspace of H that approximates H_0 in some sense, and V_1 is a shift of V_0 ($V_1 + V_0 = V_1$) that approximates H_1 in some sense.

Let us now show how V_0 and V_1 can be defined. Let T be a mesh, or a collection of finite elements, that approximates Ω. In this book, we consider only two kinds of meshes: mesh of squares (in which each finite element is a square) and mesh of triangles (in which each finite element is a triangle.)

Assume that T is also conformal in the sense that if two finite elements in it share an edge, then they must share it in its entirety, including its endpoints, which serve as vertices in both finite elements. With this additional assumption, we are ready to define the subspace V_0.

When T is a mesh of squares, V_0 is the space of functions that are continuous in T and bilinear in each square $t \in T$ and vanish at nodes in T that also lie in Γ_D. By "bilinear" we mean that, in each square, the function can be written uniquely as

$$a_0 + a_1 x + a_2 y + a_3 xy,$$

where the coefficients a_0, a_1, a_2, and a_3 depend on the particular square under consideration. Because the function is determined uniquely in each square by four degrees of freedom, it can also be determined uniquely by its values at the vertices of the square. Thus, the function is determined uniquely by its values at the nodes in T. Because it is linear along each edge, it is also continuous throughout the mesh, as required.

When T is a mesh of triangles, V_0 is the space of functions that are continuous in T and linear in each triangle $t \in T$ and vanish at nodes in T that also lie in Γ_D. In other words, in each triangle the function can be written uniquely as

$$a_0 + a_1 x + a_2 y,$$

where the coefficients a_0, a_1, and a_2 depend on the particular triangle under consideration. Because the function is determined uniquely in each triangle by three degrees of freedom, it can also be determined uniquely by its values at the vertices of the triangle. Thus, the function is determined uniquely by its values at the nodes in T. Because it is linear along each edge, it is also continuous throughout the mesh, as required.

Let us now define V_1. In fact, V_1 is defined in a similar way to V_0, except that the functions in V_1 no longer vanish in $T \cap \Gamma_D$ but rather agree with $\mathcal{F}_1$ there. More precisely, let v_1 be a particular function that is continuous in T, linear (or bilinear) in each finite element, and satisfies $v_1(i) = \mathcal{F}_1(i)$ for each node i in T that also lies in Γ_D. Then we define

$$V_1 = v_1 + V_0 = \{v_1 + v_0 \mid v_0 \in V_0\}.$$

This completes the definition of the finite-element discretization method.

Let us now construct the linear system (3.5) for the finite-element discretization method. For each node i in T, let ϕ_i be the so-called nodal basis function that is continuous in T, linear (or bilinear) in each individual finite element, and satisfies

$$\phi_i(j) = \begin{cases} 1 & \text{if } j = i \\ 0 & \text{if } j \in T, \; j \neq i. \end{cases} \tag{4.9}$$

Let N be the number of nodes in $T \setminus \Gamma_D$. Let A be the $N \times N$ matrix with elements defined by

$$a_{i,j} = a(\phi_j, \phi_i), \quad i, j \in T \setminus \Gamma_D. \tag{4.10}$$

Moreover, for every node $i \in T \setminus \Gamma_D$, define

$$b_i = f(\phi_i) - \sum_{j \in T \cap \Gamma_D} a(\phi_j, \phi_i) \mathcal{F}_1(j). \tag{4.11}$$

With these definitions, the solution $\tilde{u}$ of (4.8) can be written as

$$\tilde{u} = \sum_{j \in T \setminus \Gamma_D} x_j \phi_j + \sum_{j \in T \cap \Gamma_D} \mathcal{F}_1 \phi_j, \tag{4.12}$$

where the x_js are the components in the solution x of (3.5), in which the matrix A is as in (4.10), and the components in the right-hand side b are as in (4.11). Actually, we have from (4.9) and (4.12) that

$$x_j = \tilde{u}(j), \quad j \in T \setminus \Gamma_D.$$

Because of the assumptions used in the formulation in (4.6), A is SPD and, hence, has positive main-diagonal elements. This property allows the definition of the multigrid linear system solvers later in this book.

4.4 Bilinear Finite Elements

Let us first consider the case in which T is a mesh of squares as in Figure 4.1. Let us consider one of these squares as in Figure 4.2. Without loss of generality, assume that this square is the unit square $[0, 1] \times [0, 1]$. (Otherwise, the finite element could be mapped onto the unit square.) The four nodal basis functions are defined by:

$$\phi_{0,0}(x, y) = (1 - x)(1 - y)$$
$$\phi_{0,1}(x, y) = (1 - x)y$$
$$\phi_{1,0}(x, y) = x(1 - y)$$
$$\phi_{1,1}(x, y) = xy.$$

Each of these four nodal basis functions has the value 1 at one of the vertices of the square and 0 at the other three vertices. Because the nodal basis function is linear along each edge of the square, it can be extended continuously to the adjacent squares in the mesh T. This completes the definition of the nodal basis of V_0 in this case.

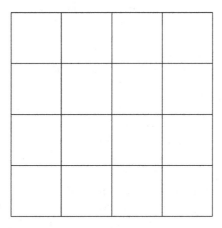

Fig. 4.1. The bilinear finite-element mesh.

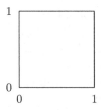

Fig. 4.2. The bilinear finite element.

When bilinear finite elements are used in (4.8), (4.6), (4.7), the linear system in (3.5) has the 9-coefficient stencil (or 9-point stencil)

$$\begin{bmatrix} NW & N & NE \\ W & C & E \\ SW & S & SE \end{bmatrix}.$$
(4.13)

(Do not confuse N that stands here for the Northern coefficient in the stencil with the integer N used often to denote the order of A.) Assuming that an $n \times n$ mesh is used, the kth equation in the linear system is written as

$$NEx_{k+n+1} + Nx_{k+n} + NWx_{k+n-1} + Ex_{k+1}$$
$$+ Cx_k + Wx_{k-1} + SEx_{k-n+1} + Sx_{k-n} + SWx_{k-n-1} = b_k,$$
(4.14)

where the coefficients NW, N, NE, W, C, E, SW, S, and SE depend on the particular node (i, j) in the mesh, and are contained in the kth row in A.

4.5 Triangulation

Uniform meshes are insufficient for most practical applications, where the domain may be nonrectangular and irregular. Although the bilinear finite-element mesh could actually use general quadrilaterals rather than squares, it is still not sufficiently flexible to approximate well a complicated domain. In particular, it is unsuitable for automatic refinement procedures that use local and adaptive refinement. This is why we turn our attention to mesh of triangles, or triangulation.

Let us show how the coefficient matrix A is constructed for a triangulation T. For this purpose, we need first to define the nodal basis functions ϕ_i for every node i in T. Let us consider first a right-angle triangle as in Figure 4.4. Clearly, for this triangle, the nodal basis functions are

$$\phi_{1,0} = x$$
$$\phi_{0,1} = y$$
$$\phi_{0,0} = 1 - x - y.$$

Indeed, each of these functions is indeed linear in the triangle and has the value 1 at one vertex and 0 at the other two vertices. This special triangle is called the *reference triangle r*, because it is used further to define the nodal basis functions in the entire mesh.

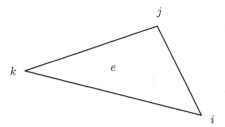

Fig. 4.3. A triangle e in the triangulation T.

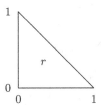

Fig. 4.4. The reference triangle r that is mapped onto each triangle e in T by the affine mapping $M_e = S_e + i$, where i is a vertex in e.

Consider a general triangle e in T as in Figure 4.3, with the vertices i, j, and k. Let M_e be the affine mapping that maps r onto e. In fact, one can write

$$M_e(x, y) = S_e \begin{pmatrix} x \\ y \end{pmatrix} + i,$$

where S_e is the 2×2 matrix that is the Jacobian of M_e. In fact, the first column in S_e is the two-dimensional vector $j - i$, and the second column is just $k - i$. This way, the vertices of r are indeed mapped to the vertices of e:

$$M_e(0, 0) = i$$
$$M_e(1, 0) = j$$
$$M_e(0, 1) = k.$$

The nodal basis functions are now defined in e by

$$\phi_i(x, y) = \phi_{0,0}(M_e^{-1}(x, y))$$
$$\phi_j(x, y) = \phi_{1,0}(M_e^{-1}(x, y))$$
$$\phi_k(x, y) = \phi_{0,1}(M_e^{-1}(x, y)).$$

Let us now use the above definitions to calculate the contribution from e to the element $A_{j,k}$ in A that couples the jth and kth nodes to each other. For this purpose, recall that, from the chain rule, we have

$$\nabla \phi_j(x, y) = S_e^{-t} \nabla \phi_{1,0}(M_e^{-1}(x, y))$$
$$\nabla \phi_k(x, y) = S_e^{-t} \nabla \phi_{0,1}(M_e^{-1}(x, y)).$$

Using (4.6) and (4.10), we have that the contribution from e to $a_{j,k}$ is

$$\int_e (\mathcal{D}(x,y)\nabla\phi_k) \cdot \nabla\phi_j dxdy + \int_e \beta(x,y)\phi_k\phi_j dxdy$$

$$= \int_r \left(\mathcal{D}(M_e(x,y))S_e^{-t}\nabla\phi_{0,1}\right) S_e^{-t}\nabla\phi_{1,0} dM_e(x,y)$$

$$+ \int_r \beta(M_e(x,y))\phi_{0,1}\phi_{1,0} dM_e(x,y)$$

$$= \int_r \left(\mathcal{D}(M_e(x,y))S_e^{-t}\begin{pmatrix}0\\1\end{pmatrix}\right) S_e^{-t}\begin{pmatrix}1\\0\end{pmatrix} |\det(S_e)| dxdy$$

$$+ \int_r \beta(M_e(x,y))\phi_{0,1}\phi_{1,0}|\det(S_e)| dxdy.$$

The contribution to $a_{j,k}$ from the triangle that lies on the other side of the edge leading from j to k is calculated in a similar way. If both j and k lie on Γ, though, then there is no such triangle; instead, one needs to add also the contribution from the boundary term in (4.6). The construction of A is also called *assembling*, because the contributions from the various triangles are assembled to construct each individual matrix element.

In the next section, we study the isotropic case, in which $\mathcal{D}$ is the identity matrix times a scalar function. For a sufficiently regular mesh, it is shown that A has some attractive properties, that are also used later in the book.

4.6 Diagonal Dominance in the Isotropic Case

Here we consider the isotropic case, in which the diffusion coefficients are the same in the x and y spatial directions. We show that, for a sufficiently regular mesh, the coefficient matrix A is a diagonally dominant M-matrix. These properties indicate that the discretization makes sense. Indeed, the M-matrix property means that A^{-1} has only nonnegative elements, and hence is suitable for approximating the positive Green function, the inverse of the original differential operator. Furthermore, the diagonal-dominance property indicates that the multigrid linear system solver should work well, as shown later in the book.

Assume that (4.6) is isotropic in the sense that

$$\mathcal{D} = \begin{pmatrix} \tilde{D} & 0 \\ 0 & \tilde{D} \end{pmatrix},$$

where $\tilde{D} \equiv \tilde{D}(x,y)$ is a bounded function in Ω [$\tilde{D} \in L_\infty(\Omega)$] satisfying $\tilde{D} \geq \tilde{D}_{\min}$ for some positive constant $\tilde{D}_{\min} > 0$. For simplicity, assume also that $\beta \equiv 0$ in (4.6), and that Dirichlet boundary conditions only are used ($\Gamma_D = \partial\Omega$, $\Gamma = \emptyset$). In this case, (4.6) takes the form

$$a(u,v) = \int_\Omega \tilde{D}\nabla u \cdot \nabla v d\Omega. \tag{4.15}$$

The coefficient matrix A obtained from a finite-element discretization of (4.15) is called the *stiffness matrix*. Let us calculate the value of a particular element in A.

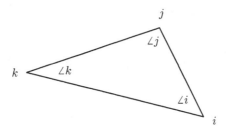

Fig. 4.5. Vertices and angles of a triangle.

Let e be some triangle in the triangulation T as in Figure 4.5, with vertices i, j, and k, and corresponding positive angles $\angle i$, $\angle j$, and $\angle k$. Assume also that the mesh is regular in the sense that the angles in the triangles are not too small or large. More specifically, assume that the angles are either acute or right:

$$\max(\angle i, \angle j, \angle k) \leq \pi/2. \tag{4.16}$$

Let $\triangle(e)$ denote the area of e. Let ϕ_i, ϕ_j, and ϕ_k be the nodal basis functions in e corresponding to i, j, and k, respectively. Let

$$\xi = (k - i)/\|k - i\|_2$$
$$\eta = \xi^{\perp}$$

be a pair of orthonormal vectors in the Cartesian plane R^2. Let $\mathcal{O}$ be the 2×2 orthogonal matrix, with the first column ξ and the second column η. Clearly, $\mathcal{O}$ maps the standard vector $(1,0)$ (the x co-ordinate) to ξ and $(0,1)$ (the y co-ordinate) to η. Furthermore, every function $f(x,y)$ can also be written as a function of ξ and η: $f(x(\xi,\eta), y(\xi,\eta))$. Using the chain rule, we have that the gradient with respect to ξ and η satisfies

$$\nabla_{\xi,\eta} = \begin{pmatrix} \partial/\partial\xi \\ \partial/\partial\eta \end{pmatrix} = \mathcal{O}^{-t} \begin{pmatrix} \partial/\partial x \\ \partial/\partial y \end{pmatrix} = \mathcal{O}\nabla.$$

Using the above, we have that the contribution from e to the element $a_{j,i}$ in A is

$$\int_e \tilde{D}\nabla\phi_i \cdot \nabla\phi_j dxdy = \int_e \tilde{D}\mathcal{O}\nabla\phi_i \cdot \mathcal{O}\nabla\phi_j dxdy$$

$$= \int_e \tilde{D}\nabla_{\xi,\eta}\phi_i \cdot \nabla_{\xi,\eta}\phi_j dxdy$$

$$= \int_e \tilde{D}(\phi_i)_\eta (\phi_j)_\eta dxdy$$

$$= -\frac{\cot(\angle k)}{\|k - i\|_2} \cdot \frac{\|k - i\|_2}{2\triangle e} \int_e \tilde{D}dxdy$$

$$= -\frac{\cot(\angle k)}{2\triangle e} \int_e \tilde{D}dxdy \leq -\cot(\angle k)\tilde{D}_{\min}/2. \tag{4.17}$$

This shows that A is an L-matrix. Now, the sum of the ith row in A (corresponding to a node $i \in T \setminus \Gamma_D$) is

$$\sum_{j \in T \setminus \Gamma_D} a(\phi_j, \phi_i) = a(\mathbf{1}, \phi_i) - \sum_{j \in T \cap \Gamma_D} a(\phi_j, \phi_i) = - \sum_{j \in T \cap \Gamma_D} a(\phi_j, \phi_i) \geq 0,$$

where $\mathbf{1}$ is the function of constant value 1 in T. This shows that A is also diagonally dominant. Thus, it follows from [118] that A is also an M-matrix.

In the following, we'll see that having a regular mesh with moderate angles is a great advantage not only for isotropic problems but also for anisotropic ones to guide and motivate the construction of suitable meshes.

4.7 Diagonal Dominance in the Anisotropic Case

In Section 3.9, we've seen that the mesh size in the strong-diffusion direction could actually be much larger than that in the weak-diffusion direction, with practically no effect on the discretization error. Here we use this principle also in the context of finite-element triangulation to stretch the triangles in the strong-diffusion direction and come up with a diagonally dominant M-matrix A.

Consider the PDE in (3.2) with the diffusion coefficients D_1 and D_2 as in Figure 4.6. The discretization method proposed in [38] (which is also supported by results from approximation theory) uses finite elements that are stretched in the strong-diffusion direction. Unfortunately, as discussed there, in order to preserve conformity also across lines of discontinuity in the diffusion coefficients, one must compromise regularity, that is, use degenerate triangles with very small angles. In Figure 4.7, we display a mesh that is both conformal and reasonably regular. More specifically, we use $\log_2 8 = 3$ layers of longer and longer triangles (along the discontinuity lines in the diffusion coefficients) to have a gradual pass from the isotropic area, where unstretched triangles are used, to the anisotropic area, where the triangles are stretched in the strong-diffusion direction. Once mapped to the isotropic coordinates [i.e., $(x, \tilde{y})$ in the lower-left subsquare and $(\tilde{x}, y)$ in the upper-right

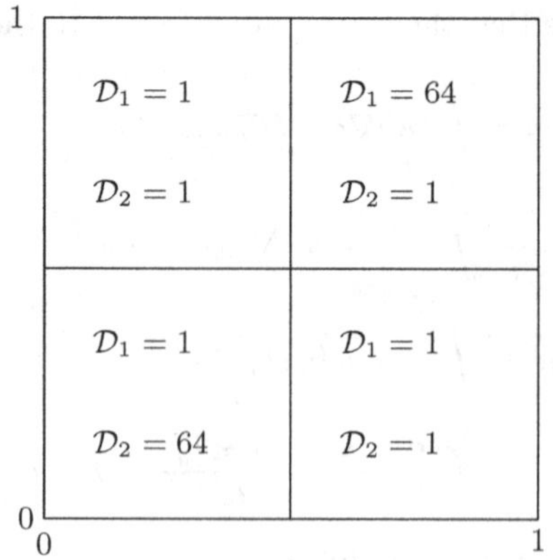

Fig. 4.6. The diffusion coefficients D_1 and D_2 for the anisotropic diffusion problem.

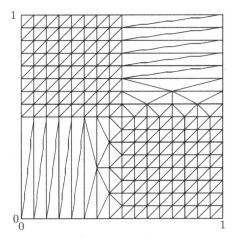

Fig. 4.7. The stretched finite-element mesh for the anisotropic diffusion equation.

subsquare, where $\tilde{x} = x/8$ and $\tilde{y} = y/8$], these triangles satisfy the conditions in Section 4.6, so the coefficient matrix A is indeed a diagonally dominant M-matrix.

The finite elements can be stretched in the strong-diffusion direction because the solution can change only slightly in this direction, so no high resolution is needed in this direction. This principle is also used in local refinement, as discussed below.

4.8 Locally Refined Meshes

The main advantage of the finite-element discretization method is in the opportunity to use finite elements of different shapes and sizes to approximate complicated domains. For this purpose, triangulations are most suitable. Indeed, one can use small triangles near the complicated and irregular boundary, where high resolution is necessary, and larger triangles elsewhere. Furthermore, if some idea about the behavior of the solution is known in advance, then the mesh can be constructed accordingly, using particularly small triangles where the solution is expected to vary sharply, and larger triangles elsewhere. This local refinement approach may save valuable computational resources by using high resolution only where necessary.

In some cases, the behavior of the solution can be deduced from the original boundary-value problem. In diffusion problems, for example, the solution may be irregular (i.e., nondifferentiable and with large variation) at corners in either the boundary of the domain or the discontinuity lines of the diffusion coefficients [112]. Local refinement in the neighborhood of these corners only can capture this irregular behavior, while keeping the number of nodes (and, hence, the computational cost) moderate.

The process of local refinement must be carried out automatically on the computer. For this purpose, a coarse mesh that provides a poor approximation is used to start. This mesh is then refined further in a refinement step or level, in which triangles are split and more edges and nodes are introduced at the neighborhoods of the places where irregularities are expected. This procedure is repeated in further refinement steps, each of which introduces more nodes, edges, and small triangles in smaller and smaller neighborhoods of the points of irregularity. The final (finest) mesh obtained from the final refinement step is then used as the required finite-element mesh that provides a good approximation to the original boundary-value problem.

4.9 The Refinement Step

Here we describe in detail a single refinement step applied to a given triangulation to improve it at some prescribed locations. Let S be a *triangulation* of Ω (a set of triangles whose interiors are disjoint from each other, whose union is a good approximation to Ω, and that are conformal in the sense that every two edge-sharing triangles share also the endpoints of this edge as their joint vertices). Assume that there exists some criterion that determines whether a triangle $s \in S$, should be refined (e.g., s should be refined if and only if the diffusion coefficient $\tilde{D}$ jumps in it in both the x and y spatial directions). The refinement step that produces the refined triangulation T from the current triangulation S is as follows:

Refinement Step 4.1

1. *Initialize T by $T = S$.*
2. *For every triangle $s \in S$ that should be refined according to the refinement criterion, connect the midpoints of its edges to each other, and include the four resulting triangles in T instead of s. (We then say that s has been fully refined.)*
3. *Repeat the following until it adds no more triangles to T.*
   - *Every triangle $s \in S$ that is also a triangle in T and currently has five or more nodes from T on its edges (its three vertices and two or more extra nodes that have been introduced in this or the former refinement steps on its edges) is refined fully as well.*
4. *If T is already sufficiently fine in terms of the refinement criterion and no more resolution is needed, then, for every triangle $s \in S$ that is also a triangle in T and currently has exactly four nodes from T on its edges (its three vertices and one extra node that has been introduced in this or the former refinement steps on one of its edges), connect this extra node to the opposite vertex in s, and include the two resulting triangles in T instead of s. (We then say that s has been half refined.) This guarantees that T is indeed conformal.*

This refinement step is illustrated in Figure 4.8. In practice, the refinement step is applied iteratively (with the substitution $S \leftarrow T$ after each iteration), until the desired resolution is achieved, and T is accepted as the required finite-element mesh.

There are also other possible refinement methods, such as those in [8] and [76]. Here is the refinement step used in a refinement method that uses only half refinement:

Refinement Step 4.2

1. *Initialize T by $T = S$.*
2. *For every triangle $s \in S$ that is also a triangle in T and should be refined according to the refinement criterion, do the following.*

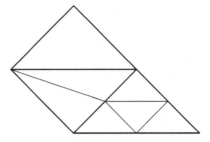

Fig. 4.8. The refined triangulation T resulting from the original triangulation S using Refinement Step 4.1.

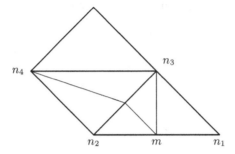

Fig. 4.9. The refined triangulation T resulting from the original triangulation S using Refinement Step 4.2. (It is assumed that the edge leading from n_1 to n_2 is refined before the edge leading from n_2 to n_3.)

a) *Connect the newest vertex of s (the vertex that has been created most recently in the refinement process) to the midpoint of the opposite edge in s.*

b) *Include the two resulting triangles in T instead of s. (We then say that s has been half refined.)*

c) *If there is a triangle $t \in T$ that also uses the edge that has just been split when s was half-refined, then it is half-refined as well in the same way: the midpoint of the edge that has just been split is connected to the opposite vertex in t, and the two resulting triangles are included in T instead of t.*

This refinement step is illustrated in Figure 4.9.

The above half-refinement of the triangle $s \in S$ uses the most newly created vertex in s. This choice increases regularity by avoiding a situation in which the same vertex is used over and over again in subsequent half refinements in subsequent refinement steps. Another possible choice that may also produce high regularity is using the vertex i in s with the maximal positive angle $\angle i$ in s. In fact, the refinement criterion can demand that a triangle $s \in S$ with an angle larger than $\pi/2$ should be refined as well to increase regularity.

As before, the Refinement Step 4.2 is repeated iteratively (with the substitution $S \leftarrow T$ after each iteration), and the mesh T at the end of the iteration is accepted as the sufficiently accurate finite-element mesh.

4.10 Adaptive Mesh Refinement

So far, we've assumed that a refinement criterion to determine whether a particular triangle $s \in S$ should be refined is available. Naturally, this criterion depends on properties of the particular application, such as the domain or the coefficients in the PDE. Clearly, this is a drawback: one would much prefer to have a general, robust, and automatic discretization method, that can be implemented once and for all. Furthermore, good refinement criteria are not always easy to deduce only from looking at the properties of the domain or PDE. A computational process that decides where to refine is clearly better.

Fortunately, such a computational process is available, and is called *adaptive refinement*. In this approach, the refinement step uses the numerical solution on the current mesh S to detect particular edges along which this numerical solution varies sharply. These edges are then split and the triangles in S that use them are half-refined, leading to the adaptively refined mesh T.

Thus, the refinement step scans not the triangles in S but rather the edges in it, to decide whether they should be split. The advantage of this approach is in the opportunity to treat boundary edges in a special way, to improve the approximation of complicated boundary segments.

The refinement step defined below uses some predetermined threshold, say 0.01.

Refinement Step 4.3

1. *Initialize T by $T = S$.*
2. *Use the triangulation S to define the coefficient matrix A from (4.10) and the right-hand side b from (4.11).*
3. *Solve (3.5) with the above A and b, and obtain the numerical solution $\tilde{u}(i) = x_i$ for every node i in S.*
4. *Let E be the set of edges in S.*
5. *Scan the edges in E one by one in some order. For every edge $e \in E$ encountered in this scanning, do the following.*
   - *Denote the endpoints of e by i and j.*
   - *If*

$$|\tilde{u}(i) - \tilde{u}(j)| \geq threshold, \tag{4.18}$$

   then do the following.
   a) *If there exist two triangles $t_i, t_2 \in T$ that share e as their joint edge in T, then do the following.*
      - *Half-refine t_1 by connecting the midpoint of e, $(i+j)/2$, to the opposite vertex in t_1 and including the two resulting triangles in T instead of t_1.*
      - *Half-refine t_2 by connecting the midpoint of e, $(i+j)/2$, to the opposite vertex in t_2 and including the two resulting triangles in T instead of t_2.*
   b) *If, on the other hand, there is only one triangle $t \in T$ that uses e as an edge in T, then we say that e is a boundary edge and do the following.*
      - *Denote the third vertex in t by k, so $t = \triangle(i, j, k)$.*

– If $(i + j)/2$ is in Ω, then connect k to $(i + j)/2$ and continue this line until it meets $\partial\Omega$ at a point m. Include in T the four triangles $\triangle(k, (i + j)/2), i,$ $\triangle(k, (i + j)/2), j,$ $\triangle(m, (i + j)/2), i,$ and $\triangle(m, (i + j)/2), j$ instead of t

– If, on the other hand, $(i + j)/2$ lies outside Ω, then connect k to $(i+j)/2$ and denote by l the point where this line meets $\partial\Omega$. Include in T the two triangles $\triangle(k, l, i)$ and $\triangle(k, l, j)$ instead of t.

This refinement step is illustrated in Figures 4.10 through 4.12 for a nonrectangular domain. As before, the refinement step is repeated iteratively (with the substitution $S \leftarrow T$ after each iteration), and the mesh T at the end of the iteration is accepted as the sufficiently accurate finite-element mesh.

The above adaptive-refinement algorithm requires the solution of a large, sparse, unstructured linear system of the form (3.5) in each refinement step. Efficient algorithms for the numerical solution of (3.5) are thus most important. This is the subject of the next part.

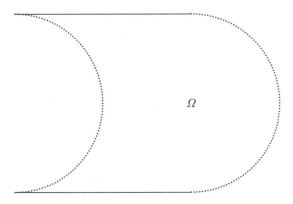

Fig. 4.10. An example of a domain Ω with boundary that is convex on the right and concave on the left.

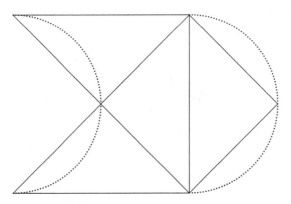

Fig. 4.11. The original mesh S that provides a poor approximation to Ω.

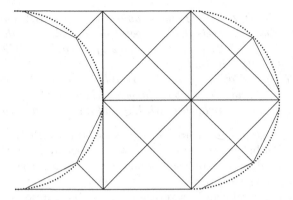

Fig. 4.12. The refined mesh T produced by Refinement Step 4.3. (It is assumed for simplicity that the refinement criterion holds for all the edges in S.)

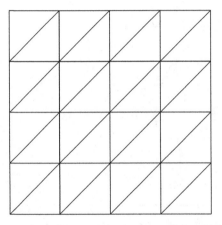

Fig. 4.13. Uniform triangulation of the square.

4.11 Exercises

1. Show that the bilinear form in (4.6) is symmetric.
2. Conclude that the coefficient matrix A resulting from the finite-difference discretization of the above bilinear form is symmetric.
3. Under the assumptions in Section 4.6, show that A is an irreducibly diagonally dominant L-matrix. (For the meaning of irreducibly diagonal dominance, see the exercises at the end of Chapter 3.)
4. Conclude from the theory in [118] that A is also nonsingular.
5. Conclude from Lemma 2.1 that A is also SPD.
6. Conclude from the theory in [118] that A is also an M-matrix.
7. Conclude from the theory in [118] that A has a unique eigenvector with positive components and positive eigenvalue, which is the minimal eigenvalue of A.

8. Compute the stiffness matrix A for the Poisson equation in a square on a uniform triangulation as in Figure 4.13. Consider various possible kinds of boundary conditions. Show that the stencil obtained is a 5-point stencil, and that it is the same as in the finite-difference and finite-volume schemes, at least at the interior of the grid.

9. Use an IMSL routine to compute the nearly singular eigenvector and its corresponding eigenvalue for the above matrix. Verify that the eigenvalue and components of the nearly singular eigenvector are indeed positive. Verify that, with Dirichlet boundary conditions, the nearly-singular eigenvector is just $v^{(1,1)}$ of Chapter 2.4.

10. Compute the stiffness matrix A for the Poisson equation in a square with Dirichlet boundary conditions on a uniform bilinear finite-element mesh as in Section 4.4. Show that the nearly singular eigenvector of A is the 2-D Fourier (sine) vector $v^{(1,1)}$ in Section 2.4. Use also an IMSL routine to verify this result numerically.

11. Show that the weak formulation of the diffusion problem in (4.6) is well-posed in the sense that it has a unique solution with derivatives in $L_2(\Omega)$. The solution can be found in Chapter 11 in [103].

12. Write a general code that computes the stiffness matrix A for a general triangulation. The solution can be found in Chapter 13 in [103].

Part III

The Numerical Solution
of Large Sparse Linear Systems
of Algebraic Equations

The discretization method produces large linear systems of algebraic equations that couple the values at the gridpoints (or mesh nodes) to each other. The numerical solution of the linear systems also provides the numerical solution to the original boundary-value problem. It is thus most important to have efficient algorithms to solve these algebraic systems, that is, efficient linear system solvers.

The coefficient matrix A in the linear system (3.5) is in general sparse in the sense that most of the elements in it are zero. An efficient linear system solver should exploit this property and avoid introducing too many new nonzero elements (fill-in). Unfortunately, the traditional Gauss elimination (LU factorization) for solving (3.5) introduces massive fill-in, and is thus highly inefficient in terms of both time and storage. Furthermore, Gauss elimination is essentially sequential, and cannot be implemented efficiently on parallel computers.

Iterative linear system solvers, on the other hand, introduce no or little fill-in, and may also be parallelized efficiently. In the first chapter in this part (Chapter 5), we start from relaxation methods, in which the unknowns are updated one by one. In each such relaxation sweep, the algebraic error is slightly reduced. When the sweeps are repeated iteratively, the algebraic error is reduced significantly, and the current values at the unknowns can be accepted as the required values in the vector of unknowns x in (3.5).

Relaxation methods are particularly efficient in removing the high-frequency modes in the algebraic error, but not in removing the low-frequency ones, which vary only slightly from one gridpoint to the next one. In order to annihilate the low-frequency error modes as well, the equation should be transferred to a coarser grid, on which the low-frequency error modes are still well approximated and can be removed more efficiently. To this end, however, one must have a good coarse-grid approximation to the original system and also good transfer operators to transfer information between the various grids. These components are supplied in the multi-grid linear system solvers described in the second chapter in this part (Chapter 6).

The multigrid method can thus be interpreted as an iterative method that uses relaxation on the fine grid until it is no longer efficient because the high-frequency error modes have already been annihilated; then, the equation is transferred to the next coarser grid, where relaxation is still efficient in reducing the lower frequencies that can still be observed on this grid. Thus, multigrid can also be thought of as an acceleration technique for the basic relaxation method. In this point of view, once the relaxation has been exhausted on the fine grid, it is reused on a coarser grid, where it is still useful. (The relaxation on the coarse grid is applied to the residual equation, in order to avoid enhancing error modes that have already been reduced on the fine grid.) Furthermore, once the relaxation has been exhausted on the coarse grid, it is reused on a yet coarser grid, and so on. Once the relaxation has been used on the entire hierarchy of the coarser and coarser grids, we say that one multigrid iteration (or V-cycle) has been completed. Since every algebraic error mode has been reduced substantially on some grid, the multigrid iteration should converge rapidly to the algebraic solution x of (3.5).

5

Iterative Linear System Solvers

In this chapter, we present several iterative methods for the solution of large sparse linear systems of algebraic equations. We start with relaxation methods such as point and block Jacobi and Gauss–Seidel (GS) and Kacmarz versions, proceed to conjugate-gradient type acceleration methods, and conclude with incomplete LU (ILU) versions.

5.1 Iterative Sparse Linear System Solvers

As we have seen above, large sparse linear systems of equations arise often from the discretization of PDEs. Having efficient and robust algorithms for solving these systems is thus most important.

Unfortunately, direct methods such as the traditional Gauss elimination method are inefficient for large sparse linear systems because of the large amount of fill-in that is introduced in the LU factorization. In fact, for both (3.7) and (4.14) this method requires $O(n^4)$ time and $O(n^3)$ storage units. Iterative methods, on the other hand, usually require only $O(n^2)$ storage units, and are thus much more attractive. Furthermore, they are usually much more efficient also in terms of time, particularly when parallel implementation is considered.

In iterative methods, one first makes an initial guess $\tilde{x}$ to approximate the solution x of the linear system (3.5); then, this initial guess is successively improved in the iteration, until it (one hopes) converges to the true solution x. Each iteration can be formulated as

$$\tilde{x} \leftarrow \tilde{x} + \mathcal{P}^{-1}(b - A\tilde{x}), \tag{5.1}$$

where "$\leftarrow$" stands for substitution, and $\mathcal{P}$, the preconditioning matrix or preconditioner, is an easily invertible matrix that approximates A in a spectral sense, as discussed later. By "invertible" we mean here that it is rather easy to solve a system of the form

$$\mathcal{P}e = r$$

(where r is a given vector and e is the vector of unknowns); of course, $\mathcal{P}^{-1}$ is never calculated explicitly, because this may be more expensive than the entire solution of the original system (3.5).

In iterative methods, it is important to estimate the error $\tilde{x} - x$. In this respect, the iteration matrix

$$\mathcal{M} = I - \mathcal{P}^{-1} A$$

(where I is the identity matrix of the same order as A) is particularly useful. This matrix governs the reduction of error in each iteration. Indeed, from (5.1) we have

$$\tilde{x} - x \leftarrow \mathcal{M}(\tilde{x} - x).$$

Thus, the l_2-norm of the iteration matrix may provide a good indication for the rate of convergence. When the number of iterations is large, the rate of convergence can actually be estimated asymptotically by the spectral radius $\rho(\mathcal{M})$ of the iteration matrix $\mathcal{M}$.

In the following, we describe several iterative methods in more concrete terms.

5.2 Relaxation Methods

An important class of iterative methods is the class of relaxation methods. In a relaxation, the vector $\tilde{x}$ that approximates the solution x is improved by scanning its components one by one and updating each of them to reduce the residual at the corresponding equation in the linear system. Some common relaxation methods are described next.

5.3 The Jacobi Relaxation Method

In the point-Jacobi relaxation method, all the components in $\tilde{x}$ are updated simultaneously by

$$\tilde{x} \leftarrow \tilde{x} + diag(A)^{-1}(b - A\tilde{x}).$$

In other words, the point-Jacobi iteration is obtained by setting the preconditioner $\mathcal{P}$ to be

$$\mathcal{P} = diag(A).$$

It is well known that the Jacobi iteration converges whenever A is an M-matrix or a diagonally dominant matrix [118]. However, the convergence may be extremely slow. For the Poisson equation, for example, the spectral radius of the iteration matrix is as large as

$$\rho(I - diag(A)^{-1}A) = 1 - 2\sin^2(\pi h/2),$$

where $h = 1/(n+1)$ is the meshsize. Therefore, the number of iterations required to reduce the l_2 norm of the error by a constant factor is $O(h^{-2})$, which is prohibitively large. Still, the Jacobi iteration can be useful thanks to its efficient parallel implementation.

5.4 The Damped Jacobi Relaxation Method

In some cases, however, one might want to use a damping factor $0 < \sigma \leq 1$ to prevent divergence:

$$\tilde{x} \leftarrow \tilde{x} + \sigma \cdot diag(A)^{-1}(b - A\tilde{x}),$$

or

$$\mathcal{P} = \sigma^{-1} diag(A).$$

This is called the damped Jacobi iteration.

5.5 The Block Jacobi Relaxation Method

The Jacobi iteration has also a "block" version:

$$\tilde{x} \leftarrow \tilde{x} + \sigma \cdot blockdiag(A)^{-1}(b - A\tilde{x}),$$

or

$$\mathcal{P} = \sigma^{-1} blockdiag(A).$$

Here "$blockdiag(A)$" is the matrix that has the same elements as A inside a chain of blocks located along its main diagonal and zeroes elsewhere. This iterative method usually converges slightly better than the above point Jacobi iteration, at the expense of inverting the individual blocks in "$blockdiag(A)$" in each iteration.

5.6 The Gauss–Seidel Relaxation Method

In the point-Jacobi relaxation, all unknowns are relaxed simultaneously. Somewhat better convergence rates can be obtained by the point-*Gauss–Seidel (GS) relaxation method* described next.

In the point-GS relaxation, the components of $\tilde{x}$ are updated one by one: for $k = 1, 2, 3, \ldots, n^2$, the unknown component $\tilde{x}_k$ is updated by

$$\tilde{x}_k \leftarrow \tilde{x}_k + (A_{k,k})^{-1} (b - A\tilde{x})_k. \tag{5.2}$$

The subscript "k" in (5.2) corresponds to the kth unknown in the vector of unknowns. Once all the unknowns are updated in this way, we say that the GS relaxation (or sweep, or iteration) is complete.

Since the unknowns are relaxed in the usual order, the point-GS iteration can be also formulated with the preconditioner being the lower triangular part of A, denoted by $\tilde{L}$:

$$\mathcal{P} = \tilde{L} \quad \text{and} \quad \mathcal{M} = I - \tilde{L}^{-1}A. \tag{5.3}$$

It is well known that the point-GS iteration converges to x whenever A is diagonally dominant or SPD [118]. However, the convergence is still rather slow, and $O(h^{-2})$ iterations are required even for the Poisson equation.

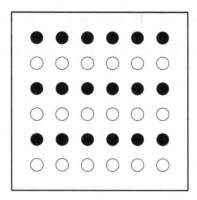

Fig. 5.1. The "zebra" coloring: odd-numbered lines are colored white, and even-numbered lines are colored black.

5.7 The Block–Gauss–Seidel Relaxation Method

In the block-GS version, the subscript "k" in (5.2) corresponds to the kth block of unknowns in $\tilde{x}$. The preconditioner for this iteration is

$$\mathcal{P} = \tilde{L},$$

where $\tilde{L}$ is now interpreted as the lower block triangular part of A. The rate of convergence is usually better than in the point-GS iteration, at the expense of inverting the block submatrices on the diagonal of $\tilde{L}$ in each iteration.

The blocks in the block-GS iteration can be chosen in different ways. For example, when the above blocks correspond to lines of gridpoints, we obtain the line-GS relaxation method. When odd-numbered lines are relaxed before even-numbered lines, we obtain the "zebra" line-GS relaxation method (Figure 5.1).

The lines in the line-GS relaxation can be either in the x- or in the y-direction in the Euclidean plane where the grid is embedded. The x-line relaxation uses lines in the x spatial direction, whereas the y-line relaxation uses lines in the y spatial direction. The alternating line relaxation is combined of two relaxation sweeps: an x-line relaxation followed by a y-line relaxation.

Similarly, the "zebra" line relaxation can also use lines in either the x or the y spatial direction. The alternating "zebra" line relaxation is composed of two relaxation sweeps: first a "zebra" relaxation with lines in the x spatial direction, and then a "zebra" relaxation with lines in the y spatial direction.

Note that the point-GS relaxation can actually be viewed as a special case of the block relaxation, with trivial blocks of order 1.

5.8 Reordering by Colors

In the point-Gauss–Seidel iteration (5.2), the unknowns must be relaxed in the standard order. The relaxation of a particular unknown depends on the updated values of unknowns that have been relaxed before. The algorithm is thus essentially sequential, and cannot be implemented efficiently in parallel.

In the first edition of this book, there is an attempt to define a parallelizable point-GS version. Unfortunately, it turns out that this version may be parallelizable only for banded matrices, but not in general.

Fortunately, parallelizable point-GS versions can be designed by using a special order of unknowns. In fact, the most attractive order is the order by colors, which allows parallel relaxation in subsets of unknowns. This kind of relaxation is referred to as "colored" relaxation; here a "color" is a maximal subset of unknowns that are decoupled from each other in A and, hence, can be relaxed simultaneously in parallel.

Here is how the colored relaxation is carried out. The original set of unknowns is split into disjoint subsets ("colors") and then relaxed color by color. Each color can be relaxed simultaneously in parallel, because the unknowns in it are decoupled from each other in A. For example, for the 5-point stencil (3.7), the red-black coloring that uses two colors as in a checkerboard provides a well-parallelizable relaxation method. The red color r is the index subset

$$r = \{(i,j) \in g \mid i+j \equiv 0 \mod 2\}, \qquad (5.4)$$

and the black color b [not to be confused with the right-hand side in (3.5)] is the index subset

$$b = \{(i,j) \in g \mid i+j \equiv 1 \mod 2\} \qquad (5.5)$$

(Figure 5.2). For 9-point stencils as in (4.13), 4-color methods as in [1] [95] are suitable. For example, these four colors can be the four index subsets in (5.8) below.

The following theorem shows that the colored relaxation is as good as the original one in terms of asymptotic convergence rate.

Theorem 5.1 *Let $\mathcal{M}$ be the iteration matrix for the point-GS method applied to a particular linear system with the coefficient matrix A. Then there exists a reordering by colors for which the iteration matrix has the same spectrum as $\mathcal{M}$. More specifically, when A is tridiagonal and the original order is the standard order, this coloring is just the odd-even splitting. Furthermore, when A corresponds to the 5-point stencil and the original order is the usual lexicographical order, this coloring is just the red-black (checkerboard) coloring.*

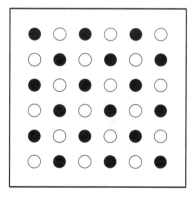

Fig. 5.2. The red-black coloring: the circles are relaxed first, and the bullets are relaxed last.

For the proof, see [1] and [95].

Let us now apply Theorem 5.1 to an example, which is also useful in Section 10.4 below. Recall that, for the point-GS iteration, the preconditioner is just the lower triangular part of A [see (5.3)]. Consider the 5-point stencil in (3.8). For this stencil, the coloring in Theorem 5.1 is just the above red-black coloring. Let r denote the set of red points and b [not to be confused with the right-hand side in (3.5)] the set of black points. With this partitioning, A has the block form

$$A = \begin{pmatrix} A_{rr} & A_{rb} \\ A_{br} & A_{bb} \end{pmatrix}.$$

Note that because the unknowns in r are decoupled from each other in A, A_{rr} is diagonal. The same is true also for the black color b, so A_{bb} is diagonal as well. These properties allow simultaneous relaxation within each color, as discussed below.

Let us first consider the first "leg" of the red-black relaxation, in which only the red points are relaxed. The inverse of the preconditioner for this leg is thus

$$(\mathcal{P}_r)^{-1} = \begin{pmatrix} (A_{rr})^{-1} & 0 \\ 0 & 0 \end{pmatrix}.$$

Therefore, the iteration matrix for the first leg is

$$S_r = I - (\mathcal{P}_r)^{-1} A = \begin{pmatrix} 0 & -(A_{rr})^{-1} A_{rb} \\ 0 & I \end{pmatrix}. \tag{5.6}$$

Similarly, in the second "leg" of the red-black relaxation only black points are relaxed. The inverse of the preconditioner for this leg is, thus,

$$(\mathcal{P}_b)^{-1} = \begin{pmatrix} 0 & 0 \\ 0 & (A_{bb})^{-1} \end{pmatrix},$$

and, hence, the iteration matrix for the second leg is

$$S_b = I - (\mathcal{P}_b)^{-1} A = \begin{pmatrix} I & 0 \\ -(A_{bb})^{-1} A_{br} & 0 \end{pmatrix}. \tag{5.7}$$

The iteration matrix for the entire red-black relaxation is, thus,

$$
\begin{aligned}
S &= S_b S_r \\
&= \begin{pmatrix} I & 0 \\ -(A_{bb})^{-1} A_{br} & 0 \end{pmatrix} \begin{pmatrix} 0 & -(A_{rr})^{-1} A_{rb} \\ 0 & I \end{pmatrix} \\
&= \begin{pmatrix} 0 & -(A_{rr})^{-1} A_{rb} \\ 0 & (A_{bb})^{-1} A_{br} (A_{rr})^{-1} A_{rb} \end{pmatrix}.
\end{aligned}
$$

Alternatively, the iteration matrix for the red-black relaxation can be computed also as a whole, without partitioning into red and black "legs." Indeed, the preconditioner for the red-black relaxation is

$$\mathcal{P} = \begin{pmatrix} A_{rr} & 0 \\ A_{br} & A_{bb} \end{pmatrix},$$

so its inverse is

$$\mathcal{P}^{-1} = \begin{pmatrix} (A_{rr})^{-1} & 0 \\ -(A_{bb})^{-1}A_{br}(A_{rr})^{-1} & (A_{bb})^{-1} \end{pmatrix}.$$

Therefore, the iteration matrix for the red-black relaxation is

$$S = I - \mathcal{P}^{-1}A$$

$$= I - \mathcal{P}^{-1}\left(\mathcal{P} + \begin{pmatrix} 0 & A_{rb} \\ 0 & 0 \end{pmatrix}\right)$$

$$= -\mathcal{P}^{-1}\begin{pmatrix} 0 & A_{rb} \\ 0 & 0 \end{pmatrix}$$

$$= \begin{pmatrix} 0 & -(A_{rr})^{-1}A_{rb} \\ 0 & (A_{bb})^{-1}A_{br}(A_{rr})^{-1}A_{rb} \end{pmatrix},$$

which is exactly the same as calculated before.

Assume now that A is lower triangular, for example, A represents the upwind discretization (see Section 3.11) of a convection equation as in (3.36) with no diffusion ($\varepsilon = 0$) and constant coefficients a_1 and a_2 there (straight entering flow). We assume that the unknowns in the linear system have been ordered in advance in such a way that A is indeed lower triangular. (We refer to this order as the *downwind* or *downstream order*, whereas the reversed order for which A is upper triangular is referred to as the *upwind* or *upstream order*.) From (5.3), the iteration matrix for the point-GS method is just the zero matrix. In this case, we say that the relaxation is done in the "downstream" direction, so convergence is achieved in one iteration only. From Theorem 5.1, the red-black relaxation must also have the zero spectrum. From the representation of the iteration matrix S above, one can see that it indeed has a zero spectrum, as its lower-right block is strictly lower triangular. Thus, the theorem indeed predicts what the above calculations reveal.

One may ask: why bother with the red-black ordering, when the standard ordering produces relaxation that converges immediately? The answer is that this only happens here by chance, because the unknowns have fortunately been ordered properly. In general, one needs to write an algorithm that will work well for any given matrix A, may it be triangular or not. Although the standard order works well for the above example, it would be rather inefficient for most other cases, including cases in which A is upper triangular or is not triangular at all. Thus, in general, the red-black ordering is preferable thanks to its efficient parallel implementation.

5.9 Four-Color Reordering

The red-black reordering is suitable for a 5-point stencil: since the points in the first subgrid r, are decoupled from each other, they can be relaxed simultaneously in parallel. The same is also true for the second subgrid, b. However, when a 9-point stencil as in (4.13) is encountered, this reordering is no longer sufficient; one should switch to a four-color reordering.

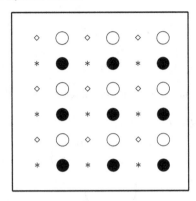

Fig. 5.3. Four-color relaxation. The asterisks are relaxed in the first stage, the bullets are relaxed in the second stage, the diamonds are relaxed in the third stage, and the circles are relaxed in the fourth stage.

The four-color partitioning uses the following four subgrids:

$$
\begin{aligned}
c_{0,0} &= \{(i,j) \in g \mid i \equiv j \equiv 0 \bmod 2\} \\
c_{0,1} &= \{(i,j) \in g \mid i \equiv j+1 \equiv 0 \bmod 2\} \\
c_{1,0} &= \{(i,j) \in g \mid i+1 \equiv j \equiv 0 \bmod 2\} \\
c_{1,1} &= \{(i,j) \in g \mid i \equiv j \equiv 1 \bmod 2\}
\end{aligned}
\tag{5.8}
$$

(Figure 5.3). Each of these subgrids can be thought of as colored by a different color; thus, each subgrid is also referred to as "color." Furthermore, the points in each subgrid are decoupled from each other in the 9-point stencil, and hence can be relaxed simultaneously in parallel.

Thus, the four-color reordering is defined by placing the first color ($c_{0,0}$) first, then the second color ($c_{1,0}$), then the third color ($c_{0,1}$), and finally the fourth color ($c_{1,1}$). The four-color point relaxation is actually a point-GS relaxation that uses this order. From [95], it follows that the iteration matrix of this relaxation method has the same spectrum as the one that uses the Cauchy (diagonal-by-diagonal) order in g. Thanks to its high degree of parallelism, the four-color relaxation is commonly used for 9-point stencils. In this book, it is often used within the multigrid cycle.

5.10 Cache-Oriented Reordering

In modern computers, the processors may be so strong that computation is no longer the most time-consuming part in the solution process. Accessing the memory may be much more time consuming, and minimizing it may be more immediate in the design of algorithms and implementations. In this respect, the cache memory is particularly useful. The access to it is relatively inexpensive, therefore one can put in it a certain amount of data (usually, a 32 × 32 or 64 × 64 array of floating-point numbers) and make the most of it as long as it is available in the cache. Domain decomposition algorithms that solve local subproblems are particularly suitable for cache-oriented implementation because the data about a particular subdomain that are placed in the cache are used fully to solve the subproblem in that subdomain. Block-relaxation methods that solve block subproblems are also attractive in terms of cache use.

Point-relaxation methods, on the other hand, are not so attractive in terms of cache access because a datum that is brought from the memory to the cache can be used only for the current relaxation before it is returned to the memory. It seems rather inefficient to bring a datum from the memory only for one relaxation. A better idea would be to use the datum that is already in the cache for as many relaxations as possible. This, however, would require a slight reordering [47].

In Figures 5.4–5.6 we illustrate the cache-oriented point relaxation. We assume a rectangular uniform grid, and decompose it into subsquares or subdomains. We also imagine a z-axis perpendicular to the domain, with z being the iteration number. One can imagine layers that are put on the grid one by one, each representing an approximate solution obtained from a point-relaxation applied to the previous approximation represented by the layer underneath it. In cache-oriented relaxation, once the data corresponding to a particular subdomain are brought from the memory to the cache, they are not only used for the current relaxation but also for subsequent relaxations, at least at those gridpoints where the information required to carry out more relaxations is available. In the present example, we assume that A is from 9-point stencil as in (4.13). The data required for relaxing a particular grid-point are, thus, the data from the previous iteration at the eight points surrounding it. The relaxation process can be thought of as building with Lego bricks with the rule that a brick must be supported by at least nine bricks underneath it.

Our aim is to build as many layers as possible in a particular subdomain without violating the above rule; this would make the most of the data about this subdomain already available in the cache. The first step is illustrated in Figure 5.4: a pyramid is built in each subdomain, based solely on the data in the lowest layer in this

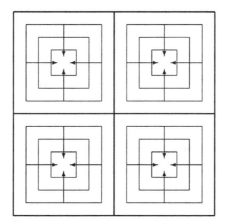

Fig. 5.4. The first step in the cache-oriented relaxation. In each of the four subsquares, the points are relaxed layer by layer. Each layer is smaller than the one underneath it, so four pyramids are built in the four subsquares. The arrows show the order in which layers are put. The height of a layer corresponds to the iteration number: the lowest layer corresponds to the initial guess, the next layer above it corresponds to the first iteration, and so on.

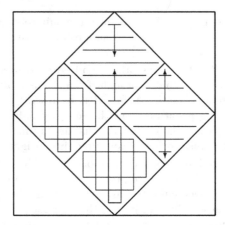

Fig. 5.5. The second step in the cache-oriented relaxation that fills the "holes" between adjacent pyramids. In each of the four subsquares, the points are relaxed layer by layer, where the lines indicate the boundary of the layer. (In the right and upper subsquares, only the horizontal boundaries are indicated.) The lowest layer corresponds to the initial guess, the next layer above it corresponds to the first iteration, and so on.

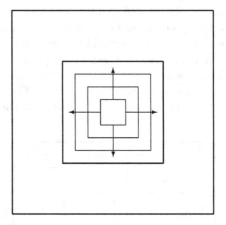

Fig. 5.6. The third and final step in the cache-oriented relaxation. An upside-down pyramid is built in the middle of the square. The layers are put in the order shown by the arrows. Each layer is larger than the one underneath it. This completes $m/2$ point-GS iterations, where m^2 is the number of points in a subsquare.

subdomain. The second step is illustrated in Figure 5.5: the "holes" in between adjacent pyramids are filled. The final step is illustrated in Figure 5.6: the "upside-down pyramid" left in the middle of the domain is also filled. Thus, at the end of the process, $m/2$ layers are constructed (where m^2 is the number of points in a subdomain). In other words, $m/2$ consecutive relaxations are done, using only few memory accesses. The extension of the algorithm in Figures 5.4 through 5.6 to more subsquares is straightforward. The cache-oriented relaxation is particularly suitable to parallel implementations, where the subdomains can be processed simultaneously in parallel.

5.11 Symmetric Gauss–Seidel Relaxation

Basic iterative methods could be prohibitively slow in terms of iteration count. Fortunately, they can often be accelerated by acceleration techniques such as the preconditioned conjugate gradient method. However, this method requires that both the coefficient matrix A and preconditioner $\mathcal{P}$ are SPD. It is thus important to construct iterative methods with a preconditioner that is SPD whenever the coefficient matrix is.

Such an iterative method is the symmetric point- (or block-) GS iteration. One iteration of this method consists of two consecutive relaxations: a standard GS relaxation, followed by another GS relaxation in the reverse order. In the following, we prove that the preconditioner for this iteration is indeed SPD whenever A is.

Let us denote the lower- (block-) triangular part of A by $\tilde{L}$, the upper- (block-) triangular part of A by $\tilde{U}$, and the (block-) diagonal part of A by $\tilde{D}$. With this notation, we have

$$A = \tilde{L} + \tilde{U} - \tilde{D}.$$

The preconditioner and iteration matrix for the GS iteration are, thus,

$$\mathcal{P} = \tilde{L} \quad \text{and} \quad \mathcal{M} = I - \tilde{L}^{-1}A.$$

When the unknowns (or blocks) are relaxed in the reverse order, we obtain the reverse GS relaxation, whose preconditioner and iteration matrix are:

$$\mathcal{P} = \tilde{U} \quad \text{and} \quad \mathcal{M} = I - \tilde{U}^{-1}A.$$

One symmetric GS iteration consists of a standard GS iteration followed by a reverse GS iteration. The iteration matrix for the symmetric GS method is thus

$$\mathcal{M} = \left(I - \tilde{U}^{-1}A\right)\left(I - \tilde{L}^{-1}A\right).$$

The inverse of the preconditioner of the symmetric GS method can thus be written as follows.

$$
\begin{aligned}
\mathcal{P}^{-1} &= (I - \mathcal{M})A^{-1} \\
&= \left(\tilde{U}^{-1}A + \tilde{L}^{-1}A - \tilde{U}^{-1}A\tilde{L}^{-1}A\right)A^{-1} \\
&= \tilde{U}^{-1} + \tilde{L}^{-1} - \tilde{U}^{-1}A\tilde{L}^{-1} \\
&= \tilde{U}^{-1} + \tilde{L}^{-1} - \tilde{U}^{-1}(\tilde{U} + \tilde{L} - \tilde{D})\tilde{L}^{-1} \\
&= \tilde{U}^{-1}\tilde{D}\tilde{L}^{-1}.
\end{aligned}
\tag{5.9}
$$

This representation yields the following standard lemma, which allows the use of the preconditioned conjugate gradient method in Section 5.12 below.

Lemma 5.1 *If A is symmetric (respectively, SPD), then the preconditioner for the symmetric GS method is symmetric (respectively, SPD) as well.*

Proof. Since A is symmetric, we have that

$$\mathcal{L} = \mathcal{U}^t.$$

Since A is symmetric (respectively, SPD), $\tilde{D}$ is symmetric (respectively, SPD) as well. The lemma follows now from (5.9). This completes the proof of the lemma.

5.12 The Preconditioned Conjugate Gradient Method

In this section, we describe the preconditioned conjugate gradient (PCG) method, which may accelerate the convergence of the basic iterative method. When accelerated by PCG, the basic iterative method actually serves as a preconditioner that reduces the amount of singularity in the original coefficient matrix and makes it more suitable for the conjugate gradient iteration.

Unfortunately, the PCG acceleration is limited to cases in which both the coefficient matrix A and the preconditioner $\mathcal{P}$ are SPD. In more general cases, other Lanczos-type acceleration methods should be used instead.

The amount of singularity in A is measured by its condition number. The smaller the condition number, the more regular A is, and the faster the PCG iteration converges to x.

For a matrix M that is SPD with respect to some inner product, define its condition number by

$$\kappa(M) = \rho(M)\rho(M^{-1}). \tag{5.10}$$

It is well known that the convergence rate of the Conjugate Gradient (CG) method in [62] for SPD problems is inversely proportional to $\kappa(A)$. For problems with large $\kappa(A)$, the convergence may thus be prohibitively slow. Fortunately, the convergence may improve considerably by using PCG with a suitable SPD preconditioner $\mathcal{P}$. Actually, PCG is obtained by applying the CG method to the preconditioned equation

$$\mathcal{P}^{-1}Ax = \mathcal{P}^{-1}b, \tag{5.11}$$

with the only change that the usual inner product $(\cdot, \cdot)_2$ used in CG is replaced by the induced inner product $(\cdot, \cdot)_{\mathcal{P}}$. With respect to this inner product, both $\mathcal{P}^{-1}A$ and $A^{-1}\mathcal{P}$ are SPD (see Lemma 2.9); therefore, we have (see also Lemma 12.7 below)

$$\kappa(\mathcal{P}^{-1}A) = \|\mathcal{P}^{-1}A\|_{\mathcal{P}}\|A^{-1}\mathcal{P}\|_{\mathcal{P}}. \tag{5.12}$$

The preconditioner $\mathcal{P}$ should thus be both easily invertible and close to A in terms of spectral analysis, especially with regard to the nearly singular eigenvectors (eigenvectors that correspond to nearly zero eigenvalues). This property would guarantee that $\kappa(\mathcal{P}^{-1}A)$ is indeed considerably smaller than $\kappa(A)$, and, hence, PCG converges much more rapidly than CG.

Let us now describe the PCG iteration in some more detail. The iteration is set to converge to six orders of magnitude accuracy.

Algorithm 5.1

1. *Let $\tilde{x}$ be the initial-guess vector.*
2. *Let $\acute{x}$ be the result of an application of the iterative method to $\tilde{x}$:*

$$\acute{x} = \tilde{x} + \mathcal{P}^{-1}(b - A\tilde{x}).$$

3. *Define also the residual vector*

$$r = b - A\tilde{x}.$$

4. *Define also the preconditioned residual vector*

$$\tilde{r} = \acute{x} - \tilde{x} = \mathcal{P}^{-1}(b - A\tilde{x}).$$

5. *Initialize also the vectors $p = r$ and $\tilde{p} = \tilde{r}$. ($\tilde{p}$ will serve as a direction vector to improve the guess $\tilde{x}$.)*
6. *Define also the scalars $\gamma_0 = \gamma = (r, \tilde{r})_2$.*
7. *Apply the iterative method to $\tilde{p}$ with zero right-hand side:*

$$\acute{p} = \tilde{p} - \mathcal{P}^{-1}A\tilde{p}.$$

8. *Compute the vector*

$$\tilde{w} = \tilde{p} - \acute{p} = \mathcal{P}^{-1}A\tilde{p}.$$

   Compute also the vector

$$w = A\tilde{p}.$$

   Compute also the scalar

$$\alpha = \gamma / (\tilde{p}, w)_2.$$

9. *Update the approximate solution $\tilde{x}$ by*

$$\tilde{x} \leftarrow \tilde{x} + \alpha\tilde{p}.$$

10. *Update also the residual by*

$$r \leftarrow r - \alpha w.$$

11. *Update also the preconditioned residual by*

$$\tilde{r} \leftarrow \tilde{r} - \alpha\tilde{w}.$$

12. *Compute also the scalar*

$$\beta = (r, \tilde{r})_2 / \gamma.$$

13. *Update γ by*

$$\gamma \leftarrow \beta\gamma.$$

14. *Update the direction vector $\tilde{p}$ by*

$$\tilde{p} \leftarrow \tilde{r} + \beta\tilde{p}.$$

15. *If $\gamma/\gamma_0 > 10^{-12}$, then go to Step 7.*

5.13 Incomplete LU Factorization (ILU)

An iterative method that is both general and well-accelerated by PCG-like methods is the *incomplete LU factorization method (ILU)* [56] [75] [119], which is based on the approximate decomposition of A as the product LU of the two sparse triangular matrices L and U. In an ILU iteration, the approximate solution $\tilde{x}$ is improved by

$$\tilde{x} \leftarrow \tilde{x} + U^{-1}L^{-1}(b - A\tilde{x}). \tag{5.13}$$

Another ILU version, the block-ILU (or line-ILU) method [50], is described in Section 11.4 below.

Next, we describe in detail a version of the ILU factorization. This version uses no fill-in at all: the triangular matrices L and U used in the ILU method (5.13) have the same sparsity pattern as the original matrix A, i.e., they can have a nonzero element only if the corresponding element in A is nonzero.

Algorithm 5.2

1. *Initialize $L = (l_{i,j})_{1 \le i,j \le N}$ to be the identity matrix.*
2. *Initialize $U = (u_{i,j})_{1 \le i,j \le N}$ by $U = A$.*
3. *For $i = 2, 3, \ldots, N$, do the following:*
   - *For $j = 1, 2, \ldots, i - 1$, do the following:*
     a) *define factor $= u_{i,j}/u_{j,j}$;*
     b) *for $k = j, j + 1, \ldots, N$, if $u_{i,k} \neq 0$, then set*

     $$u_{i,k} \leftarrow u_{i,k} - factor \cdot u_{j,k}.$$

     c) *set $l_{i,j} \leftarrow factor$.*

The resulting matrices L and U are now used in the ILU iteration (5.13).

5.14 Parallelizable ILU Version

Because each ILU iteration requires forward elimination in L followed by back-substitution in U, it is essentially sequential and cannot be implemented efficiently on parallel computers. Fortunately, one may modify the standard ILU method in such a way that it can be well parallelized. Here we introduce a parallelizable ILU version based on [96].

In the parallelizable version, L and U and the preconditioning method are defined in a slightly different way from before. For simplicity, we assume that the parallel computer contains only two processors: the first processor constructs the first $N/2$ rows in L and U, and the second processor constructs the last $N/2$ rows in L and U. The extension to $p \geq 2$ processors is straightforward; in this case, each processor constructs N/p consecutive rows in L and U.

The parallelizable ILU version is based on restarting the recursions used in ILU so that different parts of the recursions can be completed simultaneously in parallel. Let O be a positive integer denoting the amount of overlap between the parts, that is, number of matrix rows that are shared by both processors ("O" stands for "overlap"). A reasonable choice for O for 2-D problems is $O = \lceil \sqrt{N} \rceil$; another possibility is to set O to be a small number independent of N. The triangular matrices L and U are defined as follows.

Algorithm 5.3

1. *Initialize $L = (l_{i,j})_{1 \leq i,j \leq N}$ to be the identity matrix.*
2. *Initialize $U = (u_{i,j})_{1 \leq i,j \leq N}$ by $U = A$.*
3. *For $i = N/2 - O + 1, N/2 - O + 2, \ldots, N$, do the following.*
   - *For $j = N/2 - O, N/2 - O + 1, \ldots, i - 1$, do the following.*
     a) *Define factor $= u_{i,j}/u_{j,j}$;*
     b) *For $k = j, j + 1, \ldots, N$, if $u_{i,k} \neq 0$, then set*

$$u_{i,k} \leftarrow u_{i,k} - factor \cdot u_{j,k}.$$

   c) *set $l_{i,j} \leftarrow factor$.*
4. *For $i = 2, 3, \ldots, N/2$, do the following:*
   - *If $i > N/2 - O$, then for $j = N/2 - O, N/2 - O + 1, \ldots, N$, set $u_{i,j}$ back to its initial value by $u_{i,j} \leftarrow a_{i,j}$.*
   - *For $j = 1, 2, \ldots, i - 1$, do the following:*
     a) *Define factor $= u_{i,j}/u_{j,j}$;*
     b) *For $k = j, j + 1, \ldots, N$, if $u_{i,k} \neq 0$, then set*

$$u_{i,k} \leftarrow u_{i,k} - factor \cdot u_{j,k}.$$

   c) *Set $l_{i,j} \leftarrow factor$.*

The parameter O represents the amount of overlap between the two subsets of unknowns

$$\{1, 2, \ldots, N/2\} \quad \text{and} \quad \{N/2 - O, N/2 - O + 1, \ldots, N\}.$$

In each of these subsets, the incomplete factorization can take place independently of the other. Thus, each subset is assigned to a different processor. Of course, when $O = N/2 - 1$, the algorithm is equivalent to the standard ILU factorization in Section 5.13 above. In practice, however, O is much smaller. Indeed, the theory in [96] shows that O could be rather small. Although this theory is no longer applicable to the present version, it can still be interpreted to indicate that the present version should work well in terms of iteration count, particularly when used as a relaxation method within the multigrid V-cycle discussed later in this book.

On a parallel computer with two processors, steps 3 and 4 can be executed simultaneously in parallel. It is assumed that O is small relatively to N, so it can be safely assumed that by the time step 4 reaches row i (for some $N/2 - O < i \leq N/2$), the ith unknown is no longer in use in step 3, so its update in step 4 does not have a bad effect on step 3.

In practice, the rows in A are grouped in p groups that are assigned to p processors. Each group contains about $N/p + O$ rows in A, which are processed by one of the processors. Steps 3 and 4 above are replaced by p steps to process these groups and construct the corresponding rows in L and U. When implemented on a parallel computer, these steps can be executed simultaneously and independently of each other.

So far, we have shown how the matrices L and U should be constructed in a parallelizable algorithm. Next, we show that the iteration (5.13) can also be modified in the same spirit to suit parallel computers. More specifically, in (5.13) one needs to solve an equation of the form

$$Lv = r, \tag{5.14}$$

where r is a given N-dimensional vector and v is the N-dimensional vector of unknowns. The solution of (5.14), referred to as *forward elimination*, is approximated here by a parallelizable approximate forward elimination:

Algorithm 5.4

1. *Initialize v by $v = r$.*
2. *For $i = N/2 - O + 1, N/2 - O + 2, \ldots, N$, update v_i by:*

$$v_i \leftarrow r_i - \sum_{j=N/2-O}^{i-1} l_{i,j} v_j.$$

3. *For $i = 2, 3, \ldots, N/2$, update v_i by:*

$$v_i \leftarrow r_i - \sum_{j=1}^{i-1} l_{i,j} v_j.$$

Similarly, in (5.13) one also has to solve an equation of the form

$$Uv = r, \tag{5.15}$$

The solution of (5.15), referred to as *back-substitution*, is also approximated by a parallelizable approximate back-substitution:

Algorithm 5.5

1. *Initialize v by $v = diag(U)^{-1} r$.*
2. *For $i = N/2, N/2 - 1, \ldots, 1$, update v_i by:*

$$v_i \leftarrow \left(r_i - \sum_{j=i+1}^{N} u_{i,j} v_j \right) \Big/ u_{i,i}.$$

3. *For $i = N - 1, N - 2, \ldots, N/2 - O + 1$, update v_i by:*

$$v_i \leftarrow \left(r_i - \sum_{j=i+1}^{N} u_{i,j} v_j \right) \Big/ u_{i,i}.$$

Again, when implementation on a parallel computer with p processors is considered, step 2 (and 3) above is replaced by p different steps to process the p portions of restarted recursion. These steps can then be executed simultaneously and independently of each other using the p processors.

5.15 Nonsymmetric and Indefinite Problems

As in the Jacobi and symmetric GS relaxation methods, the preconditioner of ILU is SPD whenever A is. Therefore, it is suitable to serve as a preconditioner also in the PCG iteration. However, it turns out that the ILU preconditioner may be inferior to the symmetric GS preconditioner for SPD problems. The advantage of ILU is more apparent for problems in which A is not SPD, such as the convection-diffusion equation, where it is nonsymmetric, and the Helmholtz equation, where it is indefinite. Although PCG is no longer applicable, the preconditioned system

$$\mathcal{P}^{-1} A x = \mathcal{P}^{-1} b$$

can still be solved by more general Lanczos-type acceleration methods, such as the Generalized Minimal Residual (GMRES) method [88], the Quasi Minimal Residual (QMR) method [53], the Transpose-Free Quasi Minimal Residual (TFQMR) method [54], or the Conjugate Gradient Squared (CGS) method [110].

The above acceleration methods require frequent application of $\mathcal{P}^{-1} A$ to a vector. Fortunately, this can be done with no explicit calculation of $\mathcal{P}$ or $\mathcal{P}^{-1}$. Indeed, the basic iterative method calculates

$$x_{new} = x_{old} + \mathcal{P}^{-1} (b - A x_{old}),$$

so we have

$$x_{old} - x_{new} = \mathcal{P}^{-1} (A x_{old} - b).$$

Therefore, by applying the basic iteration to a given vector x_{old} (with the right-hand side b being set to the zero vector), one automatically gets the application of $\mathcal{P}^{-1} A$ to the vector x_{old}, with no need to know $\mathcal{P}$ or $\mathcal{P}^{-1}$ explicitly. Furthermore, one can also easily calculate the application of $\mathcal{P}^{-1}$ to the initial residual, which gives the preconditioned residual, used in the start of the acceleration process.

5.16 Numerical Comparison

We have tested the above ILU and parallelizable ILU versions for several nonsymmetric matrices from the Harwell–Boeing collection of sparse matrices. For these examples, the point-GS preconditioner isn't robust, and fails to converge for the "sherman2" and "pores" examples.

In Table 5.1 we report the number of ILU iterations used within CGS to reduce the l_2-norm of the preconditioned residual by 12 orders of magnitude. (In other words, the ILU version is used as a preconditioner in the CGS iteration, and the

Table 5.1. Number of ILU iterations used within CGS applied to nonsymmetric examples from the Harwell–Boeing collection of sparse matrices. The parameter "nonzeroes" denotes the number of nonzero elements in the coefficient matrix A. The parallelizable ILU method is implemented with $p = 20$ and $O = 25$.

Example	N	Nonzeroes	ILU	Parallelizable ILU
sherman2.rua	1080	23094	23	297
sherman3.rua	5005	20033	193	443
sherman4.rua	1104	3786	65	107
sherman5.rua	3312	20793	65	129
pores2.rua	1224	9613	81	1591
pores3.rua	532	3474	79	561

number of preconditioning steps is reported in the table.) It is apparent that parallelizable ILU (implemented with $p = 20$ and $O = 25$) is not as robust as standard
ILU, and requires more iterations to converge. Still, the cost of each parallelizable
ILU iteration may be reduced by parallelism. Furthermore, the iteration count in
the table may be reduced considerably by increasing O and/or decreasing p.

The power of (parallelizable) ILU is more apparent when used within the multigrid V-cycle. This is discussed later in the book (see Section 16.5).

5.17 The Normal Equations

When A is not SPD, PCG cannot be applied to the original system (3.5). However,
it can still be applied to the system of normal equations, obtained by multiplying
(3.5) throughout by A^*:

$$A^* A x = A^* b.$$

The Kacmarz iteration [115] is just the Gauss–Seidel iteration applied to this system.
Because $A^* A$ is always SPD, this iteration always converges to x [118]. Unfortunately, the convergence is usually prohibitively slow. Even when symmetric GS
iteration is used on the above normal equations and PCG acceleration is also used,
the convergence is still often too slow. This is because the condition number $\kappa(A^* A)$
[defined in (5.10)] is much larger than $\kappa(A)$. One must therefore develop preconditioning methods more powerful than symmetric GS, such as multigrid.

The above normal equations have also a slightly different version. When an SPD
preconditioner $\mathcal{P}$ is available for the non-SPD matrix A, one could multiply (3.5)
throughout by $A^* \mathcal{P}^{-1}$, to obtain:

$$A^* \mathcal{P}^{-1} A x = A^* \mathcal{P}^{-1} b.$$

Because the coefficient matrix in this system is SPD, one could apply PCG to it
with the preconditioner $\mathcal{P}$ [15].

5.18 Exercises

1. We say that the matrix A is of property-A if, in some order of unknowns, it has
   the block form

   $$\begin{pmatrix} A^{(1,1)} & A^{(1,2)} \\ A^{(2,1)} & A^{(2,2)} \end{pmatrix},$$

   where $A^{(1,1)}$ and $A^{(2,2)}$ are diagonal submatrices. Show that, if a matrix A has
   a 5-point stencil as in (3.8), then it is of property-A, and the above block form
   is obtained from the red-black reordering.
2. Write a computer code that implements the point-Jacobi iteration for tridiagonal linear systems.
3. Write the computer code that implements the point-Jacobi iteration for 5-point
   stencils.

4. Write the computer code that implements the point-Jacobi iteration for 9-point stencils.

5. Write a computer code that implements the point-GS iteration for tridiagonal linear systems.

6. Write the computer code that implements the point-GS iteration for 5-point stencils.

7. Modify the above code to use red-black ordering.

8. Write the computer code that implements the point-GS iteration for 9-point stencils.

9. Modify the above code to use four-color ordering.

10. Write the computer code that implements the line-GS iteration for 9-point stencils.

11. Modify the above code to use the "zebra" ordering.

12. Modify the above code to use the symmetric "zebra" line relaxation.

13. Write a computer code that uses the above relaxation method as a preconditioner in PCG.

14. Test your codes for the Poisson equation in a square, discretized on a uniform bilinear finite-element mesh as in Section 4.4. Which iterative method is the best?

15. Write the computer codes that implement the point-Jacobi, point-GS, and symmetric point-GS for general sparse linear systems. Furthermore, write a code that uses the latter relaxation method as a preconditioner in PCG. The solution can be found in Chapter 17 in [103].

16. Assume that A has a 9-point stencil. Show that $A^t A$ has a 25-point stencil (a 5×5-stencil).

17. Color the uniform grid in (3.6) with respect to a 25-point stencil. Use nine colors, each of which contains gridpoints that are (multiples of) three points away from each other in both the x and y spatial directions. Show that the point-GS relaxation can be done in each color in parallel; that is, all the points in a color can be relaxed simultaneously. Conclude that the 9-color point-GS relaxation can be completed in parallel in nine time units only, regardless of the size of the grid.

6

The Multigrid Iteration

In this chapter, we describe the various kinds of multigrid methods for the solution of large sparse linear systems arising from the discretization of elliptic PDEs. We highlight an interesting connection between multigrid and domain decomposition, and explain the various kinds of multigrid versions (including black-box multigrid and algebraic multigrid) in terms of domain decomposition.

6.1 The Two-Grid Method

The multigrid iterative method is a powerful tool for the numerical solution of large sparse linear systems arising from the discretization of elliptic PDEs. In a multigrid iteration (also known as multigrid cycle), the equation is first relaxed on the original (fine) grid by some relaxation method to smooth the error and make it ready to be solved for on a coarser grid. Because calculations on this grid are cheaper, a correction term can be computed there, and then transferred back to the fine grid and added to the approximate solution. Finally, the equation is relaxed again on the fine grid (usually by the same relaxation method), to smooth out any oscillatory error modes that may have contaminated the coarse-grid correction. The entire procedure can be described schematically in a diagram with the shape of the Latin letter V, hence the name "V-cycle" (Figure 6.1).

Actually, the coarse-grid problem can by itself be solved approximately recursively by one multigrid iteration, using subsequent coarser grids. This is the multigrid V-cycle (Figure 6.3). For model elliptic problems such as the Poisson equation discretized by finite differences on a uniform rectangular grid, the multigrid iteration converges to the algebraic solution x in the so-called *Poisson convergence rate*, in which the accuracy improves by one digit (the l_2-norm of the residual is reduced by factor 10) in each iteration [24]. This convergence rate is independent of the number of degrees of freedom (gridpoints) N, and is optimal for an iterative method that requires only $O(N)$ storage units and $O(N)$ arithmetic operations per iteration [7]. Good convergence rates have also been obtained for more difficult PDEs, such as highly anisotropic model equations [23] and equations with variable (discontinuous) coefficients [39]. The attempt to obtain good convergence rates for more difficult problems in complicated domains and nonuniform grids and nonsymmetric and indefinite problems is the subject of ongoing research.

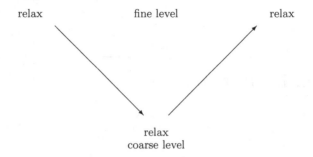

relax   fine level   relax

relax
coarse level

Fig. 6.1. The two-grid iteration has the shape of the Latin letter V: first, relaxation is used at the fine grid; then, the residual is transferred to the coarse grid to produce a correction term, which is transferred back to the fine grid; finally, relaxation is used again at the fine grid.

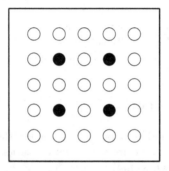

Fig. 6.2. The coarse grid: the subgrid of points that lie on even-numbered lines and even-numbered columns.

Both the two-grid and multigrid methods use a coarse grid to solve the problem (approximately) and supply a correction term. In a uniform grid, the coarse grid is constructed by taking every other point in both the x and y spatial directions in the original fine grid (Figure 6.2), so the number of coarse-gridpoints is about four times as small as the number of fine-gridpoints. More explicitly, if the original grid is the uniform grid g in (3.6), then the coarse grid is the subgrid $c \subset g$ consisting of the even-numbered points in both the x and y spatial directions:

$$c = \{(i,j) \in g \mid i \equiv j \equiv 0 \bmod 2\}. \tag{6.1}$$

The process of constructing the coarse grid c from the original fine grid g is referred to as full coarsening. (In the sequel, we also consider semicoarsening, in which the coarse grid consists of every other line of fine-gridpoints.)

The number of coarse grids used in the multigrid V-cycle is up to the user. Let us first consider the simplest case, in which only two grids are used: the original fine grid and one coarse grid. This two-grid iteration consists of three steps. First, the approximate solution $\tilde{x}$ is improved by some relaxation method:

$$\tilde{x} \leftarrow \tilde{x} + \mathcal{P}^{-1}(b - A\tilde{x})$$

(where $\mathcal{P}$ is the preconditioner of this relaxation method). After this step, it can be assumed that the subtle errors that can be viewed only on the fine grid have already been annihilated, so the error is sufficiently smooth, and ready to be solved for on the coarse grid. For this purpose, we assume that we have a restriction operator R, which transfers vectors defined on the fine grid to vectors defined on the coarse grid $[R : l_2(g) \to l_2(c)]$, and prolongation operator P, which transfers coarse-grid vectors into corresponding fine-grid vectors $[P : l_2(c) \to l_2(g)]$. We also assume that we have a matrix Q that operates on coarse-grid vector $[Q : l_2(c) \to l_2(c)]$ in the same way that A operates on corresponding fine-grid vectors. The actual definitions of R, P, and Q are specified later.

In order to use the coarse grid properly, one wouldn't like to simply transfer $\tilde{x}$ and b from the fine grid to the coarse grid, because then the benefit of the above relaxation would be lost. Instead, one should transfer the residual $b - A\tilde{x}$ from the fine grid to the coarse grid, solve for a correction term there, and prolong it back to the fine grid and add it to $\tilde{x}$:

$$\tilde{x} \leftarrow \tilde{x} + PQ^{-1}R(b - A\tilde{x})$$

(where Q is a coarse-grid approximation to the coefficient matrix A, to be specified later). The third and final step in the two-grid iteration is another relaxation of the form

$$\tilde{x} \leftarrow \tilde{x} + \mathcal{P}^{-1}(b - A\tilde{x})$$

to annihilate any subtle errors that may have been reintroduced by the coarse-grid correction. This completes the definition of the two-grid iteration, up to the specific definitions of R, P, and Q, to be discussed later in the book.

Because the coarse-grid matrix Q is of order about four times as small as the order of A, the complexity of solving the coarse-grid problem

$$Qe = R(b - A\tilde{x}) \tag{6.2}$$

(where e is the coarse-grid vector of unknowns) is much smaller than that of solving the original linear system (3.5).

The relaxation that is done before the coarse-grid correction is called prerelaxation, and the relaxation that is done after the coarse-grid correction is called postrelaxation. In practice, one could use more than one prerelaxation and one postrelaxation in a V–cycle.

Here is a more general definition of the two-grid method. Let ν_1 and ν_2 be positive integers denoting the number of prerelaxations and postrelaxations, respectively. The two-level (TL) algorithm is defined as follows.

Algorithm 6.1 $TL(x_{in}, A, b, x_{out})$:

1. *Apply to the equation (3.5) ν_1 relaxations (with x_{in} as initial guess that is updated during the relaxations).*
2. *Set $x_{out} = x_{in} + PQ^{-1}R(b - Ax_{in})$.*
3. *Apply to the equation (3.5) ν_2 relaxations (with x_{out} as initial guess that is updated during the relaxations).*

6.2 Transfer and Coarse-Grid Operators

The guidelines for how the transfer operators R and P and the coarse-grid matrix Q should be constructed follow from spectral considerations. In particular, we explain below why the transfer operators should transfer vectors that lie (almost) in the null-space of A to vectors that lie (almost) in the null-space of Q, and vice versa.

The spectrum of A (the set of eigenvalues of A) can be split into two parts: eigenvalues that are relatively large in magnitude, and eigenvalues that are relatively small in magnitude. The error in the current approximate solution $\tilde{x}$, $\tilde{x} - x$, can be decomposed according to this spectral splitting as the sum of two parts. The first part of the error, which contains eigenvectors with large eigenvalues, can be well handled by the relaxations in steps 1 and 3 in the TL method. Indeed, the preconditioner $\mathcal{P}$ of the relaxation method has probably the same effect on this part as A does, so the iteration matrix $\mathcal{M} = I - \mathcal{P}^{-1}A$ almost annihilates it. The second part, on the other hand, cannot be relaxed effectively, because $\mathcal{M}$ has almost no effect on it. Hence, it must be handled in the coarse-level correction in step 2 in the TL method.

The TL method may thus be viewed as an alternating projection method: first the error is projected onto the subspace that is almost orthogonal to the subspace spanned by eigenvectors with large eigenvalues, and then it is projected onto the subspace that is almost orthogonal to the subspace spanned by eigenvectors with small eigenvalues (in magnitude) [71],

We refer to an eigenvector with a very small eigenvalue (in magnitude) as *nearly singular eigenvector*. The restriction operator R should thus simulate well nearly singular eigenvectors of A, and transfer them (roughly) to nearly singular eigenvectors of Q. The prolongation operator P should then transfer these vectors (roughly) back to the nearly singular eigenvectors of A. Furthermore, the eigenvalue of Q for a particular eigenvector should be the same as the eigenvalue of A for the (roughly) prolonged vector. This way, the coarse-grid iteration matrix $I - PQ^{-1}RA$ will indeed annihilate the nearly singular error modes, as required.

The nearly singular eigenvectors often have rather low variation, so they can indeed be well-represented on the coarse grid. For example, when A is an M-matrix, the most nearly singular eigenvector has only positive components [118]. Furthermore, for linear systems arising from the discretization of elliptic PDEs, the nearly singular eigenvectors are the discrete approximation of the smooth nearly singular Sturm–Liouville eigenfunctions of the original differential operator. The transfer operators should thus transfer a coarse-grid discrete smooth Sturm–Liouville eigenfunction (almost) to a fine-grid discrete smooth Sturm–Liouville eigenfunction, and vice versa. In particular, the coarse-grid constant vector should be transferred (at least roughly) to the fine-grid constant vector, and vice versa. As a conclusion, P should prolong coarse-grid vectors as "smoothly" as possible, using (weighted) average to define values at fine-gridpoints that lie in between coarse-gridpoints.

6.3 The Multigrid Method

The above two-grid method requires the exact solution of the coarse-grid problem (6.2), which could be rather expensive. In practical multigrid, (6.2) is solved only

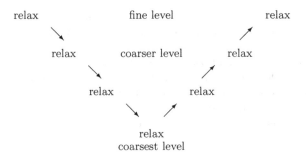

Fig. 6.3. The multigrid iteration has the shape of the Latin letter V: first, relaxation is used at the fine grid; then, the residual is transferred to the next, coarser grid, where a smaller V-cycle is used recursively to produce a correction term, which is transferred back to the fine grid; finally, relaxation is used again at the fine grid.

approximately by one multigrid iteration, using subsequent coarser grids. This way, only the problem on the coarsest grid is solved exactly. This problem is so coarse and its order so small that the cost of solving it is negligible. The only work actually required is relaxation on the grids, as is shown in Figure 6.3. Because the amount of arithmetic operations in relaxation is proportional to the number of gridpoints, the total amount of arithmetic operations in the entire multigrid V-cycle is $O(N)$.

Let us now define the multigrid iteration in detail. Let L be the positive integer denoting the number of levels used. Let ν_c be the positive integer denoting the number of relaxations used to solve approximately the coarsest-grid problem. The multilevel iteration is defined as follows.

Algorithm 6.2 $ML(x_{in}, A, b, L, x_{out})$:

1. *If $L \leq 1$, then apply to the equation (3.5) ν_c relaxations (with x_{in} as an initial guess that is updated during the relaxations), and then set $x_{out} = x_{in}$. Otherwise, proceed as follows.*
2. *Apply to the equation (3.5) ν_1 relaxations (with x_{in} as an initial guess that is updated during the relaxations).*
3. *Apply the multilevel method recursively to the coarse-grid problem by*

$$ML(\mathbf{0}, Q, R(b - Ax_{in}), L - 1, e),$$

where $\mathbf{0}$ is the zero vector defined on the coarse grid and e is the output vector defined on the coarse grid.
4. *Initialize x_{out} by*

$$x_{out} = x_{in} + Pe. \tag{6.3}$$

5. *Apply to the equation (3.5) ν_2 relaxations (with x_{out} as an initial guess that is updated during the relaxations).*

We also refer to this algorithm as the $V(\nu_1,\nu_2)$-cycle (see Figure 6.3). Note that when no pre-relaxations are used ($\nu_1 = 0$), the solution process actually starts from the coarsest grid, before going up to the finer grids to perform the ν_2 post-relaxations there. Because of its special shape, this cycle is called the *sawtooth* or F-cycle. (This cycle is actually used in the integer-division algorithm in Section 1.3 above.)

In most cases, ν_c is sufficiently large to solve the coarsest-grid problem to 6-order accuracy. The relaxation method used to solve the coarsest-grid problem is usually the same as that used for pre- and post relaxation on the finer grids. In some applications, however, we choose to use only a moderate number of Kacmarz relaxations on the coarsest grid. In the indefinite Helmholtz equation, for example, this approach may help to reduce instability due to the highly indefinite coarsest-grid problem.

6.4 Geometric Multigrid

There are two possible approaches towards the definition of the transfer and coarse-grid operators. The first one, called *geometric multigrid*, uses the original PDE and the original domain where it is defined. The second one, called *matrix-based multigrid*, uses the coefficients in A only.

In geometric multigrid, the transfer and coarse-grid operators are defined using the geometric properties of the fine and coarse grids. Most often, R and P are obtained from linear averaging. More specifically, let v be a coarse-grid vector in $l_2(c)$. By $v_{i,j}$, we denote the value of v at the point $(i,j) \in c$. Then, the prolong vector $Pv \in l_2(g)$ at a point $(i,j) \in g$ is defined by

$$(Pv)_{i,j} \equiv \begin{cases} v_{i,j} & \text{if } (i,j) \in c \\ (v_{i-1,j} + v_{i+1,j})/2 & \text{if } i+1 \equiv j \equiv 0 \bmod 2 \\ (v_{i,j-1} + v_{i,j+1})/2 & \text{if } i \equiv j+1 \equiv 0 \bmod 2 \\ (v_{i-1,j-1} + v_{i-1,j+1} + v_{i+1,j-1} + v_{i+1,j+1})/4 & \text{if } i \equiv j \equiv 1 \bmod 2. \end{cases}$$

It is assumed here that n in (3.6) is odd, so the values required in the above definition are available; otherwise, the definition should be modified slightly at the discrete boundary.

With the above definition, P indeed transfers the constant coarse-grid vector to the constant fine-grid vector, as in the guidelines in Section 6.2. Note also that P can actually be represented as a rectangular matrix, with columns that contain one entry of value 1, four entries of value 1/2, and four entries of value 1/4 each.

The restriction operator R can be defined either as simple injection

$$R = J_c$$

or (better yet) as linear averaging; that is, at each coarse-gridpoint $(i,j) \in c$, the restricted vector Rv is defined by

$$(Rv)_{i,j} = \frac{1}{4}\left(\frac{1}{4}v_{i+1,j+1} + \frac{1}{2}v_{i+1,j} + \frac{1}{4}v_{i+1,j-1} + \frac{1}{2}v_{i,j+1}\right. \tag{6.4}$$
$$\left. + v_{i,j} + \frac{1}{2}v_{i,j-1} + \frac{1}{4}v_{i-1,j+1} + \frac{1}{2}v_{i-1,j} + \frac{1}{4}v_{i-1,j-1}\right).$$

More compactly, R can be represented by the stencil

$$\frac{1}{4}\begin{bmatrix} 1/4 & 1/2 & 1/4 \\ 1/2 & 1 & 1/2 \\ 1/4 & 1/2 & 1/4 \end{bmatrix},$$

or, in other words,

$$R = \frac{1}{4}P^t.$$

Note that the row-sums of R are all equal to 1. Therefore, R indeed transfers the constant fine-grid vector into the constant coarse-grid vector, as advised in Section 6.2 above.

Finally, the coarse-grid matrix Q is obtained by rediscretizing the original PDE on the coarse grid c. Usually, the red-black point-GS method is used as a relaxation method. This completes the definition of geometric multigrid for uniform grids. In the next section, geometric multigrid is extended also to the more complicated case of unstructured meshes.

6.5 Variational Multigrid

In the following sections, we show how geometric multigrid develops naturally towards variational, matrix-based, and eventually algebraic multigrid. The first step is to extend it to more general (nonuniform) meshes.

Extending geometric multigrid to unstructured meshes requires three rather nontrivial tasks: choosing a subset of nodes to serve as a coarse grid, connecting them to form the coarse finite-element mesh, and defining the transfer operators between fine and coarse grids. Once these tasks are complete, the original PDE can be rediscretized on the coarse finite-element mesh to form the coarse-grid coefficient matrix Q.

The above algorithm can be made more concrete in the special case in which the finite-element mesh is obtained from a process of successive (local) refinement. In this case, it is natural to use the coarse mesh in the refinement process also as the coarse meshes in the multigrid algorithm for solving the linear system on the fine mesh. The coarse grid is then just the set of nodes in the coarse mesh. The prolongation operator P is defined by simple averaging:

$$(Pv)_{(i+j)/2} \equiv (v_i + v_j)/2,$$

where i and j are any two coarse-gridpoints on the same edge in the coarse mesh. This way, the prolongation operator is just the identity operator on continuous and piecewise-linear functions in the coarse finite-element function space. Indeed, the function resulting from the prolongation is linear not only in each fine finite element but also in each coarse finite element. This property can be obtained thanks to the fact that the function spaces are nested: each continuous function that is linear in every coarse finite element is also linear in every fine finite element, and thus belongs also to the function space defined on the fine mesh.

The restriction operator is now defined by

$$R \equiv P^t,$$

and the coarse-grid matrix is defined by

$$Q \equiv RAP.$$

Because R and P are just the identity operators on the corresponding function spaces, Q is actually the same as the matrix obtained from rediscretizing the original PDE on the coarse mesh. Therefore, variational multigrid can be interpreted as a special case of geometric multigrid. In the following, we show that variational multigrid can also be interpreted as a special case of the domain-decomposition approach.

6.6 Domain Decomposition and Variational Multigrid

The key factor in the multigrid algorithm is the transfer of information from fine to coarse grid and vice versa. In particular, the prolongation operator P should transform a vector v (defined on the coarse grid c) into an extended vector Pv (defined on the original grid) with energy norm as small as possible. In other words, Pv must be close to a nearly singular eigenvector of the coefficient matrix A.

Here we consider a domain-decomposition approach towards the definition of the coarse grid c and prolongation operator P. The restriction and coarse-grid operators are then defined as in variational multigrid by $R = P^t$ and $Q = RAP$. With this approach, variational multigrid can actually be viewed as a domain decomposition algorithm.

In the domain-decomposition approach, the original domain is decomposed into disjoint subdomains (Figure 6.4). The vertices (corners) of these subdomains (the bullets in Figure 6.5) are then used to form the coarse grid c. (Although this figure illustrates a case of structured grid, this is only an example; the same approach can be used also in unstructured grids.) The prolongation consists of two steps: the first step extends the values given on c also to the nodes that lie on the edges of subdomains, and the second step extends them further also to the interiors of subdomains.

For a given coarse-grid vector v, the first prolongation step produces the values of the extended vector Pv at nodes that lie on edges of subdomains (in between coarse-grid nodes). The second prolongation step uses these values to solve a discrete homogeneous Dirichlet problem in each individual subdomain (Figure 6.7). The numerical solutions of these subproblems produce Pv in the interiors of subdomains

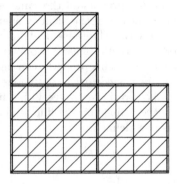

Fig. 6.4. The domain decomposition: the original domain is divided into disjoint subdomains.

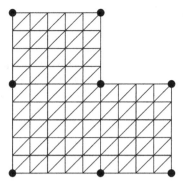

Fig. 6.5. The coarse grid c obtained from the domain decomposition contains the vertices (corners) of the subdomains (denoted by bullets).

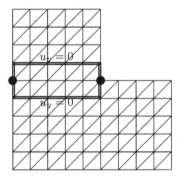

Fig. 6.6. The first prolongation step, in which the known values at the bullets are prolonged to the line connecting them by solving a homogeneous Dirichlet–Neumann subproblem in the strip surrounding it.

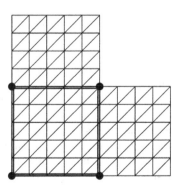

Fig. 6.7. The second prolongation step, in which the known values at the internal boundary of a subdomain (bullets and edges) are prolonged also to its interior by solving a homogeneous Dirichlet subproblem.

as well. Because these subproblems are independent of each other, they can be solved simultaneously in parallel.

The first prolongation step, which extends v to the edges in Figure 6.4, is trickier than the second step described above. Clearly, the first step should use the original values of v at the endpoints of each individual edge to solve a 1-D homogeneous Dirichlet problem in this edge. Unfortunately, it is not clear how this ODE should be defined.

In [18], it is assumed that the diffusion coefficient is constant in each individual subdomain. The ODE solved on each individual edge uses a diffusion coefficient that is the average of the original diffusion coefficients on both sides of this edge. Thus, the ODE is just the 1-D Poisson equation. This approach can be extended also to the Helmholtz equation

$$-u_{xx} - u_{yy} + \beta u = F,$$

where β is a negative parameter. In this case, the ODE along the edge is the 1-D Helmholtz equation

$$-u_{qq} + \beta u/2 = 0, \tag{6.5}$$

where $\mathbf{q}$ is the unit vector tangent to the edge. This idea is formulated algebraically in Chapter 10 below.

In the next section, the above approach is improved in such a way that there is no need to assume that the diffusion coefficient is constant in each individual subdomain. In the special case of structured grids, the resulting algorithm can be reformulated in algebraic terms only, leading to a "black-box" method that takes a matrix and a right-hand side vector as input and produces the vector of unknowns as output.

6.7 Domain Decomposition and Black-Box Multigrid

In the above, we have decomposed the prolongation operation into two steps. In the first step, values are prolonged from the vertices of subdomains to the edges of subdomains. In the second step, these values are prolonged further to the interiors of subdomains. The restriction and coarse-grid operators are then defined by the Galerkin approach: $R = P^t$ and $Q = RAP$. This completes the definition of the basic components in the multigrid iteration.

The second step of prolongation is easy to define: it solves the original PDE in the interior of subdomains, with the values calculated before serving as Dirichlet data on the edges of subdomains. The first prolongation step, on the other hand, is trickier to define: it requires the definition of a suitable ODE on each individual edge. Above, we've seen how such an ODE can be defined in some special cases. Here we turn to a more general approach for carrying out the first prolongation step. This approach has the important advantage that one no longer needs to assume that the diffusion coefficient is constant in each individual subdomain. Furthermore, in the special case of structured grids, it can be formulated algebraically, yielding a "black-box" linear-system solver that requires no human intervention whatsoever.

The more general definition of the first prolongation step is illustrated in Figure 6.6. First, the original (homogeneous) PDE is solved in a thin strip containing the edge under consideration. The boundary conditions for this subproblem are of Dirichlet type at the endpoints of the edge, where the values of the original (coarse-grid) vector v are available, and of homogeneous Neumann type elsewhere. The numerical solution of this subproblem provides the required values of Pv in the edge.

Actually, the subproblem in the strip could be reduced further into a tridiagonal system on the edge only. Consider, for example, the strip in Figure 6.8. The discrete homogeneous Neumann conditions on the top and bottom edges of this strip could be used to assume that the numerical solution of the subproblem is constant on each vertical line in the strip. The unknowns on the top and bottom edges of the strip can thus be eliminated, resulting in a reduced subproblem on the edge alone. In other words, the linear subsystem on the nodes in the strip is "lumped" into a tridiagonal system on the nodes in the edge only. The solution to this tridiagonal system produces the prolonged values of Pv in the edge (see also Chapter 14).

When structured grids as in (3.6) are considered and the coarse grid is as in (6.1), the above algorithm coincides with the black-box multigrid method in [39]. Consider, for instance, a fine gridpoint that lies in between two coarse-gridpoints on its left and its right (Figure 6.9). The above vertical lumping reduces the subsystem

$$u_y = 0 \qquad u_y = 0$$

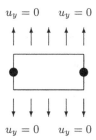

$$u_y = 0 \qquad u_y = 0$$

Fig. 6.8. Vertical lumping: the discrete homogeneous Neumann boundary conditions are used to assume that the numerical solution of the subsystem in the strip is constant along vertical lines, so the unknowns on the top and bottom edges of the strip can be eliminated.

$$u_y = 0$$

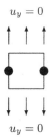

$$u_y = 0$$

Fig. 6.9. Vertical lumping in black-box multigrid: the discrete homogeneous Neumann boundary conditions are used to assume that the numerical solution of the subsystem is constant along vertical lines, so the unknowns on the top and bottom edges can be eliminated.

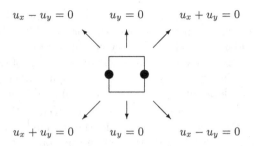

$$u_x - u_y = 0 \qquad u_y = 0 \qquad u_x + u_y = 0$$

$$u_x + u_y = 0 \qquad u_y = 0 \qquad u_x - u_y = 0$$

Fig. 6.10. Oblique and vertical lumping: the discrete homogeneous Neumann boundary conditions are used to assume that the numerical solution of the subsystem at the points on the top and bottom edges is the same as at the middle point, hence can be eliminated.

on the strip into a single algebraic equation for the middle point in this figure. The solution of this equation gives the required prolonged value at this middle point.

The main advantage of black-box multigrid is that, for problems that use structured grids, it indeed serves as black box: that is, it takes as input the coefficient matrix and right-hand-side vector, and automatically produces the required numerical solution. No human intervention is required at all. This property is particularly attractive when ones wants to write a general computer code that will solve any given problem with no presumptions or conditions.

As we'll see later in this book, black-box multigrid works well for many diffusion problems, including problems with discontinuous coefficients with discontinuity lines that do not necessarily align with the coarse grids. Still, there are some examples for which black-box multigrid fails (Section 9.4). This is probably because black-box multigrid is backed by no theory.

The multigrid version in [64], on the other hand, although inferior to black-box multigrid in most practical cases, is more robust and is also backed by theory, as we show later in this book. In this version, the lumping used to eliminate the unknowns on the top and bottom edges of the strip is not only vertical but also oblique at the corners of the strip (Figure 6.10). In fact, it can be viewed as a special case of the algorithm in Chapter 14 below, when applied to structured grids. In the following, we show how this version can be extended to unstructured grids as well.

6.8 Domain Decomposition and Algebraic Multigrid

So far, we've assumed that some domain decomposition is available, and used it to design the prolongation operator. This, however, is not always the case. In many applications, the system of algebraic equations is given with no information on the underlying grid whatsoever. In such cases, one must define the coarse grid c (which is just a subset of indices of unknowns) and the prolongation operator P (which is just a rectangular matrix with no apparent geometrical meaning) in terms of the coefficients in A alone. The restriction and coarse-grid matrices are then defined as in variational multigrid by $R = P^t$ and $Q = RAP$, which completes the definition of the basic components in the multigrid algorithm.

The original multigrid algorithm that uses no geometric information whatsoever is the Algebraic Multigrid (AMG) method in [25] and [86]. In this method, both the coarse grid c and the prolongation operator P are constructed using the elements in the coefficient matrix A alone. In [116], it is shown that the convergence factor for a particular AMG version depends only polynomially on the number of levels used.

Algebraic multigrid versions are thus characterized by the lack of any notion of PDE, domain, grid, or mesh in their definition. In fact, PDEs and grids are never mentioned or used in the construction of transfer and coarse-level matrices, let alone the coarse levels, which are in fact just subsets of unknowns, with no geometric interpretation whatsoever.

The main advantage of algebraic multigrid versions is in the opportunity to write a general computer code to solve general linear systems of equations. Because the algorithm depends on no specific property of the original PDE or discretization method, the computer code does not have to be rewritten for each particular application. Furthermore, AMG is robust for diffusion problems with discontinuous and anisotropic coefficients, at least on structured grids.

Although algebraic multigrid versions use no domain or grid, they can still benefit from the present domain-decomposition formulation to come up with a good prolongation scheme. Once the prolongation is well defined in terms of domain decomposition, it can usually be reformulated algebraically, leading to an efficient algebraic multigrid version. This is done in this section.

The idea of solving a discrete homogeneous Dirichlet–Neumann subproblem in the strip and then using the numerical solution of this subproblem to provide the required values of Pv in the edge contained in this strip is formulated in a more general way in the AMGe method in [36]. In this method, it is assumed that only the original finite-element mesh is given, with no domain decomposition. The coarse grid c is constructed algebraically, using the coefficients in the matrix A only. The set f is then defined as the set that contains all the nodes (or unknowns) that are not in c. The prolongation operator P is then defined as follows. At each node $i \in f$, $(Pv)_i$ is defined by solving the homogeneous PDE numerically in the "molecule" of finite elements that surround the node i (Figure 6.11). The boundary conditions for this subproblem are of Dirichlet type at nodes in c (where v is available) and of homogeneous Neumann type elsewhere. The numerical solution of this subproblem at i is then accepted as the prolonged value $(Pv)_i$. This defines $(Pv)_i$ at every $i \in f$, so no second prolongation step is needed.

In the AMGm method in [66], the above approach is reformulated in pure algebraic terms. The "molecules" are defined algebraically, so they can also be used in the recursive calls in the multigrid V-cycle.

The present domain-decomposition approach can also lead to a purely algebraic definition of P. Indeed, let's define the prolonged value $(Pv)_i$ at some $i \in f$. For this

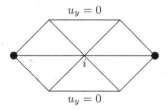

Fig. 6.11. Prolonging to node i by solving a homogeneous Dirichlet–Neumann subproblem in the "molecule" of finite elements that surround it.

purpose, instead of solving a subproblem in the molecule, we assume (momentarily) that the prolonged vector Pv should satisfy the ith equation in the homogeneous linear system:

$$\sum_{j\in c} A_{i,j}v_j + \sum_{j\in f} A_{i,j}(Pv)_j = 0.$$

Our algebraic molecule is, thus, the set of unknowns j for which $A_{i,j} \neq 0$. Of these unknowns, v_j is available for each $j \in c$, which is the algebraic analogue to Dirichlet boundary conditions. Furthermore, we use the algebraic analogue of homogeneous Neumann boundary conditions to assume that $(Pv)_j$ is the same for every $j \in f$ in the algebraic molecule, which leads to the definition

$$(Pv)_i \equiv \frac{-\sum_{j\in c} A_{i,j}v_j}{\sum_{j\in f} A_{i,j}}. \tag{6.6}$$

The above definition applies to every $i \in f$, so it actually completes the definition of P. Still, one could use the resulting values of Pv to improve the prolongation with a second step. In this step, algebraic Dirichlet conditions are used at all the unknowns $j \neq i$, using the original values v_j at unknowns $j \in c$ and the values $(Pv)_j$ calculated in the first prolongation step at unknowns $j \in f$ ($j \neq i$):

$$(Pv)_i \equiv -\frac{\sum_{j\in c} A_{i,j}v_j + \sum_{j\in f,\ j\neq i} A_{i,j}(Pv)_j}{A_{i,i}}. \tag{6.7}$$

This second prolongation step is actually an extra Jacobi relaxation, limited to unknowns in f only. It can thus be described more compactly as follows. Define the diagonal matrix D by

$$D_{i,i} \equiv \begin{cases} 1 & \text{if } i \in f \\ 0 & \text{if } i \in c. \end{cases}$$

Modify the original prolongation matrix P (which uses the first step only) by the substitution

$$P \leftarrow (I - diag(A)^{-1}DA)P$$

(where I is the identity matrix of the same order as A).

Using a second prolongation step in algebraic multigrid versions can reduce the number of iterations required for convergence in the numerical experiments at the end of this book by up to 50%. Unfortunately, it also requires much more set-up time due to the expensive matrix-times-matrix operation in the construction of P. This is why we stick to the original definition of P, and don't use the second prolongation step any more here. (The second prolongation step may be worthwhile only when a powerful parallel computer is available to carry out matrix-times-matrix operations efficiently.)

6.9 The Algebraic Multilevel Method

Because algebraic multigrid versions use no physical grids, the name *algebraic multilevel* is more adequate, and is therefore used for versions introduced here.

In the following, we introduce a prototype algebraic multilevel method in detail (Section 17.8 in [103]).

The first stage is to define the subset of indices c (the coarse level). This subset should contain indices of unknowns that are decoupled (or at most weakly coupled) in the coefficient matrix A. The subset c should also contain as many indices as possible, namely, it should be maximal in the sense that no more indices can be added to it. This way, every unknown with index that is left outside c (or, equivalently, remains in f) is strongly coupled in A to some other unknown with index in c, and therefore can have its prolonged value taken from this unknown.

It is assumed that the main-diagonal elements in A are not too small in magnitude. The algorithm to define c uses some small predetermined parameter (say 0.05) to help distinguish between strongly coupled and weakly coupled unknowns in A:

Algorithm 6.3

1. *Initialize c to contain all the unknowns:*

$$c = \{1, 2, 3, \ldots, N\}.$$

2. *For $i = 1, 2, 3, \ldots, N$, do the following.*
   - *If $i \in c$, then, for every $1 \leq j \leq N$ for which $j \neq i$ and*

$$|A_{i,j}| \geq threshold \cdot |A_{i,i}|,$$

   drop j from c.
3. *For $i = 1, 2, 3, \ldots, N$, do the following.*
   - *If $i \notin c$ and for every $j \in c$*

$$\frac{A_{i,j}}{A_{i,i}} \geq -threshold,$$

   then add i back to c.

Next, we introduce the algorithm to define the (rectangular) prolongation matrix P. This algorithm makes sure that the elements in P are nonnegative, so the prolongation to an unknown $i \in f$ actually uses a weighted average of values of unknowns in c. For those $i \in f$ for which $A_{i,i} > 0$, $A_{i,j} \leq 0$ for every $j \neq i$, and $\sum_j A_{i,j} = 0$, the definition is actually the same as in (6.6). (These conditions indeed hold for interior nodes in diffusion problems.)

The algorithm uses some small predetermined parameter (say, 0.05) to help drop matrix elements that are too small in magnitude and exclude them from P:

Algorithm 6.4

1. *Initialize P by $P = A$.*
2. *For every $i \in f$ and $j \in c$, if*

$$\frac{P_{i,j}}{P_{i,i}} > -threshold,$$

then drop $P_{i,j}$ from P and replace it by zero:

$$P_{i,j} \leftarrow 0.$$

3. *For every index $i \in c$, replace the ith row*

$$(P_{i,1}, P_{i,2}, \ldots, P_{i,N})$$

in P by the standard unit row $e^{(i)}$, of which all components vanish except of the ith one, which is equal to 1.
4. *For every $j \in f$, drop the corresponding column from P. (In this step, P becomes rectangular.)*
5. *For every $i \in f$, divide the ith row in P by its row-sum. (After this step, the row-sums in P are all equal to 1.)*

As in variational multigrid, the restriction and coarse-level matrices are defined by $R = P^t$ and $Q = RAP$. This completes the definition of the basic components in the algebraic multilevel method.

6.10 Algebraic Multigrid

In the algebraic multilevel method above, c is the maximal set of unknowns that are either completely decoupled from each other or at most coupled by positive or very small elements in A. More precisely, if i and j are in c, then either $A_{i,j}/A_{i,i} > 0$ or $|A_{i,j}/A_{i,i}| \leq$ threshold. In AMG, on the other hand, c is defined in a slightly different way. In a first stage, it is defined as a maximal set of unknowns that are only weakly coupled to each other in A. Then, in the second stage, it is modified in such a way that if i and j are two indices in f that are strongly coupled to each other in A, then there must exist a third index $k \in c$ to whom both i and j are strongly coupled as well. Otherwise, either i or j cannot remain in f, and must be added to c.

As can be seen in Algorithm 6.3, we don't bother here with the latter condition. After all, Algorithm 6.3 guarantees that both i and j are strongly coupled to some unknowns in c, so what difference does it make whether they are strongly coupled to a single unknown $k \in c$ or to two distinct unknowns in c? This is why Algorithm 6.3 is used to define c in the numerical experiments at the end of this book also for AMG.

A more important difference between AMG and the above multilevel method is in the definition of P. In the following, we give algebraic motivation to the definition of P in Algorithm 6.4 and to the modification used in AMG.

Once c is defined and its complementary subset f is defined by

$$f = \{1, 2, 3, \ldots, N\} \setminus c,$$

the coefficient matrix A can be written in the block form

$$A = \begin{pmatrix} A_{ff} & A_{fc} \\ A_{cf} & A_{cc} \end{pmatrix}. \tag{6.8}$$

Let $S(A; c)$ be the Schur complement of A with respect to this partitioning:

$$S(A; c) = A_{cc} - A_{cf} \left(A_{ff}\right)^{-1} A_{fc}.$$

With this definition, A can be decomposed in a block-LU decomposition as

$$A = \begin{pmatrix} I & 0 \\ A_{cf}\,(A_{ff})^{-1} & I \end{pmatrix} \begin{pmatrix} A_{ff} & 0 \\ 0 & S(A;c) \end{pmatrix} \begin{pmatrix} I & (A_{ff})^{-1}\,A_{fc} \\ 0 & I \end{pmatrix}.$$

In other words,

$$\begin{pmatrix} A_{ff} & 0 \\ 0 & S(A;c) \end{pmatrix} = \begin{pmatrix} I & 0 \\ -A_{cf}\,(A_{ff})^{-1} & I \end{pmatrix} A \begin{pmatrix} I & -(A_{ff})^{-1}\,A_{fc} \\ 0 & I \end{pmatrix}.$$

Thus, if we had defined

$$R = \left(-A_{cf}\,(A_{ff})^{-1} \quad I\right) \tag{6.9}$$

and

$$P = \begin{pmatrix} -\,(A_{ff})^{-1}\,A_{fc} \\ I \end{pmatrix}, \tag{6.10}$$

then we would have

$$Q = RAP = S(A;c). \tag{6.11}$$

Of course, this is not a practical approach, because $(A_{ff})^{-1}$ is prohibitively expensive to compute. Furthermore, it is dense, so R, P, and Q would be dense as well. The cure must be to approximate (A_{ff}^{-1}) along the guidelines in Section 6.2.

Let us define a matrix in which A_{ff} is replaced by its row-sum:

$$\tilde{A} = \begin{pmatrix} rs\,(A_{ff}) & A_{fc} \\ A_{cf} & A_{cc} \end{pmatrix}. \tag{6.12}$$

Because this matrix has the same effect as A on the constant N-dimensional vector, it can be used to produce sparse, easily computable matrices R, P, and Q:

$$R = \left(-A_{cf}\,rs\,(A_{ff})^{-1} \quad I\right) \tag{6.13}$$

$$P = \begin{pmatrix} -rs\,(A_{ff})^{-1}\,A_{fc} \\ I \end{pmatrix} \tag{6.14}$$

$$Q = RAP.$$

In fact, the P defined here is the same as in (6.6). This makes clear the close relation between the spectral analysis in Section 6.2, the domain decomposition approach in Section 6.8, and the present algebraic formulation.

When A is symmetric, the R defined above indeed satisfies $R = P^t$, as in variational multigrid. When A is nonsymmetric, on the other hand, the above definition of R makes more sense than the standard variational definition. Indeed, once P is redefined using $(A + A^t)/2$ instead of A, one obtains the algorithm used in Section 16.4 below, which is also in the spirit of [40].

The above definitions of R and P are based on the approximation to A in $\tilde{A}$. In $\tilde{A}$, the off-diagonal elements in A_{ff} are "lumped" (or "thrown," or added) to the main diagonal. (This is in the spirit of the Modified ILU (MILU) method in [56].) In AMG, on the other hand, the off-diagonal elements in A_{ff} are added not to the main diagonal but rather to other off-diagonal elements in A_{fc}. (This is in the spirit of other ILU versions in [51].) Here is the detailed algorithm to construct P in AMG:

Algorithm 6.5

1. *Initialize P by $P = A$.*
2. *For every $i \in f$ and $1 \le j \le N$, if $i \ne j$ and*

$$\frac{P_{i,j}}{P_{i,i}} > -threshold,$$

then drop $P_{i,j}$ from P and replace it by zero:

$$P_{i,j} \leftarrow 0.$$

3. *Define the matrix B by $B = P$.*
4. *For every index $i \in c$, replace the ith row*

$$(P_{i,1}, P_{i,2}, \ldots, P_{i,N})$$

in P by the standard unit row $e^{(i)}$, of which all components vanish except of the ith one, which is equal to 1.
5. *For every i and j that are both in f and satisfy $i \ne j$ and $P_{i,j} \ne 0$, define*

$$W_{i,j} \equiv \sum_{k \in c, \; P_{i,k} \ne 0} B_{j,k}.$$

6. *For every i and j that are both in f and satisfy $i \ne j$ and $P_{i,j} \ne 0$, add a fraction of $P_{i,j}$ to every nonzero element $P_{i,k}$ that lies in the same row and in column $k \in c$ as follows:*

$$P_{i,k} \leftarrow P_{i,k} + \frac{B_{j,k}}{W_{i,j}} P_{i,j}.$$

7. *For every $j \in f$, drop the corresponding column from P. (In this step, P becomes rectangular.)*
8. *For every $i \in f$, divide the ith row in P by its row-sum. (After this step, the row-sums in P are all equal to 1.)*

The rest of the definition is as in variational multigrid: $R = P^t$ and $Q = RAP$. In the numerical experiments at the end of this book, we show that AMG as defined here (with Algorithms 6.3 and 6.5) has practically the same performance as the above multilevel method (that uses Algorithms 6.3 and 6.4).

6.11 Semicoarsening

The geometric multigrid algorithm in Section 6.4 (with the point red-black GS relaxation) is efficient for the Poisson equation, but not for highly anisotropic equations as in (3.27). In order to work for this type of problem as well, the geometric multigrid algorithm must use alternating line relaxation or alternating "zebra" line relaxation. This way, the Poisson convergence rate can be achieved whether $\varepsilon \ll 1$, $\varepsilon = 1$, or $\varepsilon \gg 1$ in (3.27) [23].

Unfortunately, alternating relaxation is rather difficult to implement, especially in 3-D problems. Another possible geometric multigrid algorithm that works well for any ε in (3.27) uses *semicoarsening*. In this approach, the coarse grid consists of every other line of gridpoints:

$$c = \{(i,j) \in g \mid i \equiv 0 \bmod 2\} \tag{6.15}$$

(see Figure 5.1). With this semicoarsening, one can use line relaxation in the un-coarsened direction only [block relaxation, with a block being an i-line in the original grid in (3.6): the set of points (i,j) for some fixed i and all $1 \leq j \leq n$] to have a good solver also for highly anisotropic equations [23]. (Geometric multigrid with semicoarsening has also been extended to unstructured grids in [70].)

As discussed above, geometric multigrid has the crucial drawback that it cannot solve problems with discontinuous coefficients, unless the discontinuity lines align with all the coarse grids, which is highly unlikely in general problems. It is thus most important to have a matrix-based semicoarsening algorithm, which is efficient for general problems with both anisotropic and discontinuous coefficients.

The matrix-based semicoarsening algorithm [43] [108] is based on the block partitioning in (6.8), where c is now defined as in (6.15). With this partitioning, A_{ff} is block-diagonal, with $n \times n$ blocks that correspond to the i-lines in f (for odd i). More specifically, we have

$$A_{ff} = blockdiag(A^{(1,1)}, A^{(3,3)}, A^{(5,5)}, \ldots). \tag{6.16}$$

The block-rows in A_{fc}, on the other hand, each contain two nonzero $n \times n$ blocks. For example, the third block-row contains the blocks $A^{(3,2)}$ and $A^{(3,4)}$ that couple the third line with the second and fourth lines, respectively, in the original grid in (3.6).

In (6.12), A is approximated by a matrix $\tilde{A}$, which has the same effect on the constant vector as A. The matrix $\tilde{A}$ is then used to construct R and P, along the guidelines in Section 6.2. This produces the algebraic multilevel method in Section 6.8, which is applicable to general grids.

For structured grids as in (3.6), though, one can use the present semicoarsening in (6.15) to come up with a yet better approach. Instead of approximating A spectrally by $\tilde{A}$ and then using this approximation to construct R and P, why not develop spectral approximations directly to (6.9) and (6.10), which will agree with them at least for the constant vector? Indeed, because A_{ff} is block diagonal as in (6.16), $(A_{ff})^{-1}A_{fc}$ contains (in its third block row) blocks of the form $A^{(3,3)\,-1}A^{(3,2)}$ and $A^{(3,3)\,-1}A^{(3,4)}$. These blocks are dense, so they cannot be used in practice. However, they can be approximated by their corresponding row-sums, to provide a practical definition to P. For example, if $\mathbf{1}$ denotes the n-dimensional vector with all components being equal to 1, then one only needs to solve

$$A^{(3,3)}e = A^{(3,2)}\mathbf{1}$$

for an n-dimensional vector of unknowns e to have the desired approximation

$$rs\left(A^{(3,3)\,-1}A^{(3,2)}\right) = diag(e),$$

that is, the diagonal matrix of order n with the components of e on its main diagonal. Similarly, one solves

$$A^{(3,3)}e = A^{(3,4)}\mathbf{1}$$

to have

$$rs\left(A^{(3,3)}{}^{-1}A^{(3,4)}\right) = diag(e).$$

This completes the definition of the third block-row in P. Doing the same for the rest of the odd-numbered block-rows completes the definition of P for the matrix-based semicoarsening algorithm. The restriction matrix can now be defined by $R = P^t$ (for symmetric problems) or in a more general way as in Section 11.7, which is suitable also for nonsymmetric problems [with P redefined using $(A + A^t)/2$ rather than A]. The coarse-grid matrix is then defined by $Q = RAP$. The relaxation method used in the V-cycle is usually "zebra" line-relaxation [in the uncoarsened direction in the original grid (3.6)]. This completes the definition of the matrix-based semicoarsening algorithm.

In Chapter 11 we describe the matrix-based semicoarsening algorithm in more detail, with relations to the domain-decomposition and block-ILU methods. Unfortunately, the matrix-based semicoarsening method is limited to structured grids as in (3.6), and cannot be easily extended to more general, unstructured grids. This is because the above construction of the blocks in P relies heavily on the property that all the blocks in A_{ff} and A_{fc} are square submatrices of order n. This property cannot be guaranteed in unstructured grids. Extending the matrix-based semicoarsening algorithm to more general grids must therefore be left to future research.

6.12 Exercises

1. Show that, for a 5-point stencil as in (3.8), the prolongation indicated in Figure 6.10 is the same as the prolongation used in black-box multigrid, indicated in Figure 6.9.
2. Conclude that, for a 5-point stencil, the TL method that uses the prolongation in Figure 6.10 is the same as the TL method that uses black-box multigrid.
3. Show that, in the previous exercise, the coarse-grid matrix Q no longer has a 5-point stencil but rather a 9-point stencil in c.
4. Conclude that, for 5-point stencils, the ML method (with more than two levels) that uses the prolongation in Figure 6.10 is not the same as the one that uses black-box multigrid.
5. Show that the prolongation indicated in Figure 6.11 is a natural extension of the prolongation indicated in Figure 6.8 to the case of completely unstructured grids.
6. Why is this prolongation likely to transfer a nearly singular eigenvector of Q into a nearly singular vector of A?
7. Let v be a coarse-grid function defined on the bullets in Figure 6.5. Let P be the prolongation indicated in Figures 6.6 and 6.7. Show that Pv is rather small in terms of the norm induced by A (energy norm). Conclude that the matrix PJ_c is moderate in terms of the A-induced norm. (This property is essential in the analysis in Sections 12.8 and 15.8 below.)
8. Write a computer code that implements the geometric multigrid method for the Poisson equation with Dirichlet boundary conditions, discretized by finite

differences as in (3.13) on a uniform grid. Use the red-black point-GS relaxation in the V(1,1)-cycle. Do you obtain the Poisson convergence factor of 0.1?

9. Modify your code to solve problems with different kinds of boundary conditions, such as Neumann and mixed boundary conditions. Does the convergence rate deteriorate?

10. Modify your code to solve highly anisotropic problems as in Section 3.9. Does the convergence rate deteriorate?

11. Modify your code to use alternating "zebra" line-GS relaxation instead of red-black point-GS relaxation. Does the convergence rate improve?

12. Modify your code to use a coarse-grid matrix of the form $Q = RAP$, as in variational multigrid. Use it to solve isotropic diffusion problems as in (4.15) with discontinuous diffusion coefficient $\tilde{D}$ (discretized on a uniform mesh as in Figure 4.13). Test cases in which the discontinuity lines in $\tilde{D}$ align with the coarse meshes, and cases in which they don't. Are the convergence rates different in the different cases?

13. Write a computer code that implements the algebraic multilevel method in Section 6.9 for general sparse linear systems. The solution can be found in Section 17.10 in [103].

Part IV

Matrix-Based Multigrid
for Structured Grids

In the rest of this book, we consider matrix-based (or matrix-dependent) multigrid linear system solvers such as those in [39] [43] [84] [86] [105] [108] and [128]). In this family of multigrid methods, the components in the multigrid algorithm such as transfer operators and coarse-grid matrix depend only on the coefficient matrix A that couples the points in the original grid g in (3.6), but are completely independent of the particular application and its properties. Thus, they are applicable to any linear system arising from finite-difference, finite-volume, or finite-element discretization on a uniform grid, independent of the original boundary-value problem. The great advantage is that they may be implemented on the computer once and for all to be ready for future use.

Matrix-based multigrid algorithms are more general than geometric multigrid, because they also work well for linear systems arising from the discretization of PDEs with variable and even discontinuous coefficients (even when the discontinuity lines do not align with the coarse grid). Moreover, they are known to work well even for severely stretched grids. Furthermore, they are also efficient for problems in nonrectangular domains, as is illustrated in [105]. Still, they assume that the grid is uniform as in (3.6), so nonrectangular domains must be completed artificially to rectangular ones, using extra fictitious points and corresponding trivial equations. The matrix-based multigrid method may use full coarsening as in (6.1) or semicoarsening as in (6.15).

A subfamily of matrix-dependent multigrid algorithms that is not limited to uniform (or structured) grids is the family of algebraic multigrid versions [86]. In algebraic multigrid, the coefficient matrix A is used to define not only the transfer and coarse-grid operators but also the subset of unknowns to serve as a virtual (nonphysical) coarse grid. This is why we prefer to refer to this subfamily as algebraic multilevel methods rather than algebraic multigrid methods. In this part, however, we limit our attention to the rest of the matrix-based multigrid methods, which do assume that the grid is structured as in (3.6), with a 9-point stencil as in (4.13).

This part contains five chapters. In Chapter 7, we describe the Automatic Multigrid (AutoMUG) method that solves problems with 5-point stencils using coarse-grid matrices with 5-point stencils only. This method is then used in Chapter 8 to denoise digital images. In Chapter 9, we describe the black-box multigrid (BBMG) method, which is applicable also to problems with 9-point stencil. We use ideas from AutoMUG to design a BBMG version for a most difficult problem with discontinuous coefficients. In Chapter 10, we further use ideas from AutoMUG to design a BBMG version for highly indefinite Helmholtz equations. We use computational two-level spectral analysis to estimate the convergence of both BBMG and AutoMUG for this kind of problem. Finally, in Chapter 11, we describe a matrix-dependent semicoarsening method, which is efficient even for problems with highly anisotropic and discontinuous coefficients. We interpret this method as an interesting combination of domain decomposition, line ILU factorization, and variational multigrid.

The Automatic Multigrid Method

In this chapter, we describe the Automatic Multigrid (AutoMUG) iterative method for the solution of linear systems that use a 5-point stencil as in (3.8) on a uniform grid as in (3.6). The special advantage of this method is that it uses inexpensive 5-point stencils also on the coarse grids, which allows the use of the red-black point-GS relaxation in the entire V-cycle. AutoMUG has good convergence rates for problems with variable and even discontinuous coefficients, even when the discontinuity lines do not align with the coarse grid.

7.1 Properties of the AutoMUG Method

The Automatic Multigrid (AutoMUG) method for the solution of problems with 5-point stencils as in (3.8) in a structured rectangular grid is introduced in [92] [104] [104] [105], and extended also to problems with 7-point stencils in the analogous three-dimensional cubic grid in [94]. AutoMUG works well for diffusion problems with variable and even discontinuous coefficients, even when the discontinuity lines do not align with the coarse grid.

The main advantage of AutoMUG is in its straightforward and inexpensive setting. In fact, for problems with 5-point stencil as in (3.7), AutoMUG uses 5-point stencils only at all the coarse grids as well, allowing the use of efficient relaxation methods such as the red-black point-Gauss–Seidel relaxation method. The cost of the set-up stage (the computation of the transfer and coarse-grid matrices in the entire multigrid hierarchy) is, thus, particularly low: it is just one work unit (the computational work required in a point-GS relaxation). An inexpensive set-up stage is particularly attractive in time-dependent and nonlinear problems in two and three spatial dimensions, where the set-up must be repeated in every linear system encountered in the implicit time marching or Newton iteration (see Chapter 8).

There are two other multigrid methods that also use 5-point stencils at all the grids. The algorithm of this kind that is proposed in [2] uses harmonic average as in electrical engineering to produce flux-preserving 5-point stencils on the coarse grids. This algorithm, however, is limited to diffusion problems, and hasn't been extended to more difficult problems such as indefinite Helmholtz equations considered later in this book. The Cyclic-Reduction Multigrid (CR-MG) method in [30] also uses 5-point stencils only, but is inferior to AutoMUG for highly indefinite Helmholtz equations with mixed complex boundary conditions, as is apparent from the tests in [92]. For these reasons and for its straightforward and clear algebraic formulation, AutoMUG is the method with which we wish to start the detailed description of multigrid algorithms. Later on, we also use ideas from AutoMUG to improve the more general black-box multigrid method.

7.2 Cyclic Reduction

We start by describing the cyclic reduction method for the solution of linear systems with tridiagonal coefficient matrices. For this purpose, we use the multigrid framework that is also used later in more general problems with 5-point stencils.

Assume that the coefficient matrix A in (3.5) is an $n \times n$ tridiagonal matrix with only three nonzero diagonals: the main diagonal, the diagonal just above it, and the diagonal just below it (all the other diagonals vanish). For this kind of linear system, the cyclic reduction method can be formulated as a multigrid method.

Let us partition the set of indices of unknowns $\{1, 2, 3, \ldots, n\}$ into two blocks: the block of odd numbers $\{1, 3, 5, \ldots, 2\lceil n/2 \rceil - 1\}$ (denoted by "o"), and the block of even numbers $\{2, 4, 6, \ldots, 2\lfloor n/2 \rfloor\}$ (denoted by "e"). With this partitioning, the block form

$$A = \begin{pmatrix} A_{oo} & A_{oe} \\ A_{eo} & A_{ee} \end{pmatrix},$$

where A_{oo} and A_{ee} are diagonal submatrices (or blocks), and A_{oe} and A_{eo} are bidiagonal blocks (blocks with only two nonzero diagonals). This block form will lead to the definition of the matrices required in the TL method in Section 6.1:

$$P = \begin{pmatrix} -A_{oo}^{-1} A_{oe} \\ I \end{pmatrix}$$
$$R = \begin{pmatrix} -A_{eo} A_{oo}^{-1} & I \end{pmatrix}$$
$$Q = A_{ee} - A_{eo} A_{oo}^{-1} A_{oe}$$

(where I is the identity matrix of suitable order). Actually, these definitions are equivalent to those in (6.9)–(6.11) in Section 6.10 (with the subsets f and c used there interpreted as the subsets e and o used here, respectively). Fortunately, unlike the general A_{ff} in Section 6.10, here A_{oo} is diagonal so A_{oo}^{-1} is also available. This leads to a practical method in the present tridiagonal case.

Note also that

$$RA = \begin{pmatrix} 0 & Q \end{pmatrix}, \tag{7.1}$$

so we have

$$RAP = Q = RA \begin{pmatrix} 0 \\ I \end{pmatrix} = RA J_e^t. \tag{7.2}$$

This observation is used later on.

The TL method in Section 6.1 is now implemented with the above defined matrices. No prerelaxation is used, and only one postrelaxation is used ($\nu_1 = 0$

and $\nu_2 = 1$). The relaxation method in the TL method is a half sweep in the point-GS method, in which only the odd-numbered unknowns are updated. The iteration matrix for this prerelaxation is [see (5.6)]

$$S_o = \begin{pmatrix} 0 & -(A_{oo})^{-1} A_{oe} \\ 0 & I \end{pmatrix}.$$

On the other hand, from (7.1) we have that the iteration matrix of the coarse-grid correction is

$$I - PQ^{-1}RA = I - P \begin{pmatrix} 0 & I \end{pmatrix} = \begin{pmatrix} I & A_{oo}^{-1} A_{oe} \\ 0 & 0 \end{pmatrix}.$$

The iteration matrix of the TL method is just the product of these two iteration matrices:

$$S_o \left(I - PQ^{-1}RA \right) = (0)$$

(the zero matrix). In other words, the algebraic error is annihilated in only one iteration: the error in e is annihilated by the coarse-grid correction, and the error in o is annihilated in the post-relaxation. (Actually, one could have the same effect also with the alternative definition $P = J_e^t$, but this definition would be more difficult to extend to the 2-D case below.) Thus, the TL method with the present definitions is actually a direct tridiagonal-system solver. Furthermore, because Q is tridiagonal as well, the same method itself can also be used recursively to apply Q^{-1} to $R(b - Ax_{in})$ in the TL algorithm, which actually leads to the ML method in Section 6.3. This completes the definition of the Cyclic Reduction method.

As a matter of fact, tridiagonal systems can be also solved efficiently by Gaussian elimination (see Section 2.2), because no fill-in is produced. However, the cyclic reduction method is much easier to implement on parallel and vector computers [79]. Furthermore, it can be extended to problems with 5-point stencils, arising, for example, from the discretization of two-dimensional diffusion equations.

7.3 The Two-Dimensional Case

The main point in the cyclic reduction method is that the coarse-grid matrix Q is tridiagonal as well as A, which allows recursion in the multigrid V-cycle. Unfortunately, for 5-point stencils as in the finite-difference and finite-volume discretization of 2-D diffusion problems, the situation is more complicated: in order to have the same stencil as in A, Q must be modified, as discussed below.

Assume that A has the 5-point stencil arising from the finite-difference or finite-volume discretization of a 2-D diffusion equation on a uniform grid as in (3.6). In this case, A can be written as the sum

$$A = X + Y, \tag{7.3}$$

where X is a tridiagonal matrix representing the discrete approximation to the derivatives in the x spatial direction (including the discrete boundary conditions associated with this direction), and Y contains the discrete derivatives in the

y spatial direction (including the discrete boundary conditions associated with this spatial direction). As in Section 7.2 above, one can define matrices P_X and R_X from X, and P_Y and R_Y from Y. More precisely, P_X and R_X are block-diagonal matrices, with blocks that are defined as in Section 7.2 in the relevant x-lines in the grid in (3.6), and P_Y and R_Y are obtained in a similar way from Y in the relevant y-lines. Since the prolongation operator is naturally the product of P_X and P_Y and the restriction operator is the product of R_X and R_Y, a natural definition of the coarse-grid matrix would be

$$R_Y R_X X P_X P_Y + R_X R_Y Y P_Y P_X. \tag{7.4}$$

Unfortunately, this matrix is no longer of a 5-point stencil but rather of a 9-point stencil in the coarse grid c in (6.1). In order to preserve the 5-point structure also in the coarse grid matrix, one should recall from Section 6.2 that a good coarse-grid matrix should act in much the same way as the original matrix A on the constant vector. Thus, the matrices R_Y and P_Y in the first term in (7.4) could be replaced by their row-sum. More precisely, because the coarse-grid matrix is later restricted to c, and because the prolongation in the y spatial direction leaves the values in c unchanged, P_Y could actually be replaced by the identity matrix rather than $rs(P_Y)$. [This can also be seen from (7.2).] Thus, the first term in (7.4) can be well approximated by $rs(R_Y)R_X X$ (for gridpoints in c). When the second term in (7.4) is approximated in a similar way, one gets

$$rs(R_Y)R_X X + rs(R_X)R_Y Y.$$

When this matrix is restricted to c, one gets the coarse-grid matrix

$$Q = J_c \left(rs(R_Y)R_X X + rs(R_X)R_Y Y \right) J_c^t, \tag{7.5}$$

which has a 5-point stencil in c, as required. Yet better, because prolongation matrices should naturally also be used in the definition of Q, one could modify the above definition by replacing $rs(R_Y)$ by $rs(P_Y)$ and $rs(R_X)$ by $rs(P_X)$:

$$Q = J_c \left(rs(P_Y)R_X X + rs(P_X)R_Y Y \right) J_c^t. \tag{7.6}$$

For the Poisson equation, this definition is different from the previous one only at boundary points in c. Later on, we show more clearly why (7.6) is better than (7.5). In the next section, we give a more complete definition of the AutoMUG method.

7.4 Definition of the AutoMUG Method

Consider the PDE

$$-(D_1 u_x)_x - (D_2 u_y)_y + \beta u = \mathcal{F} \tag{7.7}$$

in the unit square, with suitable boundary conditions. Assume that A is obtained from the finite-difference or finite-volume discretization of (7.7) on the uniform grid g in (3.6), so it has the 5-point stencil in (3.7). Furthermore, it can be written in the form (7.3), where the matrix X contains the discrete first term and half of the last term in (7.7) and has the 3-point stencil

$$\begin{bmatrix} 0 & 0 & 0 \\ W & C_x & E \\ 0 & 0 & 0 \end{bmatrix}, \tag{7.8}$$

and the matrix Y contains the discrete second term and half of the third term in (7.7) and has the 3-point stencil

$$\begin{bmatrix} 0 & N & 0 \\ 0 & C_y & 0 \\ 0 & S & 0 \end{bmatrix} \tag{7.9}$$

[where $C_x + C_y = C$ is the middle element in the stencil of A, and the Northern element N in (7.9) is not to be confused with the integer N used often to denote the order of A]. Note that X also incorporates the boundary conditions at the left and right edges, and Y also incorporates the boundary conditions at the lower and upper edges.

We first define the partial (1-D) prolongation and restriction matrices that are used later to define the transfer matrices P and R. The definitions are as in Section 7.2, except that here the matrices are square rather than rectangular:

$$P_X = 2I - diag(X)^{-1}X$$
$$P_Y = 2I - diag(Y)^{-1}Y$$
$$R_X = 2I - Xdiag(X)^{-1}$$
$$R_Y = 2I - Ydiag(Y)^{-1}.$$

The prolongation operator P is now defined by

$$P = P_Y P_X J_c^t.$$

This means that the prolongation is done in two steps: first in the x spatial direction, and then in the y spatial direction. Consider, for instance, a coarse-grid vector $v \in l_2(c)$. The prolonged vector $Pv \in l_2(g)$ is defined in two steps: first, it is defined in every other x-line in g, or, more specifically, at each point $(i,j) \in g$ for which i is even. This is done using the 3-point stencil (7.8) at (i,j) as follows.

$$(Pv)_{i,j} \equiv \begin{cases} v_{i,j} & \text{if } j \text{ is even} \\ -\dfrac{Wv_{i,j-1} + Ev_{i,j+1}}{C_x} & \text{if } j \text{ is odd.} \end{cases} \tag{7.10}$$

Next, Pv is prolonged further to each point $(i,j) \in g$ with odd i. This is done using the 3-point stencil (7.9) at (i,j) as follows.

$$(Pv)_{i,j} \equiv -\frac{S(Pv)_{i-1,j} + N(Pv)_{i+1,j}}{C_y}. \tag{7.11}$$

This completes the definition of Pv in the entire grid g.

Next, the restriction matrix in the AutoMUG method is defined by

$$R = J_c R_X R_Y.$$

This means that the restriction is also done in two steps: first in the y spatial direction, and then in the x spatial direction. To make things clearer, consider a fine-grid vector $v \in l_2(g)$. The restricted vector $Rv \in l_2(c)$ is defined in two steps. First, it is defined in every other x-line in the original grid g [or at each point $(i, j) \in g$ with even i]. This is done by three different 3-point stencils: the stencil (7.9) at (i, j), the corresponding stencil

$$\begin{bmatrix} 0 & N_S & 0 \\ 0 & C_{y,S} & 0 \\ 0 & S_S & 0 \end{bmatrix} \tag{7.12}$$

at $(i-1, j)$ (indicated by the subscript "S"), and the corresponding stencil

$$\begin{bmatrix} 0 & N_N & 0 \\ 0 & C_{y,N} & 0 \\ 0 & S_N & 0 \end{bmatrix} \tag{7.13}$$

at $(i+1, j)$ (indicated by the subscript "N"). Here is how these three 3-point stencils are used to define $R_Y v$ at points (i, j) with even i:

$$(R_{Yv})_{i,j} \equiv -\frac{Sv_{i-1,j}}{C_{y,S}} + v_{i,j} - \frac{Nv_{i+1,j}}{C_{y,N}}.$$

Then, the final value of Rv at each point $(i, j) \in c$ is defined by three other 3-point stencils: the stencil (7.8) at (i, j), the corresponding stencil

$$\begin{bmatrix} 0 & 0 & 0 \\ W_W & C_{x,W} & E_W \\ 0 & 0 & 0 \end{bmatrix} \tag{7.14}$$

at $(i, j-1)$ (denoted by the subscript "W"), and the corresponding stencil

$$\begin{bmatrix} 0 & 0 & 0 \\ W_E & C_{x,E} & E_E \\ 0 & 0 & 0 \end{bmatrix} \tag{7.15}$$

at $(i, j+1)$ (denoted by the subscript "E"). Here is how these three 3-point stencils are used to define Rv in c:

$$(Rv)_{i,j} \equiv -\frac{W(R_{Yv})_{i,j-1}}{C_{x,W}} + (R_{Yv})_{i,j} - \frac{E(R_{Yv})_{i,j+1}}{C_{x,E}}.$$

This completes the definition of Rv in c.

Note that R is the transpose of a prolongation matrix derived from A^t rather than A.

The definition of the coarse-grid matrix Q uses the observation made in (7.2) that

$$J_c R_X X J_c^t = J_c R_X X P_X J_c^t \quad \text{and} \quad J_c R_Y Y J_c^t = J_c R_Y Y P_Y J_c^t.$$

Furthermore, it uses the principle in Section 6.2 that Q should have about the same effect as A on the constant vector. This means that positive matrices may be replaced by their row-sum. This leads to the definition in (7.5), or, better yet, the definition in (7.6):

$$Q = J_c \left(rs(P_Y) R_X X + rs(P_X) R_Y Y \right) J_c^t. \tag{7.16}$$

This means that Q has the following stencil at the point $(i,j) \in c$:

$$\frac{-S + C_y - N}{C_y} \begin{bmatrix} 0 & 0 & & 0 \\ -\dfrac{W_W W}{C_{x,W}} & C_x - \dfrac{E_W W}{C_{x,W}} - \dfrac{W_E E}{C_{x,E}} & -\dfrac{E_E E}{C_{x,E}} \\ 0 & 0 & & 0 \end{bmatrix}$$

$$+ \frac{-W + C_x - E}{C_x} \begin{bmatrix} 0 & -\dfrac{N_N N}{C_{y,N}} & 0 \\ 0 & C_y - \dfrac{S_N N}{C_{y,N}} - \dfrac{N_S S}{C_{y,S}} & 0 \\ 0 & -\dfrac{S_S S}{C_{y,S}} & 0 \end{bmatrix} \tag{7.17}$$

[where the subscripts S, N, W, and E indicate stencils that are evaluated at the points $(i-1,j)$, $(i+1,j)$, $(i,j-1)$, and $(i,j+1)$ in g, respectively].

The stencil of Q in (7.17) is the sum of two 3-point stencils: a 3-point stencil that couples the points in c in the x spatial direction, and another 3-point stencil that couples the points in c in the y spatial direction. The coefficient of the first term comes from $rs(P_Y)$ and uses elements from (7.9), and the coefficient of the second term comes from $rs(P_X)$ and contains elements from (7.8). It is now clear why (7.6) is better than (7.5): it guarantees that the coefficients in (7.17) use elements from the original stencil of A at the relevant point $(i,j) \in c$, which makes sense.

The stencil of Q in (7.17) is a 5-point stencil in c, so recursion can be used as in the ML algorithm in Section 6.3. The relaxation method used in the V-cycle is usually the red-black point-GS method. This completes the definition of the AutoMUG iterative method.

The AutoMUG method can also be easily extended to 3-D elliptic PDEs, discretized by finite differences or finite volumes on a uniform cubic grid [94]. In the next section, we also define a simplified AutoMUG version, which is easier to implement and analyze.

7.5 The AutoMUG(q) Version

The 5-point stencil in (7.17) that couples the coarse-gridpoints contains two terms: a 3-point stencil in the x spatial direction, and a 3-point stencil in the y spatial direction in c. Thanks to the fact that (7.6) is used rather than (7.5), the coefficients of these terms can be simplified and slightly modified, to produce another AutoMUG version that is easier to implement and analyze. This is done below.

Let us focus on the first term in (7.17). Before the 3-point stencil that couples the coarse-gridpoints in the x spatial direction, there is the scalar coefficient

$$\frac{-S + C_y - N}{C_y} = 2 - \frac{S + C_y + N}{C_y}. \tag{7.18}$$

For matrices as in (7.3) with Y with nearly zero row-sum, this coefficient is only slightly different from 2. Such matrices indeed arise often in the finite-difference and finite-volume discretization of diffusion problems and PDEs as in (7.7). Thus, it makes sense to simplify the original AutoMUG version by replacing the coefficient in (7.18) by the scalar

$$2 + q,$$

where q is a small parameter that approximates the row-sum of $-diag(Y)^{-1}Y$. For diffusion equations, for example, in which the row-sums of Y are almost always zero, it makes sense to use $q = 0$. Similarly, the coefficient of the second term in (7.17) (the 3-point stencil that couples the coarse-gridpoints in the y spatial direction) is replaced by $2 + q$ as well. In summary, instead of (7.16), we have

$$Q = (2 + q)J_c(R_X X + R_Y Y)J_c^t.$$

The same parameter q is also used in the recursion in the ML algorithm. This is the AutoMUG(q) version, which is easier not only to implement but also to analyze. In particular, the AutoMUG(0) method is used in the applications in Chapter 8 below.

A prototype version of AutoMUG(q) that uses multiple coarse grids rather than the standard V-cycle is analyzed in some model cases [92] and [94]. Unfortunately, this theory is inapplicable to the present AutoMUG and AutoMUG(q), which use the common V-cycle. Indeed, the corollaries in [92] and [94] that attempt to prove convergence of AutoMUG(0) assume that the iteration matrix of the post-relaxation is PJ_c. This assumption makes sense only when two different coarse grids are used, c as in Figure 6.2 in the first coarse-grid correction, and

$$d = \{(i,j) \in g \mid i \equiv j \equiv 1 \bmod 2\} \tag{7.19}$$

as in Figure 7.1 in the second coarse-grid correction, but not in the present V-cycle, which uses only c to produce a single coarse-grid correction term.

Still, AutoMUG works well for various kinds of examples with 5-point stencils [105], including diffusion problems with discontinuous coefficients, even when the discontinuity lines do not align with the coarse grid. In the next chapter, AutoMUG(0) is used to solve such problems in the field of image processing.

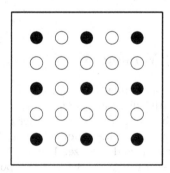

Fig. 7.1. The coarse grid d: the subgrid of points that lie in odd-numbered lines and odd-numbered columns.

7.6 Exercises

1. Count the number of arithmetic operations required to solve a tridiagonal linear system of order N by constructing the LU decomposition (Gauss elimination). Note that both L and U are bidiagonal (no fill-in), so the forward elimination and back-substitution is efficient. (The answer is a function of N.)

2. Assume that an arithmetic operation requires α seconds, where α is a small positive parameter. Calculate the sequential time required to solve the above problem. (See the definitions in the exercises at the end of Chapter 1.)

3. Count the number of arithmetic operations required in the cyclic-reduction method for solving the above problem.

4. Consider a parallel computer of p processors as in the exercises at the end of Chapter 1. Use your answer from the previous exercise to calculate the parallel computation time.

5. Calculate also the communication time on that computer.

6. Calculate also the parallel time on that computer.

7. Calculate also the speedup and average speedup on that computer.

8. As in the exercises at the end of Chapter 1, assume now that the concrete parameters $\alpha = 10^{-6}$, $\gamma = 10^{-3}$, $\delta = 10^{-7}$, and $N = 10^9$ are used. How do the speedup and average speedup behave as functions of p (number of processors)?

9. Show that, if A is of 5-point stencil as in (3.8), then the coarse-grid matrix Q in AutoMUG is of 5-point stencil as well.

10. Show by induction that, if A is of 5-point stencil, then so are all the coarse-grid matrices used in the ML algorithm with AutoMUG.

11. Conclude that the red-black point-GS relaxation can be used in all the levels in the V-cycle in AutoMUG.

12. Consider the Poisson equation (3.12), discretized as in (3.13) on the uniform grid g in (3.6). Show that the stencil of Q in (7.6) (at least at inner points in the coarse grid c) is the same as that of A (at inner points in the original grid g), and is different from that of the geometric coarse-grid matrix in Section 6.4 by factor 4 only.

13. Show that the above factor is canceled with the factor $1/4$ in (6.4) (which is missing in the restriction matrix $J_c R_X R_Y$ used in AutoMUG), so in this case AutoMUG is the same as geometric multigrid at inner gridpoints.

14. Show that if the matrices X and Y in (7.3) are nonsingular, then the corresponding matrices that approximate them on the coarse grid in AutoMUG are nonsingular as well.

15. Show that if the matrices X and Y in (7.3) are diagonally dominant L-matrices, then the corresponding matrices that approximate them on the coarse grid in AutoMUG are diagonally dominant L-matrices as well.

16. Show that if the matrices X and Y in (7.3) are M-matrices, then the corresponding matrices that approximate them on the coarse grid in AutoMUG are M-matrices as well. The solution can be found in [105].

17. Write the computer code that implements the AutoMUG(0) method.

18. Define an isotropic diffusion problem [as in (4.15)] and discretize it on a uniform grid using a 5-point stencil (using finite volumes or uniform finite-element

mesh as in Figure 4.13). Apply your AutoMUG(0) code to it. Test also discontinuous diffusion coefficients with discontinuity lines that don't align with the coarse grid.

19. Verify that the coarse-grid matrices obtained in the previous exercise are symmetric and diagonally dominant. Conclude from Lemma 2.1 that they are also SPD. Conclude from the theory in [118] that they are also M-matrices. Use an IMSL routine to compute their nearly singular eigenvectors, and verify that their eigenvalues and components are all positive.

8

Applications in Image Processing

In this chapter, we use the multigrid method in the field of image processing. In particular, automatic multigrid is used to remove noise from grayscale and color digital images.

8.1 The Denoising Problem

Two important problems in image processing that are closely related to each other are denoising and object segmentation. In the denoising problem, one needs to clarify a given digital image that is contaminated with random noise by removing the noise without spoiling the true features in the image. In the object segmentation problem, on the other hand, one has to find particular objects in the given digital image. In both cases, one has to filter out irrelevant features in the image while leaving the relevant features unharmed.

The algorithms that solve these problems are based on nonlinear PDEs, discretized on the uniform grid that represents the digital image and solved by some iterative method. The original denoising algorithm for grayscale images has been introduced in [81]. Improved versions of this algorithm have been also introduced in [10] and [11]. Extensions to color images in red-green-blue (RGB) form have been also proposed in [89] and [109]. Similarly, algorithms for object segmentation have been proposed in [33] (for grayscale images) and [55] (for RGB color images).

The above extensions to RGB color images suffer from the drawback that the PDE also contains mixed derivatives, so the coefficient matrices resulting from the linearization and discretization are no longer M-matrices. The present extension to color images, on the other hand, uses M-matrices with efficient 5-point stencils only.

In existing works in denoising and object segmentation, little attention is paid to computational costs. When large digital images are considered, the amount of calculations required in the above algorithms may be prohibitively large, because it may increase superlinearly with the problem size (the number of pixels in the image). Indeed, hundreds of iterations are needed in the above denoising algorithms [89], and hundreds of CPU seconds are consumed by the object-segmentation algorithm in [55] (even for small 64×64 images).

The main advantage of the present algorithm is its efficiency: it requires only ten iterations to denoise rather large 512×512 grayscale and color images. The algorithm uses M-matrices with 5-point stencils for both grayscale and color images, allowing the use of the efficient AutoMUG linear system solver. The algorithm to denoise RGB color images can be viewed as a natural extension of the algorithm to denoise grayscale images. Both algorithms can probably be modified to solve the object-segmentation problem as well, by adding to the PDE the extra terms introduced in the PDE in [33].

8.2 The Denoising Algorithm for Grayscale Images

We first consider grayscale digital images, which are actually 2-D arrays of pixels that take numerical values to indicate the gray level (amount of light) at the individual points in the image. Let z be the grid function (vector) that is defined on the above array and contains the gray levels in the given noisy image. The denoising algorithm that removes the noise is a variant of the algorithm in [81], and is defined by the sequence of linear PDEs listed below. To this end, we assume for simplicity that the digital image is square, so it can be embedded in the unit square like the grid in (3.6). First, let $\tilde{z}$ and $u^{(0)}$ be functions (defined in the unit square) that agree with z on the gridpoints corresponding to the pixels in the digital image. Then, for $i = 1, 2, \ldots$, solve

$$u^{(i)} - \alpha \nabla \left(\frac{\nabla u^{(i)}}{1 + \frac{|u_x^{(i-1)}|^2 + |u_y^{(i-1)}|^2}{k_i}} \right) = \tilde{z} \qquad (8.1)$$

in the unit square $(0, 1) \times (0, 1)$, with homogeneous Neumann boundary conditions. As before, the first appearance of the Nabla operator "∇" stands for the divergence operator, and the second one stands for the gradient operator. Also, α and k_i are positive parameters to be specified later.

Each linear PDE in (8.1) is discretized by finite volumes (Section 3.12) on a uniform $n \times n$ grid of meshsize $h = 1/n$, where n is the size of the image. (We assume for simplicity a square image of $n \times n$ pixels.) The derivatives $u_x^{(i-1)}$ and $u_y^{(i-1)}$ in (8.1) are discretized by second-order central differencing of $u^{(i-1)}$ at midpoints that lie in between gridpoints. For example, for some function f, the derivatives f_x and f_y are approximated by

$$(f_y)_{i+1/2,j} \doteq (f_{i+1,j} - f_{i,j})/h$$
$$(f_x)_{i+1/2,j} \doteq 0$$
$$(f_y)_{i,j+1/2} \doteq 0$$
$$(f_x)_{i,j+1/2} \doteq (f_{i,j+1} - f_{i,j})/h.$$

With this discretization, the coefficient matrices are diagonally dominant M-matrices with a 5-point stencil, so the AutoMUG method can be used to solve the linear systems. Each linear system obtained from the discretization of (8.1) is solved approximately by one V(1,1)-cycle of the AutoMUG(0) version. Thanks to the inexpensive set-up in AutoMUG(0), the total cost of this solve (solution process) is only 4–5 *work units* (where a work unit is the work required in a point relaxations on a 5-point stencil). The initial guess in the AutoMUG(0) cycle in the ith iteration in

(8.1) is taken from the previous iteration, $u^{(i-1)}$. The output of the algorithm, the final iteration [say, $u^{(10)}$] in the discrete grid, contains the required denoised image.

The parameters α and k_i in (8.1) have not yet been specified. Usually, they depend on the size of the image and the amount of noise. In particular, α should be small enough to avoid introducing extra blur, but not too small, to allow effective denoising. In the present example with uniformly-distributed random noise (with magnitude of at most 25% of the maximal gray level) at each and every pixel, a good choice seems to be $\alpha = .000025$ for a 512×512 image.

The parameters k_i are rough estimates to the average value of $|u_x^{(i-1)}|^2 + |u_y^{(i-1)}|^2$ over the gridpoints. Thus, the k_is should decrease, to reflect the decrease in the noise during the iteration. In the present examples, we have used $k_i = 10000$ for $i \leq 2$, $k_i = 1000$ for $3 \leq i \leq 5$, and $k_i = 100$ for $6 \leq i \leq 10$.

8.3 The Denoising Algorithm for Color Images

In this section, we extend the above denoising algorithm also to RGB color images. Let z be a noisy color image in the RGB form: $z \equiv (z^{(r)}, z^{(g)}, z^{(b)})$ contains the three color vectors that contain the intensities of the red, green, and blue colors in the pixels in the entire noisy image. The denoising algorithm requires the solution of the following system of nonlinear PDEs in the unit square with homogeneous Neumann boundary conditions. The unknown functions in this system, $\mathcal{R}(x,y)$, $G(x,y)$, and $B(x,y)$, combine to form the unknown vector function $u \equiv (\mathcal{R}, G, B)$. The square grid associated with the pixels in the image is embedded in the unit square. (The meshsize in the grid is $h = 1/n$, and the number of gridpoints, n^2, is the same as the number of pixels in the digital image.) The right-hand sides in the equations are interpreted as continuous functions that agree with the color vectors $z^{(r)}$, $z^{(g)}$, and $z^{(b)}$ in the grid. Here is the system of PDEs:

$$\mathcal{R} - \alpha \nabla \left(F(T(u)) \nabla \mathcal{R} \right) = z^{(r)} \tag{8.2}$$

$$G - \alpha \nabla \left(F(T(u,h)) \nabla G \right) = z^{(g)} \tag{8.3}$$

$$B - \alpha \nabla \left(F(T(u,h)) \nabla B \right) = z^{(b)}, \tag{8.4}$$

where

$$k^2(u) \equiv \int_0^1 \int_0^1 \left(\mathcal{R}_x^2 + \mathcal{R}_y^2 + G_x^2 + G_y^2 + B_x^2 + B_y^2 \right) dx\,dy, \tag{8.5}$$

$$T(u) \equiv \begin{pmatrix} 1 + \dfrac{\mathcal{R}_x^2 + G_x^2 + B_x^2}{k^2(u)} & \dfrac{\mathcal{R}_x \mathcal{R}_y + G_x G_y + B_x B_y}{k^2(u)} \\[4mm] \dfrac{\mathcal{R}_x \mathcal{R}_y + G_x G_y + B_x B_y}{k^2(u)} & 1 + \dfrac{\mathcal{R}_y^2 + G_y^2 + B_y^2}{k^2(u)} \end{pmatrix}, \tag{8.6}$$

and $F()$ is a function defined on the set of 2×2 matrices. (The precise definition of $F()$ is specified below.)

Note that $T(u) \equiv T(u(x,y))$ and $F(T(u)) \equiv F(T(u(x,y)))$ are actually functions of the spatial variables x and y. Thus, (8.2)–(8.4) is a system of three coupled PDEs with variable nonlinear coefficients. This system is now linearized in the Newton iterative method, in which each iteration is the solution of a linearized system, in which $T(u)$ uses the u value from the previous iteration:

$$u^{(0)} \equiv (\mathcal{R}^{(0)}, G^{(0)}, B^{(0)}) = z$$

$$\mathcal{R}^{(i)} - \alpha \nabla \left(F(T(u^{(i-1)})) \nabla \mathcal{R}^{(i)} \right) = z^{(r)} \tag{8.7}$$

$$G^{(i)} - \alpha \nabla \left(F(T(u^{(i-1)})) \nabla G^{(i)} \right) = z^{(g)} \quad (i = 1, 2, 3, \ldots) \tag{8.8}$$

$$B^{(i)} - \alpha \nabla \left(F(T(u^{(i-1)})) \nabla B^{(i)} \right) = z^{(b)}. \tag{8.9}$$

Let us now specify the function $F()$ used above. In fact, this function can be defined in two possible ways. The first way, in the spirit of [109], uses $F(K) = K^{-1}$ (where K is a 2×2 nonsingular matrix). With this approach, it is natural to discretize (8.7)–(8.9) on a uniform finite-element mesh. Unfortunately, with either square finite elements as in Section 4.4 or triangular finite elements as in Section 4.5, there is no guarantee that the coefficient matrices are M-matrices. Indeed, because the eigenvalues of $T(u)$ could be distinct (at least at some points in the domain), the problem may well be anisotropic, so the conditions in Section 4.6 are not satisfied in general.

The second way, on the other hand, produces coefficient matrices that are diagonally-dominant M-matrices and use a 5-point stencil only. In the spirit of [55], $F(K)$ is defined to be the inverse of the determinant of K:

$$F(K) = \det(K)^{-1} \tag{8.10}$$

(where K is a 2×2 nonsingular matrix). Note that, with this choice, $F()$ is no longer a matrix but rather a scalar. Using finite volumes as in Section 3.12 or uniform finite-element triangulation would thus produce diagonally dominant M-matrices (see Section 4.6). This is why the definition in (8.10) is used here.

Using discrete homogeneous Neumann boundary conditions, the function $k^2(u)$ in (8.5) is discretized on the uniform grid in (3.6) by forward finite differencing as follows:

$$k^2(u) \doteq \sum_{i=1}^{n} \sum_{j=1}^{n-1} (\mathcal{R}_{i,j+1} - \mathcal{R}_{i,j})^2 + (G_{i,j+1} - G_{i,j})^2 + (B_{i,j+1} - B_{i,j})^2$$

$$+ \sum_{i=1}^{n-1} \sum_{j=1}^{n} (\mathcal{R}_{i+1,j} - \mathcal{R}_{i,j})^2 + (G_{i+1,j} - G_{i,j})^2 + (B_{i+1,j} - B_{i,j})^2. \tag{8.11}$$

Each of the three linear PDEs in (8.7)–(8.9) is discretized by the finite-volume discretization method in Section 3.12 on the uniform grid in (3.6). The derivatives in $T(u)$ in (8.6) are discretized by second-order central differencing at midpoints of the form $(i + 1/2, j)$ and $(i, j + 1/2)$ as follows:

$$(G_Y)_{i+1/2,j} \doteq (G_{i+1,j} - G_{i,j}) / h \tag{8.12}$$

$$(G_x)_{i+1/2,j} \doteq (G_{i+1,j+1} + G_{i,j+1} - G_{i+1,j-1} - G_{i,j-1}) / (4h) \tag{8.13}$$

$$(G_y)_{i,j+1/2} \doteq (G_{i+1,j+1} + G_{i+1,j} - G_{i-1,j+1} - G_{i-1,j}) / (4h) \tag{8.14}$$

$$(G_x)_{i,j+1/2} \doteq (G_{i,j+1} - G_{i,j}) / h, \tag{8.15}$$

and similarly for $\mathcal{R}$ and B.

Note that when (i, j) in (8.12) through (8.15) is a boundary point, some points in the right-hand sides in (8.12) through (8.15) may lie outside the grid, which is of course impossible. Fortunately, thanks to the discrete homogeneous Neumann boundary conditions, the right-hand sides in (8.12) through (8.15) can be modified by replacing the nonexisting point by the nearest point that lies within the grid.

With the present discretization, (8.7) through (8.9) are actually three independent linear systems, each of which has a coefficient matrix which is a diagonally dominant M-matrix and uses a 5-point stencil. Thus, the AutoMUG method can be used to solve each of these algebraic systems separately. Actually, each algebraic system is solved approximately by one V(1,1)-cycle of the AutoMUG(0) method. The initial guess for this V-cycle is taken from the previous iteration in Newton's method [$u^{(i-1)}$ in (8.7)–(8.9)].

The choice of the parameter α depends on the size of the digital image and the amount of noise. In the present experiments, we have used uniformly-distributed random noise (with magnitude of at most 25% of the maximal color intensity) at every pixel and every color. In this case, the optimal choices seem to be $\alpha = .000025$ for a 512×512 image and $\alpha = .0000025$ for a 864×1152 image. Ten Newton iterations are used, and the output $u^{(10)}$ seems to be well denoised.

8.4 Numerical Examples

In the numerical examples, we use digital images of the model Lena, used often in tests in image processing. The special features in these images make them particularly challenging and suitable for scientific comparison, which is the only reason for using them here [77].

We first take a 512×512 grayscale image of Lena and add to each pixel in it uniformly distributed random noise with magnitude of at most 25% of the maximum light intensity in the image. The present denoising algorithm in Section 8.2 uses $\alpha = .000025$, $k_i = 10000$ for $i \leq 2$, $k_i = 1000$ for $3 \leq i \leq 5$, and $k_i = 100$ for $6 \leq i$. Ten iterations are used in (8.1), each of which is solved approximately by a single AutoMUG(0) iteration. The AutoMUG(0) method uses a V(1,1)-cycle (with ten levels) with the red-black point-GS relaxation method.

Although the AutoMUG method fails to solve the linear system in Section 7.1.3 in [97], this is because the problem there uses $k_i = 1$, which leads not only to high local anisotropy but also to poor denoising. By refusing to solve the problem, the AutoMUG method actually indicates that something is wrong with the system of equations and it may be too irregular. Indeed, the large k_is used here guarantee not only good denoising but also good performance of the multigrid method.

The noisy image is shown below in the file "noisy.pgm" in Figure 8.1. The image that is the output of the denoising algorithm is shown in the file "denoised.pgm" in Figure 8.2. The result exhibits good denoising with almost no blur or any other side effects.

Next, we turn to RGB color images. We take a 512×512 color image of Lena and add to each pixel and each color in the RGB form of it uniformly distributed random noise with magnitude of at most 25% of the maximum color intensity. The present denoising algorithm in Section 8.3 uses $\alpha = .000025$. Ten Newton (fixed-point) iterations in (8.7) through (8.9) are used. Each of the three individual linear systems

Fig. 8.1. The noisy grayscale (noncolor) image that is the input to the present denoising algorithm.

Fig. 8.2. The denoised grayscale (noncolor) image that is the output of the present denoising algorithm.

Fig. 8.3. The noisy RGB color image that is the input to the present denoising algorithm.

Fig. 8.4. The denoised RGB color image that is the output of the present denoising algorithm.

Fig. 8.5. The result of the Wiener filter with a 10-by-10 stencil applied to the noisy RGB color image.

in each Newton iteration is solved approximately by one AutoMUG(0) iteration. The AutoMUG(0) method uses a V(1,1)-cycle (with ten levels) with the red-black point-GS relaxation method. (The AutoMUG method may give slightly better results.)

The noisy image is shown in the file "noisy.tif" in Figure 8.3. The image that is the output of the denoising algorithm is shown in the file "denoised.tif" in Figure 8.4. The result exhibits good denoising with almost no blur or other side effects.

For comparison, we also show the result of the Wiener filter in the Matlab library applied separately to each noisy color in the RGB form of the original image in "noisy.tif." This filter is based on statistical characteristics of the input image. At each pixel, the value is calculated as a weighted average of values in a subsquare of size 10×10 pixels surrounding it. (This size is the minimal size required to remove some noise in this example.) The output of this Wiener filter is shown in the file "Wiener10.tif" in Figure 8.5. It turns out that this filter introduces a considerable amount of blur. A comparison between Figures 8.4 and 8.5 shows the advantage of the present denoising algorithm in terms of sharpness of the output image.

8.5 Exercises

1. Assume that the real function f and its derivative f' are available. Assume also that f has a root $\tilde{x}$, for which $f(\tilde{x}) = 0$. The Newton (or Newton–Raphson) iteration to find $\tilde{x}$ is as follows. Let $x^{(0)}$ be an initial approximation (initial guess), which is not too far from $\tilde{x}$. For $i = 0, 1, 2, \ldots$, define

$$x^{(i+1)} = x^{(i)} - f(x^{(i)})/f'(x^{(i)}).$$

For sufficiently large i, $|x^{(i+1)} - \tilde{x}|$ is sufficiently small, and $x^{(i+1)}$ is accepted as a sufficiently good approximation to $\tilde{x}$. Write the computer code that implements the Newton iteration.

2. The fixed-point problem is as follows: given a function $g(x)$, find a point $\tilde{x}$ for which

$$g(\tilde{x}) = \tilde{x}.$$

The Picard (fixed-point) iteration for this problem is defined by

$$x^{(i+1)} = g(x^{(i)}).$$

Write the computer code that implements this iteration.

3. Show that the Newton iteration can be formulated as a Picard iteration for the function

$$g(x) = x - f(x)/f'(x).$$

4. Show that (8.2)–(8.4) can be written in the form

$$T(u) = F, \tag{8.16}$$

where F is a given function, u is the unknown function, and T is a nonlinear differential operator.

5. Extend the above Newton iteration to solve (8.16), or to find the "root" u that solves the equation

$$T(u) - F = 0. \tag{8.17}$$

6. For every function v, let $\mathcal{L}_v$ be the linearization of T at v. Use the above exercises to show that Newton's iteration for the solution of (8.17) can be written in the form

$$u^{(i+1)} = u^{(i)} - \mathcal{L}_{u^{(i)}}^{-1}\left(T(u^{(i)}) - F\right) = u^{(i)} - \left(u^{(i)} - \mathcal{L}_{u^{(i)}}^{-1}F\right) = \mathcal{L}_{u^{(i)}}^{-1}F.$$

7. Conclude that (8.7) through (8.9) indeed form the Newton iteration for the solution of (8.2) through (8.4).

8. Show that the coefficient matrix in each Newton iteration in the denoising algorithms is a symmetric, strictly diagonally dominant L-matrix.

9. Conclude from Lemma 2.1 that it is also SPD.

10. Conclude from the theory in [118] that it is also an M-matrix.

11. Use an IMSL routine to compute its nearly singular eigenvector. Verify that its eigenvalue and components are all positive.

12. Use your AutoMUG(0) code from the exercise at the end of the previous chapter in a denoising code. Apply it to grayscale and color images, and assess its performance in terms of noise and blur.

13. Pick one of the linear systems at some Newton iteration. Verify that the coarse-grid matrices used in the AutoMUG(0) method for this system are all symmetric and strictly diagonally dominant. Conclude that they are also SPD M-matrices.

The Black-Box Multigrid Method

In this chapter, we describe the black-box multigrid (BBMG) method for the numerical solution of structured linear systems with a 9-point stencil. This matrix-based multigrid method is suitable for diffusion problems with variable and even discontinuous coefficients, even when the discontinuity lines don't align with the coarse grid. Furthermore, we show how it can be modified to solve efficiently diffusion problems with strong-diffusion areas that interact with each other.

9.1 Definition of Black-Box Multigrid

Here we describe the black-box multigrid (BBMG) method [39]. This method is more general and robust than AutoMUG, because it is applicable not only to 5-point stencils but also to 9-point stencils. Still, one can use ideas from AutoMUG to come up with an optimal version of BBMG [97].

Assume that the linear system (3.5) is from 9-point stencil as in (4.13) on the uniform $n \times n$ grid g in (3.6). The main part in the definition of BBMG is the definition of the prolongation operator $P : l_2(c) \rightarrow l_2(g)$, where the coarse grid c is as in (6.1). For this purpose, we first introduce the so-called "lumped" stencil; here lumping means summing the coefficients in the original stencil (4.13) in the spatial direction that is perpendicular to the direction along which prolongation is carried out. More precisely, for a vector $v \in l_2(c)$, $Pv \in l_2(g)$ is defined as follows. For every coarse-gridpoint $(i, j) \in c$, the value of Pv is, of course, the same as the corresponding value of v:

$$(Pv)_{i,j} \equiv v_{i,j}.$$

Consider now a gridpoint $(i, j) \in f = g \setminus c$. Assume first that i is even and j is odd, so (i, j) lies in between the two coarse-gridpoints $(i, j - 1)$ and $(i, j + 1)$. In this case, the prolongation is in the x spatial direction, so the lumped stencil is obtained from lumping (summing) in the y spatial direction in the original stencil at (i, j):

$$\begin{bmatrix} 0 & 0 & 0 \\ NW + W + SW & N + C + S & NE + E + SE \\ 0 & 0 & 0 \end{bmatrix} \tag{9.1}$$

(see Figure 6.9). This leads to the definition

$$(Pv)_{i,j} \equiv -\frac{(NW + W + SW)v_{i,j-1} + (NE + E + SE)v_{i,j+1}}{N + C + S} \quad (9.2)$$

(as in Section 6.7).

When i is odd and j is even, (i, j) lies in between the coarse-gridpoints $(i - 1, j)$ and $(i + 1, j)$. In this case, the lumping is done in the x spatial direction, and the lumped stencil is

$$\begin{bmatrix} 0 & NW + N + NE & 0 \\ 0 & W + C + E & 0 \\ 0 & SW + S + SE & 0 \end{bmatrix}.$$

Thus, the prolonged value is

$$(Pv)_{i,j} \equiv -\frac{(NW + N + NE)v_{i+1,j} + (SW + S + SE)v_{i-1,j}}{W + C + E}. \quad (9.3)$$

This completes the first prolongation step in Section 6.7.

Finally, when both i and j are odd, the value of Pv is defined simply from the original stencil:

$$\begin{aligned} (Pv)_{i,j} \equiv - & (NWv_{i+1,j-1} + NEv_{i+1,j+1} + SWv_{i-1,j-1} \\ & + SEv_{i-1,j+1} + N(Pv)_{i+1,j} + S(Pv)_{i-1,j} \\ & + W(Pv)_{i,j-1} + E(Pv)_{i,j+1})/C \end{aligned} \quad (9.4)$$

[as in (6.7)]. Of course, when (i, j) is a boundary point, some of the coefficients in the original stencil vanish, so no fictitious points are needed. This completes the definition of the prolongation operator P. As in variational multigrid, R and Q are defined by $R = P^t$ and $Q = RAP$. (For nonsymmetric systems, an improved definition is proposed in [40], see Section 16.4.)

For highly anisotropic problems, alternating ("zebra") line relaxation (Section 5.7) should be used within the V-cycle [41]. For problems with variable strong-diffusion direction, however, the convergence may deteriorate. Alternating line-relaxation cannot help in this case, because the difficulty is with the coarse-grid correction (see Section 15.1), not the relaxation method. In such problems, one should probably turn to matrix-based semicoarsening (Chapter 11) or algebraic multilevel method (Chapter 15). In the present applications, however, we are mostly interested in isotropic problems, so we use BBMG with the 4-color point-GS relaxation (Section 5.9).

In the next section, the above definition of the prolongation matrix P is studied and improved, in the spirit of AutoMUG.

9.2 Improvements in Diffusion Problems

In the special case in which A is of 5-point stencil, one would naturally like the prolongation operator P defined above to agree with the prolongation operator used in the AutoMUG method. Fortunately, this is indeed the case for diffusion problems as in (3.2) discretized by finite differences or finite volumes, at least at interior gridpoints that lie in between two coarse-gridpoints. Indeed, because a 5-point stencil is used, we have

$$NW = NE = SW = SE = 0.$$

Therefore, (9.2) and (9.3) are simplified to read

$$(Pv)_{i,j} \equiv \begin{cases} -\dfrac{Wv_{i,j-1} + Ev_{i,j+1}}{N+C+S} & \text{if } i \equiv j+1 \equiv 0 \bmod 2 \\[2mm] -\dfrac{Nv_{i+1,j} + Sv_{i-1,j}}{W+C+E} & \text{if } i+1 \equiv j \equiv 0 \bmod 2. \end{cases}$$

The row-sums of both X and Y at the interior point (i,j) are zero, thus we also have

$$C_x = C - C_y = C + N + S$$
$$C_y = C - C_x = C + W + E.$$

Therefore, we have from (7.10) and (7.11) that the prolongation for AutoMUG is the same as for BBMG for interior points (i,j) with $i+j \equiv 1 \bmod 2$.

Let us now extend the above also to boundary gridpoints in (3.6). For this, however, we need to assume that n in (3.6) is odd, so no boundary point lies in between two coarse-gridpoints. Consider, for example, a boundary point (i,j) that lies on the rightmost column of gridpoints, that is, $j = n$ and i is even. For this point, we have $E = 0$; however, thanks to the fact that the contribution from the discrete boundary condition is incorporated in C_x, we still have

$$C_x = C - C_y = C + N + S,$$

which again implies that the prolongation in BBMG is the same as in AutoMUG. Similarly, consider a gridpoint at the upper row in the grid; that is, $i = n$ and j is even. For this point, $N = 0$ and the discrete boundary conditions are incorporated in C_y. Thus, we again have

$$C_y = C - C_x = C + W + E,$$

which implies that the prolongation in BBMG is the same as in AutoMUG. The same result is also true for boundary gridpoints at the lower row (where $i = 1$ and j is even) and the leftmost column of points in the grid (where $j = 1$ and i is even).

It only remains to show that the prolongation in BBMG is the same as in AutoMUG also for points (i,j) of which both i and j are odd. For this, however, we need an additional assumption.

Lemma 9.1 *Assume that A in (3.5) is obtained from finite-difference or finite-volume discretization of the diffusion equation (3.2), so it can be written as in (7.3). Assume also that n in (3.6) is odd, and that $diag(X)^{-1}X$ and $diag(Y)^{-1}Y$ commute with each other. Then the prolongation matrix for BBMG is the same as the prolongation matrix for AutoMUG.*

Proof. From the latter assumption, we have that P_X and P_Y defined in Section 7.4 commute with each other as well. Using also the assumption that n is odd, we have from the above discussion that the prolongation in BBMG is the same as in AutoMUG for every gridpoint (i, j) with $i + j \equiv 1$ mod 2. Thus, we have that the prolongation matrix P in BBMG satisfies

$$\begin{aligned}
P &= diag(A)^{-1} \left(diag(X) P_X P_Y + diag(Y) P_Y P_X \right) J_c^t \\
&= diag(A)^{-1} \left(diag(X) + diag(Y) \right) P_Y P_X J_c^t \\
&= P_Y P_X J_c^t.
\end{aligned}$$

This completes the proof of the lemma.

Let us now consider an implementation that uses the coarse grid d in (7.19) (containing the odd-numbered gridpoints) rather than c in (6.1). In order to have the above property that the prolongation in BBMG is the same as in AutoMUG even at boundary points that lie in between two coarse-grid boundary points, we must replace C in (9.2) and (9.3) by

$$\tilde{C} \equiv -(NW + N + NE + W + E + SW + S + SE). \tag{9.5}$$

(Note that some of the terms on the right-hand side might be zero.) In other words, $\tilde{C}$ is obtained from C by subtracting the row-sum, or the contribution from the discrete boundary conditions. As a result, we have for a boundary point on the rightmost column of gridpoints ($j = n$, i is even, $E = 0$, and the discrete boundary conditions are incorporated in C_x) that

$$C_y = -(N + S) = \tilde{C} + W,$$

so the prolongation in BBMG is the same as in AutoMUG. Similarly, for a gridpoint on the upper row in the grid ($i = n$, j is even, $N = 0$, and the discrete boundary conditions are incorporated in C_y), we have

$$C_x = -(W + E) = \tilde{C} + S,$$

so the prolongation in BBMG is again the same as in AutoMUG. The same result holds for gridpoints on the lower row (where $i = 1$ and j is even) and the leftmost column of gridpoints (where $j = 1$ and i is even). In summary, we have

Lemma 9.2 *Assume that A in (3.5) is obtained from finite-difference or finite-volume discretization of the diffusion equation (3.2), so it can be written as in (7.3). Assume also that n in (3.6) is odd, and that the coarse grid d in (7.19) is used rather than c in (6.1). Assume also that*

$$J_c diag(X)^{-1} X diag(Y)^{-1} Y J_d^t = J_c diag(Y)^{-1} Y diag(X)^{-1} X J_d^t. \tag{9.6}$$

Assume also that C in (9.2)–(9.3) is replaced by $\tilde{C}$ defined in (9.5) at boundary gridpoints that lie in between two boundary coarse-gridpoints ($i = 1$ and j even, $i = n$ and j even, $j = 1$ and i even, or $j = n$ and i even). Then the prolongation matrix for BBMG is the same as the prolongation matrix for AutoMUG.

Proof. Let P be the prolongation matrix for BBMG. Thanks to the above modification that uses $\tilde{C}$ rather than C at boundary points, we have

$$J_b P = J_b P_Y P_X J_d^t,$$

where b is as in (5.5). Furthermore, from (9.6), we have

$$J_c P_X P_Y J_d^t = J_c P_Y P_X J_d^t,$$

which implies that

$$\begin{aligned} J_c P &= J_c diag(A)^{-1} \left(diag(X) P_X P_Y + diag(Y) P_Y P_X \right) J_d^t \\ &= J_c diag(A)^{-1} \left(diag(X) + diag(Y) \right) P_Y P_X J_d^t \\ &= J_c P_Y P_X J_d^t. \end{aligned}$$

Clearly, we also have

$$J_d P = J_d P_Y P_X J_d^t = I.$$

In summary, we have

$$P = P_Y P_X J_d^t,$$

as required. This completes the proof of the lemma.

It is thus advisable to replace C in (9.2) and (9.3) by $\tilde{C}$ in (9.5) also in the recursive calls to BBMG and in general 9-point stencils, at least at boundary points that lie in between two coarse-grid boundary points. In Table 9 in [97], it is indeed shown that this can improve the convergence rate for a diffusion example with discontinuous coefficients.

9.3 Using the Right-Hand Side

In Section 6.7, the first and second prolongation steps are defined by solving suitable homogeneous subproblems. A slightly better approach is to solve inhomogeneous subproblems instead, where the right-hand side is taken from the residual. This way, the prolonged coarse-grid term (Pe in the ML method in Section 6.3) may approximately satisfy the residual equation, hence provide a good correction term to add to x_{in}.

This improvement is made by adding to the three right-hand sides in (9.2)–(9.4), the extra term

$$\frac{(b - A x_{in})_k}{C}, \tag{9.7}$$

where $k = (i - 1)n + j$ is the index of the row in A corresponding to the (i, j)th gridpoint. With this improvement, P is actually no longer a rectangular matrix but rather a square matrix, as is formulated algebraically in Section 12.2 below.

The cost of the modification is negligible because the residual $b - A x_{in}$ is already available. It may reduce slightly the number of multigrid iterations required for convergence in diffusion problems, although not in indefinite Helmholtz equations.

BBMG works well for many diffusion problems with variable and even discontinuous coefficients, even with discontinuity lines that don't align with the coarse grid. Furthermore, it can also be extended to 3-D diffusion problems, discretized on uniform cubic grids with 27-point stencil [41]. In the next section, we study a particularly difficult example, for which BBMG works well after further modification.

9.4 Improvement for Problems with Discontinuous Coefficients

Geometric and variational multigrid work well for diffusion problems with discontinuous coefficients, provided that the discontinuity lines align with all the coarse grids. Unfortunately, this significant limitation excludes most realistic cases.

Matrix-based multigrid methods such as BBMG and AutoMUG, on the other hand, work well even when the discontinuity lines don't align with the coarse grid. This allows one to write a general computer code that is independent of the particular application.

Usually, BBMG is superior to AutoMUG thanks to its variational properties (Section 6.7). Still, one could learn from AutoMUG to develop further improvements for BBMG as well. Here we describe such an improvement in a particularly difficult case, in which standard BBMG fails to converge.

Consider the isotropic diffusion equation

$$-(\tilde{D}u_x)_x - (\tilde{D}u_y)_y = \mathcal{F}, \quad 0 < x, y < 62. \tag{9.8}$$

Here the diffusion coefficient $\tilde{D}$ is defined as in Figure 9.1:

$$\tilde{D}(x,y) \equiv \begin{cases} \dfrac{1000+1}{2} \cdot \dfrac{1}{1000} & \text{if } (x,y) \in \Theta \equiv \{(x,y) \mid |x-\xi| + |y-\xi| \le 1\} \\ 1000 & \text{if } (x,y) \in ((0,\xi) \times (0,\xi)) \cup ((\xi,62) \times (\xi,62)) \setminus \Theta \\ 1 & \text{if } (x,y) \in ((0,\xi) \times (\xi,62)) \cup ((\xi,62) \times (0,\xi)) \setminus \Theta \\ 0 & \text{if } (x,y) \notin (0,62) \times (0,62), \end{cases} \tag{9.9}$$

where $0 < \xi < 62$ is a parameter denoting the location of the discontinuity in $\tilde{D}$. Thus, the strong-diffusion regions at the lower-left and upper-right subsquares are separated by a thin strip of width $\sqrt{2}$, denoted by Θ. This strip prevents any strong diffusion between these subsquares, so they can interact only weakly; this property should also be observed in the discrete system and its coarse-grid approximation.

The above problem can be viewed as an approximation of the so-called checkerboard problem, which is almost the same, except that Θ is set to be the empty set $\emptyset$, so $\tilde{D}(x,y)$ takes the value

$$\begin{array}{ll} 1000 & (x,y) \in ((0,\xi) \times (0,\xi)) \cup ((\xi,62) \times (\xi,62)) \\ 1 & (x,y) \in ((0,\xi) \times (\xi,62)) \cup ((\xi,62) \times (0,\xi)) \\ 0 & (x,y) \notin (0,62) \times (0,62). \end{array} \tag{9.10}$$

Unfortunately, both the finite-difference and finite-volume discretization methods are inadequate for the original checkerboard problem. Indeed, they produce strong interaction between the lower-left and upper-right subsquares, and thus cannot possibly yield a good approximation to the original PDE, in which the interaction

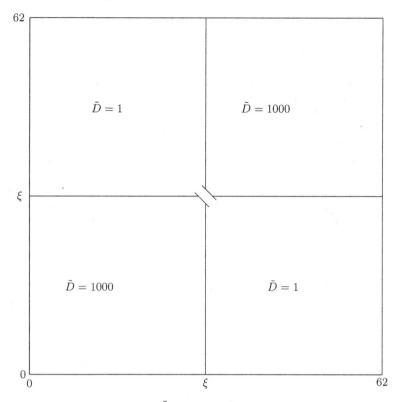

Fig. 9.1. The diffusion coefficient $\tilde{D}$ in the present example. The distance between the regions of strong diffusion at the middle of the domain is $\sqrt{2}$.

is only weak. This inadequacy is even more pronounced on the coarse grids, where the meshsize is larger; this is why AutoMUG diverges and BBMG converges with the rather nonoptimal convergence rate of 0.5. Fortunately, the present PDE that uses (9.9) rather than (9.10) is more suitable for the finite-volume discretization, which produces adequate discretization with no strong interaction between the lower-left and upper-right subsquares, as required [2]. Furthermore, with AutoMUG, this property is preserved also on the coarse grids, leading to an excellent (Poisson) convergence rate, even when the discontinuity lines don't align with the coarse grid. Below we learn from AutoMUG also how BBMG should be modified to produce such a good convergence rate as well.

Let us first complete the boundary-value problem by specifying also the boundary conditions to be of the third kind:

$$u_n = 0 \ x = 0 \text{ or } y = 0$$
$$\tilde{D}u_n + 0.5u = 0 \ x = 62 \text{ or } y = 62,$$

where $\mathbf{n}$ denotes the outer normal vector. The equation is discretized on a uniform 63×63 grid, using the finite-volume discretization method (Section 3.12) with the meshsize $h = 1$.

We test two cases: (a) $\xi = 31$, in which the discontinuity line aligns with all the coarse grids, and (b) $\xi = 30$, in which the discontinuity line does not align with any of the coarse grids. It turns out that both AutoMUG and the multigrid method displayed in Figure 6.10 (which uses lumping towards the center only) converge rapidly for both $\xi = 30$ and $\xi = 31$. This is in agreement with the theory in Chapter 12, which implies that the method displayed in Figure 6.10 should be robust, no matter whether the discontinuity lines align with the coarse grid or not.

Other matrix-based multigrid algorithms, such as the matrix-based semicoarsening method in Section 6.11 and BBMG, may fail for a certain type of ξ, probably due to lack of theory. For BBMG, in particular, the situation is as follows. With $\xi = 30$, it converges nicely with the Poisson convergence rate. With $\xi = 31$, on the other hand, it still converges rapidly as long as only two levels are used. This is indeed expected, since the finite-volume scheme uses a 5-point stencil on the original (fine) grid, so there is actually no difference between the lumping in Figures 6.9 and 6.10, and the theory in Chapter 12 applies. Unfortunately, when more than two levels are used, a 9-point stencil must be used on the coarse grids, and BBMG practically stagnates.

The reason for this stagnation is seen in Figure 9.2, which displays the strong coupling between gridpoints near the junction point $(31, 31)$ in the coefficient

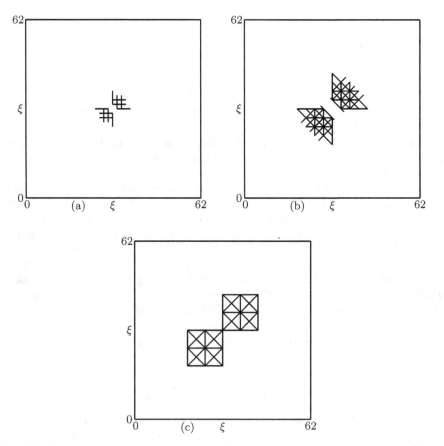

Fig. 9.2. Strong coupling between strong diffusion regions at the junction point $(\xi, \xi) = (31, 31)$ for black-box multigrid in (a) the first (finest) level, (b) the second level, and (c) the third level in standard BBMG. Clearly, the third level produces an inappropriate approximation to the original problem due to the interaction between strong diffusion regions.

matrices in the first, second, and third levels in BBMG. In the first level (a), the finite-volume scheme uses a 5-point stencil with nearest-neighbor coupling only. Thanks to the strip Θ in (9.9), there is no strong interaction between the lower-left and upper-right subsquares, as required. This property is preserved also in the second level (b), although the coarse-grid matrix $Q = RAP$ already uses a 9-point stencil with oblique coupling between coarse-gridpoints. Unfortunately, the coefficient matrix on the third level (c) involves strong coupling also between the lower-left and upper-right subsquares, hence approximates poorly the original system, and leads to stagnation of the BBMG iteration.

Because BBMG is superior to both AutoMUG and the method in Figure 6.10 in many other problems, it is particularly important to fix it for the present example as well. In fact, improper strong interaction between regions of strong diffusion may appear on some coarse grid in BBMG and slow down its convergence in many examples. In order to fix BBMG, however, we need first to understand better the source of the trouble.

As can be seen in Figure 9.2(b), the coarse-grid matrix $Q = RAP$ involves strong oblique coupling between coarse-gridpoints to the North-East (or South-West) of the junction point $(31, 31)$. Due to the vertical and horizontal lumping used in BBMG, these oblique strong couplings produce horizontal and vertical strong couplings in the prolongation from the junction point in the third level. This produces an improper strong coupling between the lower-left and upper-right subsquares through the junction point in the third level. In order to prevent this, one must modify the lumping in BBMG, and lump not only vertically or horizontally (as in Figure 6.9) but also obliquely (as in Figure 6.10), at least when the lumped elements are unusually large in magnitude.

Suppose that, for some point $(i, j) \in f$ with even i and odd j, the North-Eastern element in the stencil, NE, is relatively very large in magnitude:

$$|NE| > 10|E|.$$

In this case, it makes no sense to use vertical lumping, because the Eastern element is too small in magnitude to take the extra magnitude of NE. It makes more sense to use oblique lumping towards the main element in the stencil, C. This is done by replacing C and NE in (9.2) by

$$\tilde{C} \equiv C + NE \quad \text{and} \quad \tilde{NE} \equiv 0,$$

respectively. Similarly, assume that

$$|NE| > 10|N|$$

at a point $(i, j) \in f$ with odd i and even j. In this case, NE is too large (in magnitude) to be lumped horizontally, and must be lumped obliquely onto the central element in the stencil, C. This is done by replacing C and NE in (9.3) by

$$\tilde{C} \equiv C + NE \quad \text{and} \quad \tilde{NE} \equiv 0,$$

respectively. The same approach is also used for the other corner elements in the 9-point stencil. This improved version of BBMG doesn't suffer from the above-mentioned stagnation, and actually converges with the Poisson convergence rate.

9.5 Exercise

1. Show that the prolongation used in BBMG is indeed as indicated in Figure 6.9.
2. Show that, for a 5-point stencil as in (3.8), this prolongation is equivalent to the one indicated in Figure 6.10.
3. Show that, even for a 5-point stencil, the coarse-grid matrix Q is no longer of 5-point stencil but rather of 9-point stencil.
4. Show that, if A is of 9-point stencil, then the prolongation used in BBMG is no longer equivalent to the prolongation indicated in Figure 6.10.
5. Show that, if A is of 9-point stencil, then Q is of 9-point stencil as well.
6. Show by induction that, if A is of 9-point stencil, then all the coarse-grid matrices used in the ML algorithm are of 9-point stencil as well.
7. Conclude that the 4-color point-GS relaxation can be used at all the levels in the V-cycle.
8. Show that the ith column in the coarse-grid matrix Q can be calculated by

$$Qe^{(i)} = R\left(A\left(Pe^{(i)}\right)\right),$$

where $e^{(i)}$ is the ith column in the identity matrix of the same order as Q.
9. Color the coarse grid c with nine colors as in the last exercise at the end of Chapter 5. Let $C^{(j)}$ ($1 \le j \le 9$) be the coarse-grid vector with the value 1 at coarse-gridpoints in the jth color and 0 elsewhere. Show that

$$QC^{(j)} = R\left(A\left(PC^{(j)}\right)\right)$$

gives immediately all the columns $Qe^{(i)}$ with i corresponding to any gridpoint in the jth color. Conclude that Q can be calculated by $3 \cdot 9 = 27$ matrix-vector operations, regardless of the size of the grid.
10. Write the computer code that implements the BBMG method for problems of 9-point stencil on a uniform grid. Use the 4-color relaxation method in the V-cycle.
11. Apply your code to the Poisson equation, discretized by bilinear finite elements as in Section 4.4. Do you obtain the Poisson convergence factor of 0.1?
12. Apply your code to highly anisotropic equations as in Section 3.9. Does the convergence rate deteriorate?
13. Modify your code by using the alternating "zebra" line-GS relaxation instead of the 4-color relaxation. Does the convergence rate improve?
14. Apply your code to the diffusion problem with discontinuous coefficient as in (9.9) with $\xi = 31$. Print the coarse-grid matrices and the graphs of strong coupling around the middle point (ξ, ξ). Verify that you indeed obtain the same pictures as in Figure 9.2: the lower-left and upper-right subsquares are only weakly coupled in the first and second level, but strongly coupled in the third level.
15. Repeat the above exercise, only this time use the fix in Section 9.4. Verify that this time the lower-left and upper-right subsquares are only weakly coupled in the third level as well.
16. What is the convergence factor in your code in the previous exercise? Is it as small as the Poisson convergence factor?

The Indefinite Helmholtz Equation

In this chapter, we apply BBMG and AutoMUG versions to the indefinite Helmholtz equation, discretized on a uniform rectangular grid. Using ideas from AutoMUG, the transfer operators in the BBMG version are also designed to agree with the nearly singular eigenfunction of the original PDE. We also show how to compute the spectrum of the iteration matrix in some model cases. This analysis helps in the design of the multigrid method.

10.1 Multigrid for the Indefinite Helmholtz Equation

Here we consider the indefinite Helmholtz equation (3.14) (with $\beta < 0$), discretized by finite differences as in Section 3.8 on a uniform grid as in (3.6). This chapter contains results from [90], [91], and [97].

It is well-known that geometric and variational multigrid methods (Sections 6.4 and 6.5) do not work well for indefinite Helmholtz equations. This is because the nearly singular eigenfunctions are no longer smooth as in diffusion problems, so they are not preserved well by standard prolongation operators. Furthermore, the coarse-grid matrices obtained from rediscretizing the differential operator in (3.14) may be completely different from A with respect to these nearly singular eigenfunctions. Indeed, from (3.19) we have that the eigenvalue of A with respect to a nearly singular eigenvector is

$$4h^{-2}\left(\sin^2(\pi k h/2) + \sin^2(\pi l h/2)\right) + \beta \sim \pi^2(k^2 + l^2) + \beta,$$

provided that

$$\pi k h \leq \sqrt{|\beta|}h \ll 1.$$

Consider now some coarse grid of meshsize $H > h$, and assume that (3.14) is discretized on it using (3.18) with h replaced by H. The resulting coefficient matrix also has the same eigenvalue

$$4H^{-2}\left(\sin^2(\pi k H/2) + \sin^2(\pi l H/2)\right) + \beta \sim \pi^2(k^2 + l^2) + \beta,$$

provided that

$$\pi k H \leq \sqrt{|\beta|} H \ll 1. \tag{10.1}$$

Unfortunately, this places a severe limitation on how large H could be when geometric or variational multigrid is used. In other words, very few coarse grids can actually be used in the V-cycle.

Note that this limitation is related to the coarse-grid correction, which is responsible for reducing the nearly singular error modes, and has nothing to do with the particular relaxation method used within the V-cycle, which is responsible for reducing the rest of the error modes. Thus, no relaxation method could help here; the only way to increase the number of coarse grids that can be used is by a clever design of the transfer and coarse-grid matrices.

In [26], a projection method that "filters out" nearly singular error components is proposed. Although this method has no limit on the number of coarse grids that can be used, it is limited to the slightly indefinite 1-D equation with constant coefficients. A related oblique-projection method is also proposed in [15]. This method requires an exact solve on a coarse grid of meshsize H satisfying

$$H \leq |\beta|^{-1/2}, \tag{10.2}$$

at least for the $|\beta| \leq 150$ tested there. For larger $|\beta|$, (10.2) may be insufficient, and yet smaller H may be required.

Here we are particularly interested in matrix-based methods, which require no special treatment as in the above filtering or projection, and hence avoid significant changes to the original algorithm and its implementation. In fact, we will see below that a minor change to the original algorithm could produce transfer and coarse-grid operators that agree with the nearly singular eigenfunction of the original PDE. Because this change is given in algebraic terms, it applies to all sorts of boundary conditions and to problems with variable coefficients as well.

10.2 Improved Prolongation

In [67], the prolongation operator is designed to agree with the nearly singular eigenfunctions of the original differential operator. However, it is assumed there that it is known in advance that the eigenfunctions are of the form $\exp(ikx)$. Although this is a fair assumption for model cases, it no longer holds for more general cases with different kinds of boundary conditions and variable coefficients. This is why we propose here a matrix-based approach to define P. As is indicated in Section 6.6, P is designed to agree with the nearly singular eigenfunction with $k = l$; indeed, for this function we have $\pi^2 k^2 = \pi^2 l^2 = -\beta/2$, so (6.5) holds with the direction q being interpreted as either x or y. Once the restriction and coarse-grid operators are defined by $R = P^t$ and $Q = RAP$, Q preserves the effect of A on this nearly singular eigenfunction, leading to an appropriate coarse-grid approximation.

In order to be general with respect to different kinds of boundary conditions and variable coefficients, this approach must be given an algebraic formulation, in which P is defined in terms of the elements in A alone. This way, the same procedure can be used recursively to construct the transfer operators to and from the next (coarser) grid, and so on, until the entire grid hierarchy used in the V-cycle is created.

10.3 Improved Black-Box Multigrid

Here we consider matrix-based multigrid methods for the solution of the indefinite Helmholtz equation. We assume that the 5-point stencil (3.18) is used on the uniform grid in (3.6), and that full coarsening as in (6.1) or (7.19) is used.

The method in [30] has the advantage that it uses a 5-point stencil also on the coarse grids. However, it turns out that this method is inferior to AutoMUG, particularly for highly indefinite equations with Neumann or mixed boundary conditions [92]. In the sequel, we use ideas from AutoMUG to develop an improved BBMG version, which is more general and efficient indefinite Helmholtz equations. To this end, we need to improve the definition of the prolongation matrix P in BBMG.

Consider a point (i, j) with even i and odd j or odd i and even j. In the sequel, use the stencil at (i, j) to define $\tilde{C}$. This $\tilde{C}$ is then used instead of C in (9.2) and (9.3).

Consider the interior gridpoint that is nearest to (i, j). [If (i, j) is an interior gridpoint, then this gridpoint is (i, j) itself.] Let K be the row-sum of A at the row corresponding to this gridpoint. More precisely, if the stencil at this point is denoted by

$$
\begin{bmatrix}
NW' & N' & NE' \\
W' & C' & E' \\
SW' & S' & SE'
\end{bmatrix},
$$

then K is defined by

$$K = NW' + N' + NE' + W' + C' + E' + SW' + S' + SE'.$$

In the stencil in (3.18), for example, we have just $K = \beta$.

Furthermore, define the relative diffusion in the x and y spatial directions by

$$
D_x = \frac{W' + E'}{W' + E' + N' + S'}
$$
$$
D_y = \frac{N' + S'}{W' + E' + N' + S'}.
$$

We are now ready to define the required $\tilde{C}$. First, define $\tilde{C}$ by

$$
\tilde{C} \equiv
\begin{cases}
-(NW + N + NE + W + E + SW + S + SE) & \text{if } (i,j) \text{ is a} \\
 & \text{boundary point} \\
 & \text{that lies in between} \\
 & \text{two coarse-gridpoints} \\
C - K & \text{otherwise,}
\end{cases}
$$

At this stage, $\tilde{C}$ can be thought of as the central element in the stencil resulting from a different problem, in which β is set to zero and Neumann boundary conditions are imposed on each boundary point that lies in between two coarse-gridpoints, Now, $\tilde{C}$ is modified further by adding to it a fraction of K:

$$
\tilde{C} \leftarrow
\begin{cases}
\tilde{C} + D_x K & \text{the prolongation to } (i,j) \text{ is horizontal} \\
\tilde{C} + D_y K & \text{the prolongation to } (i,j) \text{ is vertical.}
\end{cases}
\tag{10.3}
$$

This new $\tilde{C}$ is now used instead of C in (9.2) and (9.3).

Assuming that the term β in (3.18) is divided proportionally between the terms X and Y in (7.3) [$D_x\beta$ is added to C_x in (7.8) and $D_y\beta$ is added to C_y in (7.9)], the prolongation in BBMG [at the point (i, j)] is the same as in AutoMUG. Indeed,

1. If (i, j) is a boundary gridpoint that lies in between two coarse-gridpoints, then
   - If the prolongation to (i, j) is horizontal, then

$$\tilde{C} + N + S = -(W + E) + D_x K = C_X,$$

   - And, if the prolongation to (i, j) is vertical, then

$$\tilde{C} + W + E = -(N + S) + D_y K = C_y;$$

2. And if, on the other hand, (i, j) is an interior gridpoint or a boundary point that doesn't lie in between two coarse-gridpoints, then
   - If the prolongation to (i, j) is horizontal, then

$$\tilde{C} + N + S = C - D_y K + N + S = C - C_y = C_X,$$

   - And, if the prolongation to (i, j) is vertical, then

$$\tilde{C} + W + E = C - D_x K + W + E = C - C_x = C_Y.$$

Therefore, Lemmas 9.1 and 9.2 still hold.

Note that the above results hold only for the 5-point stencil in (3.18), and not necessarily for more general stencils. Still, it is useful in computing the spectrum of the iteration matrix of BBMG in some model cases, which leads to an a priori estimate of the convergence properties of the BBMG iteration.

10.4 Computational Two-Level Analysis

Here we show how the spectrum of the iteration matrix in AutoMUG and BBMG can be computed in advance in some cases. For this purpose, we need to assume that only two levels are used (as in the TL algorithm in Section 6.1): the fine grid g in (3.6) (with odd n) and the coarse grid c in (6.1). This is why this analysis is called computational two-level analysis.

We also assume that a 5-point stencil is used as in (3.18), and that the matrices X and Y in (7.3) commute with each other and have constant main diagonals:

$$diag(X) = \mathbf{x}I \quad \text{and} \quad diag(Y) = \mathbf{y}I,$$

for some constants $\mathbf{x}$ and $\mathbf{y}$. Thanks to the above modification of BBMG, we have from Lemma 9.1 that the prolongation matrix in BBMG is the same as that in AutoMUG. Because both X and Y are symmetric, this is also true for the restriction matrix $R = P^t$. These properties are helpful in the analysis below.

Two-level analysis has been introduced in [113] for a geometric two-grid method for diffusion problems with constant coefficients and periodic boundary conditions. Here, it is extended also to matrix-based two-grid methods for definite and indefinite problems (with either Dirichlet or periodic boundary conditions). In this context, it can provide valuable information about the suitable design and convergence properties of multigrid methods also in more general cases.

The spectrum of the iteration matrix computed in the two-level analysis gives a useful indication about the convergence properties of the iterative method. Clearly, if all the eigenvalues are considerably smaller than one in magnitude, then the iterative method converges rapidly. Furthermore, even when there are few eigenvalues with large magnitude, an outer acceleration method can be used to annihilate them and yield rapid convergence.

The present assumptions are required only for the sake of the analysis. Both AutoMUG and the present BBMG version are applicable also in more general cases, including variable coefficients and multilevel implementation of the ML algorithm in Section 6.3. This is indeed apparent from the numerical experiments.

The computational two-level analysis method uses the partitioning of the original grid g in (3.6) into the four subgrids in (5.8). Let v be a common eigenvector of X and Y with the eigenvalues x_v and y_v, respectively. Define the rectangular matrix

$$V = 2 \left(J^t_{c_{0,0}} J_{c_{0,0}} v \mid J^t_{c_{0,1}} J_{c_{0,1}} v \mid J^t_{c_{1,0}} J_{c_{1,0}} v \mid J^t_{c_{1,1}} J_{c_{1,1}} v \right).$$

Define the orthogonal symmetric Haar matrix by

$$\hat{H} = \frac{1}{2} \begin{pmatrix} 1 & 1 & 1 & 1 \\ 1 & -1 & 1 & -1 \\ 1 & 1 & -1 & -1 \\ 1 & -1 & -1 & 1 \end{pmatrix}.$$

Define also the rectangular matrix

$$U \equiv (u_1 \mid u_2 \mid u_3 \mid u_4) = V\hat{H}.$$

Clearly, $u_1 = v/2$ is a common eigenvector of X and Y. Because X and Y are of property-A, it follows from Section 7.1 in [125] that u_2, u_3, and u_4 are also common eigenvectors of X and Y. In fact,

$$\begin{aligned}
Xu_1 &= x_v u_1 & Yu_1 &= y_v u_1 \\
Xu_2 &= (2\mathbf{x} - x_v)u_2 & Yu_2 &= y_v u_2 \\
Xu_3 &= x_v u_3 & Yu_3 &= (2\mathbf{y} - y_v)u_3 \\
Xu_4 &= (2\mathbf{x} - x_v)u_4 & Yu_4 &= (2\mathbf{y} - y_v)u_4.
\end{aligned}$$

Furthermore, u_1, u_2, u_3, and u_4 alias with each other on each of the subgrids in (5.8). In other words, these four eigenvectors coincide on each of the subgrids in (5.8) up to multiplication by -1. For example, if Dirichlet boundary conditions are used and β in (3.18) is constant, then, for any pair of integer numbers $1 \le k, l \le \lfloor (n+1)/2 \rfloor$, the above eigenvectors could be just the four aliasing two-dimensional Fourier modes defined in Section 2.4:

$$\begin{aligned}
u_1 &= v^{k,l} \\
u_2 &= v^{n+1-k,l} \\
u_3 &= v^{k,n+1-l} \\
u_4 &= v^{n+1-k,n+1-l}.
\end{aligned}$$

Note that, because $\hat{H}$ is orthogonal and symmetric, we also have

$$V = U\hat{H}.$$

The symbols used below are small matrices that represent the action of larger matrices in the invariant subspace spanned by the columns of V. Below we derive the symbols of the matrices used in the two-level iteration matrix, including the symbols of the iteration matrices S_r and S_b of the first and second parts in the red-black point-GS relaxation (Section 5.8):

$$\hat{X} = \hat{H} \begin{pmatrix} x_v & & & \\ & 2\mathbf{x} - x_v & & \\ & & x_v & \\ & & & 2\mathbf{x} - x_v \end{pmatrix} \hat{H}$$

$$\hat{Y} = \hat{H} \begin{pmatrix} y_v & & & \\ & y_v & & \\ & & 2\mathbf{y} - y_v & \\ & & & 2\mathbf{y} - y_v \end{pmatrix} \hat{H}$$

$$\hat{A} = \hat{X} + \hat{Y}$$

$$\hat{P} = \begin{pmatrix} 1 \\ 1 - \mathbf{x}^{-1}x_v \\ 1 - \mathbf{y}^{-1}y_v \\ (1 - \mathbf{x}^{-1}x_v)(1 - \mathbf{y}^{-1}y_v) \end{pmatrix}$$

$$\hat{R} = (\hat{P})^t$$

$$\hat{Q} = \hat{R}\hat{A}\hat{P}$$

$$\hat{S}_b = \begin{pmatrix} 1 & 0 & 0 & 0 \\ \dfrac{\mathbf{x} - x_v}{\mathbf{x} + \mathbf{y}} & 0 & 0 & \dfrac{\mathbf{y} - y_v}{\mathbf{x} + \mathbf{y}} \\ \dfrac{\mathbf{y} - y_v}{\mathbf{x} + \mathbf{y}} & 0 & 0 & \dfrac{\mathbf{x} - x_v}{\mathbf{x} + \mathbf{y}} \\ 0 & 0 & 0 & 1 \end{pmatrix}$$

$$\hat{S}_r = \begin{pmatrix} 0 & \dfrac{\mathbf{x} - x_v}{\mathbf{x} + \mathbf{y}} & \dfrac{\mathbf{y} - y_v}{\mathbf{x} + \mathbf{y}} & 0 \\ 0 & 1 & 0 & 0 \\ 0 & 0 & 1 & 0 \\ 0 & \dfrac{\mathbf{y} - y_v}{\mathbf{x} + \mathbf{y}} & \dfrac{\mathbf{x} - x_v}{\mathbf{x} + \mathbf{y}} & 0 \end{pmatrix}$$

$$\hat{S} = \hat{S}_b \hat{S}_r.$$

Let M_B be the iteration matrix of the present BBMG version, with two levels (as in Section 6.1) and the point red-black GS relaxation. Because the above symbols represent the action of the corresponding operators in the column space of V, we have

$$M_B V = V \hat{M}_B, \tag{10.4}$$

where

$$\hat{M}_B \equiv (\hat{S})^{\nu_2} (I - \hat{P}(\hat{Q})^{-1}\hat{R}\hat{A})(\hat{S})^{\nu_1}. \tag{10.5}$$

The spectra of the 4×4 matrices in (10.5) can be computed numerically, using, for example, an IMSL routine. This can be done for every pair of integers $1 \le k, l \le \lfloor n/2 \rfloor$ in (3.19), yielding the entire spectrum of the iteration matrix M_B. It is also verified numerically that the spectral radius of M_B indeed agrees with the convergence rate in the code that implements the two-level iteration, as required.

The AutoMUG(q) method can also be analyzed in a similar way. The only difference is that $\hat{Q}$ should be redefined by

$$\hat{Q} = (2 + q)\hat{J}_c(\hat{R}_X\hat{X} + \hat{R}_Y\hat{Y})\hat{J}_c^t, \tag{10.6}$$

where

$$\hat{J}_c = (1, 0, 0, 0)$$
$$\hat{R}_X = 2I - \mathbf{x}^{-1}\hat{C}$$
$$\hat{R}_Y = 2I - \mathbf{y}^{-1}\hat{Y}.$$

Fortunately, when n in (3.6) is odd, the coarse grid c in (6.1) contains no boundary points, so the original AutoMUG method is equivalent to AutoMUG(q) for a carefully chosen parameter q. For (3.18), in particular,

$$q = 1 - \frac{2}{2 + \beta h^2/2}$$

should be used. Indeed, the spectral radius of the iteration matrix of AutoMUG(q), calculated by the two-level analysis, coincides with the convergence factor of the corresponding AutoMUG iteration.

The computational two-level analysis can also be adapted for implementations that use coarse grid different from c. In fact, each subgrid in (5.8) could actually serve as a coarse grid, with only slight changes in the above analysis. The opportunity to use the two-level analysis also in these cases is particularly important in algorithms that use more than one coarse-grid correction. In fact, the computational two-level analysis shows that each coarse-grid correction should be based on a different coarse grid, namely, a different subgrid in (5.8). For example, the multigrid algorithm in [58] for highly anisotropic equations should indeed use two different coarse grids from (5.8) to compute the two correction terms. Similarly, the algorithm in [44] that uses four correction terms should indeed calculate each term from a different coarse grid in (5.8). The results of the computational two-level analysis in these cases can be found in [91].

For the indefinite Helmholtz equation, it is also possible to use more than one coarse-grid correction term. In this case, however, the different correction terms don't have to be calculated on different coarse grids; in fact, they can all use the same coarse grid c in (6.1). Still, the computational two-level analysis is helpful in predicting the convergence properties of the various possible algorithms and designing the optimal algorithm.

10.5 Multiple Coarse-Grid Corrections

The nearly singular eigenfunctions of the Helmholtz equation (3.14) (in the unit square with Dirichlet boundary conditions) are the functions

$$\sin(\pi kx)\sin(\pi ly)$$

with integers k and l satisfying

$$\pi^2(k^2 + l^2) \doteq -\beta.$$

(These eigenfunctions are the continuous counterparts of the 2-D Fourier modes $v^{k,l}$ in Section 2.4.) The pairs $(k,l) \in Z^2$ satisfying this condition are displayed in Figure 3.2. The prolongation matrix P in the present BBMG version agrees with the nearly singular eigenfunction for which

$$|k| = |l| \doteq \frac{\sqrt{|\beta|}}{\sqrt{2\pi}}.$$

Indeed, thanks to (10.3) and the fact that $D_x = D_y = 1/2$ in (3.18), the prolongation is the discrete analogue to (6.5), with the direction q being interpreted as either x or y.

It is also possible to modify P to agree with other nearly singular eigenfunctions. This can be done by replacing (10.3) by

$$\tilde{C} \leftarrow \begin{cases} \tilde{C} + 2\alpha_m D_x K & \text{the prolongation to } (i,j) \text{ is horizontal} \\ \tilde{C} + 2(1 - \alpha_m)D_y K & \text{the prolongation to } (i,j) \text{ is vertical,} \end{cases} \tag{10.7}$$

where α_m is a parameter specified later.

Clearly, when $\alpha_m = 1/2$, the original method in (10.3) is obtained. However, one could elect to use also other kinds of α_m, to produce prolongation operators that agree with other kinds of nearly singular eigenfunctions. For example, one could use three different coarse-grid corrections, each of which uses a different α_m in (10.7): the first one uses $\alpha_1 = \sin^2(\pi/12)$ to produce a prolongation operator that agrees with nearly singular eigenfunctions with

$$|k|/|l| \doteq \tan^2(\pi/12),$$

the second one uses $\alpha_2 = \sin^2(\pi/4) = 1/2$ [which is equivalent to (10.3)], and the third one uses $\alpha_3 = \sin^2(5\pi/12)$ to produce a prolongation operator that agrees with nearly singular eigenfunctions with

$$|k|/|l| \doteq \tan^2(5\pi/12)$$

(see Figure 10.1). Thus, each coarse-grid correction uses different prolongation matrix P, restriction matrix $R = P^t$, and coarse-grid matrix $Q = RAP$, and

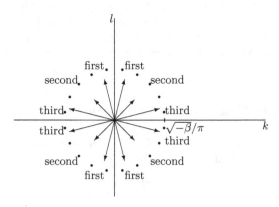

Fig. 10.1. The wave numbers $(k.l)$ of the nearly singular eigenfunctions $v^{(k,l)}$ that are handled in the first, second, and third coarse-grid corrections.

continues to use the same α_m also in further recursive calls that use coarser and coarser grids. Together, the three coarse-grid corrections may reduce significantly the error modes corresponding to a number of nearly singular eigenfunctions, hence may lead to a multigrid algorithm that converges more rapidly (in terms of iteration count) than the original algorithm that uses (10.3) alone.

There are three possible ways to combine the above three coarse-grid correction terms. In the additive approach, these terms are added one by one to x_{in} in the TL method in Section 6.1 (or the ML method in Section 6.3), without recalculating the residual in between. In other words, all three coarse-grid correction terms are calculated from the same residual. This leads to the following two-level iteration matrix:

$$(\hat{S})^{\nu_2}\left(I - \sum_{m=1}^{3} \hat{P}_m(\hat{Q}_m)^{-1}\hat{R}_m\hat{A}\right)(\hat{S})^{\nu_1}, \tag{10.8}$$

where the subscript $m = 1, 2, 3$ indicates the first, second, or third coarse-grid correction.

Unfortunately, the computational two-level analysis shows that the spectrum of this iteration matrix is even worse than that of M_B in (10.4). Therefore, we turn to the multiplicative approach, in which the residual is recomputed after adding each coarse-grid correction term to x_{in}. With this approach, the iteration matrix has the symbol

$$(\hat{S})^{\nu_2}\Pi_{m=1}^{3}(I - \hat{P}_m(\hat{Q}_m)^{-1}\hat{R}_m\hat{A})(\hat{S})^{\nu_1}. \tag{10.9}$$

Here the sum $\sum$ in (10.8) is replaced by the product Π. Unfortunately, the spectrum is still worse than that of M_B. The reason for this failure is probably threefold.

1. The nearly-singular eigenvectors of the three different coarse-grid matrices may combine to produce a large number of large eigenvalues in the iteration matrix.
2. Each coarse-grid correction term may contain error components from nearly-singular eigenfunctions that had already been annihilated by a previous coarse-grid correction, thus spoiling the work that has just been done on that coarse grid.
3. Each coarse-grid correction may also enhance error modes that correspond to eigenvectors of A that are not nearly singular at all. These error modes may lead to divergence of the entire two-grid iteration, unless reduced by an extra relaxation after each coarse-grid correction.

We must therefore turn to a third approach, in which relaxations are also performed in between the different coarse-grid corrections. This is actually equivalent to using three consecutive V-cycles: the first uses α_1 to define the transfer and coarse-grid matrices, the second uses α_2, and the third uses α_3. The symbol of the iteration matrix is then

$$\Pi_{m=1}^{3}[(\hat{S})^{\nu_2}(I - \hat{P}_m(\hat{Q}_m)^{-1}\hat{R}_m\hat{A})(\hat{S})^{\nu_1}]. \tag{10.10}$$

According to the computational two-level analysis, this iteration matrix has indeed a better spectrum than M_B in (10.4).

Actually, the spectrum of the iteration matrix can be improved further by using more than three coarse-grid corrections. The number '3' used in (10.10) could then increase to, say, 5, with parameters $\alpha_1, \alpha_2, \ldots, \alpha_5$ chosen in such a way that the circle in Figure 10.1 is covered with a yet better resolution. This way, the nearly singular eigenfunctions are handled better, and the spectral radius of the iteration matrix is smaller than with only three coarse-grid corrections.

Consider, for example, the highly indefinite equation (3.14) in the unit square with Dirichlet boundary conditions, $\beta = -790$, and the discretization in (3.18). on a uniform grid with meshsize $h = 1/64$. In this case, the spectral radius of the iteration matrix that uses five different coarse-grid corrections [whose symbol is as in (10.10), only with '3' replaced by '5' and $\nu_1 = \nu_2 = 1$] is as small as 10^{-3}. Indeed, it is also confirmed numerically that the five consecutive two-grid iterations that use the parameters $\alpha_1, \alpha_2, \ldots, \alpha_5$ to form the different transfer and coarse-grid matrices reduce the residual by a factor of 10^{-3}, which indicates that all error modes, including those corresponding to nearly-singular eigenvectors, have indeed been annihilated, as required.

When the meshsize in the fine grid is doubled to $h = 1/32$, the system is more indefinite and difficult. In this case, 17 coarse-grid corrections are required to have the same convergence rate as before. Indeed, the iteration matrix that uses 17 coarse-grid corrections with 17 different parameters $\alpha_1, \alpha_2, \ldots, \alpha_{17}$ (that provide a uniform coverage of the circle in Figure 10.1) has spectral radius as small as 10^{-3}. Again, this result is verified numerically: 17 consecutive two-grid iterations that use the parameters $\alpha_1, \alpha_2, \ldots, \alpha_{17}$ indeed reduce the residual by factor 10^{-3} as well.

The above results are most interesting from a theoretical point of view, because they show that we are indeed on the right track in our efforts to have a uniform coverage to the circle in Figure 10.1. Unfortunately, they have little practical value: the above algorithm that uses multiple coarse-grid corrections (or multiple V-cycles) is too expensive, hence inferior to the original algorithm that uses only one coarse-grid correction as in (10.3) in terms of overall operation count. Furthermore, one should also bear in mind the extra set-up cost involved in constructing the different transfer and coarse-grid matrices for each coarse-grid correction. Moreover, the above numerical results are limited to problems with Dirichlet boundary conditions and two-grid iterative methods only; with more general boundary conditions (such as mixed complex boundary conditions that are often used in practice) or more than two levels, the convergence rate is no longer as good as before. This is why we stick here to our original BBMG version that uses one coarse-grid correction only, with (10.3) rather than (10.7).

10.6 The Size of the Coarsest Grid

Geometric multigrid is rather unsuitable for indefinite Helmholtz equations. Indeed, the requirement in (10.1) limits significantly the number of coarse grids that could be used. Furthermore, geometric multigrid is inapplicable to problems with discontinuous coefficients, unless the discontinuity lines align with all the coarse grids.

The AutoMUG and BBMG versions described above, on the other hand, are much more suitable for the indefinite Helmholtz equation, with either constant or variable and even discontinuous coefficients, even when the discontinuity lines do not align with the coarse grid. Furthermore, since the coarse-grid matrices are no longer defined by rediscretizing the original PDE, the strong requirement in (10.1) is no longer necessary. Still, there is a limit on the meshsize in the coarsest grid, as discussed below.

The main point in the above AutoMUG and BBMG versions is that the prolongation operator P preserves the nearly singular eigenfunction, as do also the restriction operator $R = P^t$ and the coarse-grid matrix $Q = RAP$. However, the nearly singular eigenfunction can be well approximated on a coarse grid only if it contains sufficiently many points. More specifically, it must contain at least two points per oscillation. Since a nearly singular eigenfunction can oscillate at most $k \doteq |\beta|^{1/2}/\pi$ times in the unit interval, the meshsize in the coarsest grid must satisfy

$$H \leq k^{-1}/2 \doteq \pi|\beta^{-1/2}|/2. \tag{10.11}$$

Indeed, for $\beta = -790$, the maximal meshsizes for which the computational two-level analysis still gives good results are $h = 1/32$ and $H = 1/16$. In fact, for these values, most of the spectrum of the iteration matrix is considerably smaller than one in magnitude; only few, moderate, isolated eigenvalues exceed one in magnitude, which implies that outer acceleration can be used to annihilate the corresponding error modes, as is indeed evident from the numerical experiments below.

10.7 Numerical Examples

Here we test the present AutoMUG and BBMG versions for the indefinite Helmholtz equation in the unit square, discretized on the $n \times n$ uniform grid in (3.6) by finite differences, as in (3.18). We start with a slightly indefinite equation with $\beta = -20$ and Dirichlet boundary conditions.

The details of the multigrid iteration are as follows. Four levels are used in a V(1,1)-cycle ($\nu_1 = \nu_2 = \nu_c = 1$ and $L = 4$ in the ML algorithm in Section 6.3). The red-black point-GS relaxation method (Section 5.8) is used within AutoMUG, and the 4-color point-GS relaxation method (Section 5.9) is used within BBMG. The initial error in the iteration is random, so it contains components from all eigenvectors of A.

As expected, AutoMUG and the present BBMG version [that uses (10.3)] turn out to be superior to the standard BBMG method in Chapter 9. The results in Table 10.1 are given in terms of the convergence factor "cf", defined by

$$cf = \frac{\|Ax^{(last)} - b\|_2}{\|Ax^{(last-1)} - b\|_2}, \tag{10.12}$$

where $x^{(i)}$ denote the ith multigrid iteration, and "last" (the index of the last iteration) is so large that the l_2 norm of the residual is reduced by about six orders

Table 10.1. Convergence factors (cf) for four-level V(1,1)-cycles for the slightly indefinite Helmholtz equation in the unit square with $\beta = -20$ and Dirichlet boundary conditions.

n	Standard BBMG	Present BBMG	AutoMUG
31	> 1	0.063	0.131
63	0.431	0.064	0.096

of magnitude. Because the equation is only slightly indefinite, no acceleration is needed; therefore, the convergence is linear, and cf indeed represents the rate of convergence. In this book in general, cf is reported only when no acceleration is used.

Next, we test the highly indefinite Helmholtz equation (3.14) in the unit square with $\beta = -790$. Neumann boundary conditions are imposed on three edges, and mixed complex boundary conditions of the form

$$u_n + 10\sqrt{-1}u = 0$$

(where **n** is the outer normal vector) are imposed on the fourth edge. The finite-volume discretization method (Section 3.12) is used, so the stencil is as in (3.18) at interior gridpoints of the $n \times n$ uniform grid. The initial error is again random. The V$(0, 1)$-cycle is used ($\nu_1 = 0$ and $\nu_2 = 1$ in the ML algorithm in Section 6.3). Because the equation is highly indefinite, the iteration matrix has a few moderate and isolated eigenvalues that exceed one in magnitude, as is indicated by the computational two-level analysis. To annihilate error modes corresponding to these eigenvalues, the multigrid iteration is accelerated by the Conjugate Gradient Squared (CGS) method in [110]. (We have found CGS to be as efficient as TFQMR in [54] for the present example, and more efficient than several versions of GMRES.)

Because CGS is applied to the preconditioned system (5.11) rather than the original system (3.5), it makes sense to estimate the convergence rate in terms of the preconditioned residuals rather than the original ones. Furthermore, since CGS is a nonlinear process, the convergence rate is also nonlinear, and should be averaged. For these reasons, we use here not the convergence factor cf but rather the preconditioned convergence factor pcf, defined by

$$\text{pcf} = \left(\frac{\|\mathcal{P}^{-1}(Ax^{(last)} - b)\|_2}{\|\mathcal{P}^{-1}(Ax^{(0)} - b)\|_2} \right)^{1/last}, \tag{10.13}$$

where $\mathcal{P}$ is the multigrid preconditioner and "*last*" is the number of multigrid iterations used within CGS to reduce the l_2 norm of the preconditioned residual by about six orders of magnitude. (The l_2-norm of the preconditioned residual is available in CGS for no extra cost.)

There is also another advantage to using pcf here rather than cf. Because the preconditioned system is better conditioned than the original system, the preconditioned residual may approximate the error better than the residual itself. Hence, the norm of the preconditioned residual is a better convergence estimate than the norm of the residual itself. Indeed, it is also verified that the l_2 and l_∞ norms of the error decrease by at least four orders of magnitude during the CGS iteration.

Although the finest grid in the present numerical experiments is not sufficiently fine in terms of the adequacy criterion in Section 3.8, there is no problem to add more fine grids to the multigrid hierarchy, provided that the coarsest grid remains the same, as is indeed apparent from Tables 10.2 and 10.3. Furthermore, one could actually use a higher-order discretization method on the finest grid, as in [106]. At the end of this chapter, we will also solve the Helmholtz equation adequately.

In Table 10.2, the coarsest grid is of size 31×31, which is in agreement with (10.11). With this implementation, the present BBMG version exhibits good convergence, and AutoMUG also seems attractive thanks to its inexpensive time and storage requirements.

Table 10.2. Preconditioned convergence factors (pcf) for V(0,1)-cycles accelerated by CGS for the highly indefinite Helmholtz equation in the unit square with $\beta = -790$ and Neumann mixed boundary conditions.

n	Levels	Standard BBMG	Present BBMG	AutoMUG
63	2	.614	.509	.610
127	3	.588	.482	.592
255	4	.594	.488	.648

Table 10.3. Preconditioned convergence factors (pcf) for V(0,1)-cycles accelerated by CGS for the highly indefinite Helmholtz equation in the unit square with $\beta = -790$ and Neumann mixed boundary conditions. The coarsest-grid problem is solved approximately by 10-point Kacmarz relaxations.

n	Levels	Standard BBMG	Present BBMG	AutoMUG
31	2	.900	.862	.901
63	3	.947	.865	.949
127	4	>.99	.863	.963

In practical applications, in which a very large grid may be required, it is particularly important to use as many coarse grids as possible. This is why we attempt to use one more coarse grid. In Table 10.3, the coarsest grid is of size 15×15 rather than 31×31. This is still OK according to (10.11); however, the problem on the coarsest grid is so indefinite that solving it exactly would lead to divergence, due to nearly singular eigenvectors of the coarsest-grid matrix, which produce extremely large eigenvalues in the iteration matrix, which can never be annihilated by outer acceleration. For this reason, the coarsest-grid problem is only solved approximately by ten point Kacmarz relaxations ($\nu_c = 10$ in the ML algorithm in Section 6.3). The advantage of the present BBMG version is apparent from Table 10.3.

Finally, we also use the present BBMG version in the 9-point stencil resulting from the bilinear finite-element discretization (Section 4.4). We use $\beta = -1000$, Dirichlet boundary conditions on the right edge, homogeneous Neumann boundary conditions on the lower and upper edges, and mixed complex boundary conditions on the left edge. The boundary conditions are set in such a way that the exact solution is $\exp(-\sqrt{|\beta|}\sqrt{-1}x)$. Since the exact solution is available, one can choose a grid on which the discretization is indeed adequate. It turns out that, on a 100×100 grid, the discretization error is at most 0.02, which is rather good. The discrete system is solved by the present BBMG version, implemented in a V(1,1)-cycle that uses three levels. The coarsest grid is of size 25×25, which is in agreement with (10.11). Twenty Kacmarz relaxations are used to solve the coarsest-grid problem approximately. Outer CGS acceleration is also used. The preconditioned convergence factor for the present BBMG version is about 0.085.

Furthermore, we turn to a yet more indefinite equation with $\beta = -4000$. In this case, the fine grid must be increased to a size of 400×400 to keep the maximal discretization error as small as 0.06. (This is in agreement with the adequacy criterion in Section 3.8.) The present BBMG version uses four levels, so the coarsest grid is of size 50×50, which is OK in terms of (10.11). Forty Kacmarz relaxations are used to solve the coarsest-grid problem approximately. As before, CGS outer acceleration is used as well. With this implementation, the preconditioned convergence factor is 0.9. Although this rate is far worse than the Poisson convergence rate, it is still acceptable for such a large and highly indefinite problem.

10.8 Exercises

1. Show that the improved prolongation in Section 10.3 can be interpreted from a domain-decomposition point of view, using the 1-D Helmholtz equation (6.5) on the edges of subdomains to carry out the first prolongation step.
2. Which nearly singular eigenfunction of the indefinite Helmholtz equation is best approximated by this prolongation?
3. Show that each prolongation in Section 10.5 can also be interpreted from a domain-decomposition point of view. How should the 1-D indefinite Helmholtz equation (6.5) be modified to determine the prolonged values at the edges of the subdomains? (Distinguish between edges that are in the x spatial direction and edges that are in the y spatial direction.)
4. Which nearly singular eigenfunction of the indefinite Helmholtz equation is approximated well by each coarse-grid correction in Section 10.5?
5. Why is the multiple coarse-grid algorithm impractical? How does one coarse-grid correction spoil the work done by the previous one?
6. What is the lower bound for the number of points that can be used in the coarsest grid in the indefinite Helmholtz equation? Does this bound depend on the number of points used in the finest grid?
7. Write the computer code that implements the computational two-level analysis for the Poisson equation with Dirichlet boundary conditions, discretized as in (3.13). Note that the eigenvectors of A are just the 2-D Fourier (Sine) modes $v^{k,l}$ in Section 2.4. Verify that the spectral radius of the iteration matrix is indeed as small as the Poisson convergence factor.
8. Apply your code also to the indefinite Helmholtz equation with Dirichlet boundary conditions, discretized as in (3.18). Verify that most of the eigenvalues of the iteration matrix are much smaller than one in magnitude and only few isolated eigenvalues are larger, so long as the condition in (10.1) is met. Conclude that, with outer acceleration, the multigrid method should work well.
9. Extend the computational two-level analysis also to a coefficient matrix A with a 9-point stencil as in (4.14) on a uniform $n \times n$ grid with odd n, provided that the stencil is constant in the grid (independent of the gridpoint). This is done as follows.
   a) Assume that A can be written as

   $$A = X + Y + UZ$$

   rather than (7.3), where U has the stencil

   $$\begin{bmatrix} 0 & 0 & 0 \\ W_U & 0 & E_U \\ 0 & 0 & 0 \end{bmatrix}$$

and Z has the stencil

$$\begin{bmatrix} 0 & N_Z & 0 \\ 0 & 0 & 0 \\ 0 & S_Z & 0 \end{bmatrix}.$$

Note that the central element in the stencil in (4.14) is still

$$C = \mathbf{x} + \mathbf{y},$$

where $\mathbf{x}$ is the constant main-diagonal element in X and $\mathbf{y}$ is the constant main-diagonal element in Y.

b) Assume that X, Y, U, and Z commute with each other (e.g., they have constant stencils).

c) Let u_v and z_v be the eigenvalues of U and Z with respect to the common eigenvector v.

d) The symbol of U is

$$\hat{U} = \hat{H} \begin{pmatrix} u_v & & & \\ & -u_v & & \\ & & u_v & \\ & & & -u_v \end{pmatrix} \hat{H}.$$

e) The symbol of Z, $\hat{Z}$, is defined in a similar way.

f) $\hat{A}$ is defined by

$$\hat{A} = \hat{X} + \hat{Y} + \hat{U}\hat{Z}.$$

g) The symbol of the iteration matrix of the first "leg" in the 4-color point-GS relaxation is defined by

$$\hat{S}_{0,0} = \begin{pmatrix} 0 & \dfrac{\mathbf{x} - x_v}{\mathbf{x} + \mathbf{y}} & \dfrac{\mathbf{y} - y_v}{\mathbf{x} + \mathbf{y}} & \dfrac{-u_v z_v}{\mathbf{x} + \mathbf{y}} \\ 0 & 1 & 0 & 0 \\ 0 & 0 & 1 & 0 \\ 0 & 0 & 0 & 1 \end{pmatrix}.$$

h) The symbols of the other three "legs" in the 4-color point-GS relaxation, $\hat{S}_{1,0}$, $\hat{S}_{0,1}$, and $\hat{S}_{1,1}$, are defined in a similar way.

i) The symbol of the entire 4-color relaxation is defined by

$$\hat{S} = \hat{S}_{1,1} \hat{S}_{0,1} \hat{S}_{1,0} \hat{S}_{0,0}.$$

j) Because of the lumping used in BBMG, P_X is defined by

$$P_X = I - (\tilde{C} + S + N)^{-1}(X - \mathbf{x}I + rs(Z)U),$$

where S and N are as in (4.14), $\tilde{C}$ is as in (10.3), and $rs(Z)$ is the constant row-sum of Z.

k) The symbol of P_X is defined by

$$\hat{P}_X = I - (\tilde{C} + S + N)^{-1}(\hat{X} - \mathbf{x}I + rs(Z)\hat{U})$$

(where I here is the identity matrix of order 4).

l) $\hat{P}_Y$ is defined in a similar way.

m) The prolongation matrix P is defined by

$$P = S_{0,0}(P_X + P_Y - I)J_c^t.$$

n) The symbol of P is defined by

$$\hat{P} = \hat{S}_{0,0}(\hat{P}_X + \hat{P}_Y - I)\hat{J}_x^t.$$

o) The restriction matrix R and its symbol are defined in a similar way.

10. Show that the symbol of the first "leg" in the 4-color relaxation can also be written as

$$\hat{S}_{0,0} = \begin{pmatrix} 0 & -\hat{A}_{1,2}/\hat{A}_{1,1} & -\hat{A}_{1,3}/\hat{A}_{1,1} & -\hat{A}_{1,4}/\hat{A}_{1,1} \\ 0 & 1 & 0 & 0 \\ 0 & 0 & 1 & 0 \\ 0 & 0 & 0 & 1 \end{pmatrix},$$

where $\hat{A}_{1,1}$, $\hat{A}_{1,2}$, $\hat{A}_{1,3}$, and $\hat{A}_{1,4}$ are the elements in the first row in $\hat{A}$.

11. Apply your two-level analysis code to the Poisson equation with Dirichlet boundary conditions, discretized as in Section 4.4. (Note that the eigenvectors of A are just the 2-D Fourier modes $v^{k,l}$ in Section 2.4.) Verify that the spectral radius of the two-level iteration matrix is indeed as small as the Poisson convergence factor.

12. Apply your two-level analysis code also to the indefinite Helmholtz equation with Dirichlet boundary conditions, discretized as in Section 4.4. Verify that most of the eigenvalues of the iteration matrix of the two-level method are indeed small in magnitude and only few isolated eigenvalues are larger, provided that the condition in (10.1) is met. Conclude that, with outer acceleration, the method should work well.

13. Apply your two-level analysis code also to highly anisotropic equations as in Section 3.9. What happens to the spectrum of the two-level iteration matrix?

14. Modify your two-level analysis code to use "zebra" line-GS rather than 4-color relaxation. For this purpose, write $\hat{A}$ in the block form

$$\hat{A} = \begin{pmatrix} \hat{A}_{bb} & \hat{A}_{bw} \\ \hat{A}_{wb} & \hat{A}_{ww} \end{pmatrix},$$

where $\hat{A}_{bb}$, $\hat{A}_{bw}$, $\hat{A}_{wb}$, and $\hat{A}_{ww}$ are 2×2 submatrices of $\hat{A}$, b stands for the black horizontal lines $(c_{0,0} \cup c_{1,0})$, and w stands for the white horizontal lines $(c_{0,1} \cup c_{1,1})$. Now, define

$$\hat{S}_b = \begin{pmatrix} (0) & -\hat{A}_{bb}\hat{A}_{bw} \\ (0) & I \end{pmatrix}$$

$$\hat{S}_w = \begin{pmatrix} I & (0) \\ -A_{ww}^{-1}\hat{A}_{wb} & (0) \end{pmatrix}$$

$$\hat{S} = \hat{S}_w\hat{S}_b,$$

where (0) is the zero matrix and I is the identity matrix of order 2.

15. Modify your two-level analysis code to use alternating "zebra" line-GS relaxation.
16. Apply your modified two-level analysis code to the above highly anisotropic equations. Does the convergence rate improve?
17. Apply your BBMG code from the exercises at the end of Chapter 9 (implemented with two levels) to the above problems, and verify that the convergence factor is indeed the same as the spectral radius of the corresponding two-level iteration matrix, computed by the corresponding two-level analysis code.

11

Matrix-Based Semicoarsening Method

In this chapter, we describe in detail the matrix-based multigrid method that uses semicoarsening (Section 6.11). In particular, we show how this method can be obtained as an interesting combination of domain decomposition, line-ILU, and variational multigrid.

11.1 The Semicoarsening Approach

So far, we have used full coarsening, in which the coarse grid c in (6.1) is coarser than the original grid g in (3.6) in both the x and y spatial directions. In this chapter, we consider semicoarsening, in which the coarse grid c is coarser than g only in one spatial direction, say, the y spatial direction as in (6.15).

The matrix-based semicoarsening method works well for many practical diffusion problems with anisotropic and discontinuous coefficients, discretized on a uniform grid (e.g., by finite volumes or bilinear finite elements). Furthermore, the method can be extended to 3-D problems with a 27-point stencil as well. Unfortunately, the method is limited to structured grids only, and is unlikely to be extended to more general grids.

Here we introduce the matrix-based semicoarsening method as a combination of domain decomposition, line- (or plane in 3-D) ILU, and variational multigrid. Each component contributes to one aspect in the method: domain decomposition is the basis of the geometry (or topology) used in the method, line-ILU provides the algebraic framework to define the prolongation matrix P, and variational multigrid provides the usual definitions of restriction and coarse-grid matrices via the usual Galerkin approach $R = P^t$ and $Q = RAP$. The combination of these three components gives the matrix-based semicoarsening method, which benefits from each of them and works better than each of them alone.

We start by describing the geometric background that leads to the semicoarsened grid.

11.2 Flow of Information in Elliptic PDEs

Hyperbolic PDEs are characterized by characteristic directions, along which the boundary conditions flow into the domain. This is why discontinuities in the boundary conditions are preserved along the characteristic lines.

Elliptic PDEs, on the other hand, are characterized by the lack of any such characteristic direction. In fact, the solution of the elliptic PDE at each interior point depends not only on the characteristic line leading from the boundary to it, as in the hyperbolic case, but also on the entire boundary. In other words, the boundary conditions spread instantly into the entire domain to determine the solution in it uniquely, with no need to travel along characteristic lines. (This is also why any discontinuity in the boundary conditions smoothes out in the interior of the domain.) The solution process involves mutual interaction between each two points in the domain; the problem is global in nature, and cannot divide into independent subproblems.

The elliptic PDE could actually be viewed as the limit case of the corresponding time-dependent parabolic PDE, obtained by adding the extra term u_t (where t is the time variable). The discrete analogue of this time-dependent process is the iterative solution of the algebraic system obtained from the discretization of the original elliptic PDE. Indeed, during the iteration, the discrete boundary conditions "travel" with the index of the iteration (the discrete "time"), until they reach the entire grid to determine the numerical solution in it uniquely.

It is also interesting to see how information flows from the boundary to the interior of the grid when a direct (rather than an iterative) linear system solver is used. Assuming that the gridpoints are ordered row by row from bottom to top, the forward elimination in the lower-triangular matrix L (the first factor in the LU factorization of A) carries information from the bottom row upwards, whereas the back-substitution in the upper-triangular matrix U (the second factor in the LU factorization of A) carries information backward, from the top row downwards. The massive fill-in introduced in L and U indicates the interaction is indeed global: each row influences the solution at all the other rows. Figure 11.1 displays the mutual interaction among the rows (gridlines) in the solution process.

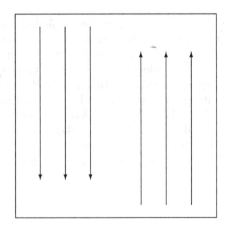

Fig. 11.1. Flow of information in the solution process of an elliptic PDE.

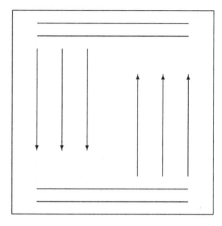

Fig. 11.2. Flow of information in the line forward elimination and line back-substitution in the line-ILU iteration.

Because of their global nature, elliptic PDEs are not easy to solve in parallel. One must use creative thinking to identify those parts of the problem (local subproblems) that can benefit from parallelism, and distinguish them from those parts (global subproblem) that must be solved in one processor only. Unfortunately, the flow of information in Figure 11.2 is inherently sequential, and cannot be implemented in parallel. Fortunately, the situation can be improved by using domain decomposition, and applying the LU decomposition to a reordered system.

11.3 Multilevel Line Reordering

In the domain-decomposition approach, the domain is divided into narrow strips (subdomains). The strips are separated from each other by the so-called interface: each strip has its own internal boundary to separate it from its neighbor strips above and below. The linear system is reordered in such a way that the interface unknowns (the unknowns corresponding to gridpoints on the interface) are ordered last, and the interior unknowns (the unknowns corresponding to gridpoints in the interiors of subdomains) are ordered first. The transfer of information from bottom to top is done in three stages. In the first stage, information is sent from the interior of each subdomain to its internal boundaries (Figure 11.3). This is done by forward elimination of the interior unknowns. This "local" transfer of data can be done in all the subdomains simultaneously in parallel. In the second stage, the interface lines share their information among themselves. This is done by solving a low-order system (the Schur complement) for the interface unknowns. This stage is global in nature, and cannot be well parallelized. In the third stage, information is sent from the interfaces back to the subdomain interiors (Figure 11.4). This is done by back substitution to the interior unknowns. This "local" stage can again be done in all the subdomains simultaneously in parallel. The second stage above, which is of "global" nature, can also be solved recursively by the same method itself, allowing some more parallelism. This completes the solution process; the details are described in Section 11.5 below.

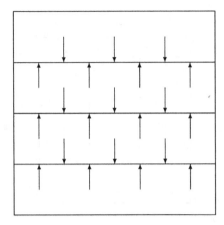

Fig. 11.3. Flow of information from the subdomain interiors to the interfaces in the first phase of the domain-decomposition method.

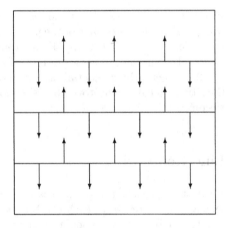

Fig. 11.4. Flow of information from the interfaces back to the subdomain interiors in the third phase of the domain-decomposition method.

In the special case in which A is block-tridiagonal and each strip in Figures 11.3 and 11.4 contains a single horizontal line of gridpoints, the domain decomposition actually forms a "zebra" coloring as in Figure 5.1, in which the subdomain interiors are colored by, say, white, and the interface lines are colored by, say, black. The black gridlines are further reordered recursively in the solution of the Schur complement in the second stage above. Thus, the above domain decomposition is actually a multilevel reordering of gridlines. In fact, it is equivalent to the Cyclic Reduction method in Section 7.2 (with matrix elements replaced by blocks corresponding to individual gridlines).

The transfer of information about the individual gridlines is also done in a multilevel scheme, rather than by marching across gridlines as in Figure 11.2. Indeed, the information from two white lines (say, the first and third lines from the bottom)

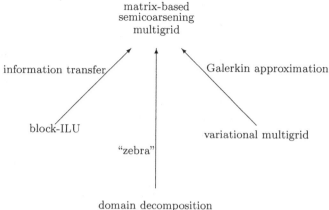

Fig. 11.5. The matrix-based semicoarsening multigrid method is a combination of three components: (a) geometry/topology as in domain decomposition to define the semicoarsening, (b) algebra as in line-ILU to define P and R, and (c) functional analysis as in variational multigrid to define $Q = RAP$.

is first transferred to the black line that lies in between them (the second line from the bottom). Then, it is transferred further recursively to the next black line (the fourth line from the bottom). Once the information about all the gridlines reaches the middle line, it is unpacked and redistributed to them. This way, the interaction among all the gridlines is completed in $\log_2 n$ steps (where n is the number of lines), rather than n steps as in Figure 11.2.

So far, we have used direct linear system solvers based on the LU factorization of A. Unfortunately, these solvers don't preserve the sparsity of A. Indeed, the Schur complement corresponding to the interface unknowns may be far less sparse than the original matrix A. Therefore, it makes sense to replace the Schur complement with some sparse matrix that approximates it in a spectral sense. This change may provide a good iterative method for the solution of the original linear system.

One option is to replace the Schur complement with a matrix obtained from rediscretizing the original PDE on the black lines only. This would yield a geometric multigrid with semicoarsening (Section 6.11). Here we are more interested in the more powerful multigrid, method that uses the low-order matrix $Q = RAP$, where R is the operator that transfers information from the white to the black lines and P is the operator that transfers information from the black lines back to the white lines. The precise way to define R and P in such a way that they remain sparse follows from the line-ILU iterative method described next.

11.4 Block-ILU Factorization

The block-ILU iterative method [50] uses the same pattern of transfer of information as in Figure 11.2, However, unlike in the standard LU factorization, here the triangular factors L and U are incomplete, so $A \neq LU$. The main advantage in block-ILU is that L and U are as sparse as the original matrix A.

In the sequel, we assume that the coefficient matrix A has a 9-point stencil as in (4.13) on a uniform grid as in (3.6). In other words, A is block-tridiagonal, with tridiagonal blocks (submatrices of order n).

The block-ILU (or line-ILU) method is motivated by the theory by Meurant about banded matrices (matrices with only few nonzero diagonals just above and below the main diagonal). This theory shows that the elements in the inverse of a banded matrix decay rapidly as their distance from the main diagonal increases. As a consequence, the inverse of a banded matrix, although dense in general, can be well approximated by a banded matrix that agrees with it on, say, three principal vectors (denoted by z_1, z_2, and z_3), defined below.

The block-ILU method scans the gridline by line from bottom to top, and performs an approximate block-LU decomposition (or approximate block Gauss elimination). Clearly, the exact (complete) block-LU decomposition produces dense block pivots (submatrices that lie on the main diagonal) and, hence, also dense factors L and U. This direct algorithm is, therefore, prohibitively expensive in terms of both time and storage. This is why the block-ILU method replaces these block pivots by sparse blocks, which produce sparse incomplete factors L and U, leading to a more efficient iterative algorithm. More precisely, the block-ILU method avoids the dense pivot blocks by replacing them with *pentadiagonal submatrices* (submatrices with at most five nonzero diagonals: the main diagonal, the two diagonals just above it, and the two diagonals just below it).

Let us now describe the block-ILU factorization in detail. First, we write the coefficient matrix A as

$$A = \text{block-tridiag}(Y_{i,i-1}, Y_{i,i}, Y_{i,i+1})_{1 \le i \le n},$$

where the $Y_{i,j}$s are tridiagonal matrices of order n that correspond to horizontal lines in the $n \times n$ grid g. In the following, the Z_is are the pentadiagonal approximate block pivots, and the F_is are the tridiagonal blocks in the incomplete factor U.

The Z_i's and F_i's are defined by induction. First, Z_1 is defined by $Z_1 = Y_{1,1}$. Then, for $i = 2, 3, 4, \ldots, n$, the tridiagonal matrix F_{i-1} of order n is defined by

$$F_{i-1}z_1 = (Z_{i-1})^{-1}Y_{i-1,i}z_1$$
$$F_{i-1}z_2 = (Z_{i-1})^{-1}Y_{i-1,i}z_2$$
$$F_{i-1}z_3 = (Z_{i-1})^{-1}Y_{i-1,i}z_3,$$

where

$$z_1 = (1, 0, 0, 1, 0, 0, 1, 0, 0, \ldots) \tag{11.1}$$
$$z_2 = (0, 1, 0, 0, 1, 0, 0, 1, 0, 0, \ldots) \tag{11.2}$$
$$z_3 = (0, 0, 1, 0, 0, 1, 0, 0, 1, 0, 0, \ldots) \tag{11.3}$$

are the n-dimensional principal vectors. Actually, F_{i-1} is not exactly tridiagonal, because it may contain two extra nonzero elements, $(F_{i-1})_{1,3}$ and $(F_{i-1})_{n,n-2}$. Thus, F_{i-1} contains exactly $3n$ elements, which are determined uniquely by the $3n$ equations in (11.1)–(11.3). As a result, F_{i-1} may be viewed as a spectral approximation to $Z_{i-1}^{-1}Y_{i-1,i}$ in the sense that it agrees with it on the three principal vectors z_1, z_2, and z_3.

The calculation of the right-hand sides in (11.1)–(11.3) requires applications of Z_{i-1}^{-1} to vectors. Fortunately, as can be shown by induction, Z_{i-1} is banded, so its

LU decomposition can be calculated with no extra fill-in, and used in the required calculations.

The matrix F_{i-1} is now used to define the pentadiagonal approximate block pivot Z_i:

$$Z_i \equiv Y_{i,i} - Y_{i,i-1}F_{i-1}.$$

More precisely, because F_{i-1} is not exactly tridiagonal, Z_i is not exactly pentadiagonal. In fact, it may contain two extra nonzero elements, $(Z_i)_{1,4}$ and $(Z_i)_{n,n-3}$. Fortunately, these elements produce no extra fill-in in the LU decomposition of Z_i.

The incomplete factors L and U are defined now by

$$L = \text{block-bidiag}(Y_{i,i-1}, Z_i, (0))$$
$$U = \text{block-bidiag}((0), I, F_i),$$

where (0) is the zero matrix of order n. Finally, the easily invertible preconditioner in the block-ILU method is defined by

$$\mathcal{P} = LU.$$

This preconditioner is now used in the iteration (5.1), which can be further accelerated by a Lanczos-type acceleration method as in Section 5.12.

We now turn back to the direct solver that uses Gauss elimination and complete LU decomposition of A. We describe in detail the domain-decomposition reordering in Section 11.3, which is used later also in block-ILU and, ultimately, in the matrix-based semicoarsening method.

11.5 The Domain-Decomposition Direct Solver

Here we describe in detail the domain-decomposition solver in Section 11.3. This solver is based on Gauss elimination (or LU decomposition) on a reordered system. Although this is a direct solver, the idea of reordering is relevant in iterative methods such as block-ILU and semicoarsening, as is discussed below.

The domain-decomposition approach may be viewed as a "divide-and-conquer" approach. The problem is divided into two parts: a "fine" part, which contains "local" subproblems that are independent of each other and can be solved in the individual subdomains simultaneously, and a "coarse" part, consisting of a low-order "global" system that couples the interfaces between the subdomains to each other (see also Section 1.18).

The main advantage of the domain-decomposition approach lies in the opportunity to implement the first part efficiently in parallel. Unfortunately, The second part is global in nature, hence more difficult to parallelize. Nevertheless, in some cases, it can be divided recursively into fine and coarse parts, allowing more parallelism.

In this section, we describe a domain-decomposition solver introduced in [35], which uses strips as in Figure 11.6 as subdomains. In Sections 11.6–11.7 below, we show how the present semicoarsening method emerges from the domain-decomposition method as a special case.

In the sequel, we assume that the coefficient matrix A has a 9-point stencil as in (4.13) on a uniform grid as in (3.6). We also assume that the grid is divided

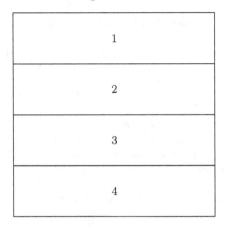

Fig. 11.6. The domain decomposition that uses four strips denoted by 1, 2, 3, and 4.

into strips as in Figure 11.6, with the interfaces between subdomains aligning with the grid. This decomposition induces the following partitioning of unknowns in the linear system (3.5), in which the unknowns are divided into two subsets: c, the subset of unknowns that lie on the interface between subdomains, and f, the rest of the unknowns. The unknowns are reordered so that the unknowns in f come before the unknowns in c. This partitioning induces the following block form of A,

$$A = \begin{pmatrix} A_{ff} & A_{fc} \\ A_{cf} & A_{cc} \end{pmatrix}$$

and similarly for other matrices of the same order.

Within c and f, the unknowns are ordered in the usual order, that is, line by line, with the gridpoints in each line ordered left to right. For the example in Figure 11.6, the order is thus as follows.

1. Unknowns in the interior of subdomain 1
2. Unknowns in the interior of subdomain 2
3. Unknowns in the interior of subdomain 3
4. Unknowns in the interior of subdomain 4
5. Unknowns on the interface between subdomains 1 and 2
6. Unknowns on the interface between subdomains 2 and 3
7. Unknowns on the interface between subdomains 3 and 4

This 7-block partitioning induces the following block form,

$$A = \begin{pmatrix} A_{1,1} & & & & A_{1,5} & & \\ & A_{2,2} & & & A_{2,5} & A_{2,6} & \\ & & A_{3,3} & & & A_{3,6} & A_{3,7} \\ & & & A_{4,4} & & & A_{4,7} \\ A_{5,1} & A_{5,2} & & & A_{5,5} & & \\ & A_{6,2} & A_{6,3} & & & A_{6,6} & \\ & & A_{7,3} & A_{7,4} & & & A_{7,7} \end{pmatrix} \qquad (11.4)$$

and similarly for other matrices of the same order.

Define the matrices P, R, and Q as follows [compare with (6.9)–(6.11)].

$$P = \begin{pmatrix} I & -(A_{ff})^{-1}A_{fc} \\ 0 & I \end{pmatrix} \tag{11.5}$$

$$= \begin{pmatrix} I & & & & -(A_{1,1})^{-1}A_{1,5} & & & \\ & I & & & -(A_{2,2})^{-1}A_{2,5} & -(A_{2,2})^{-1}A_{2,6} & & \\ & & I & & & -(A_{3,3})^{-1}A_{3,6} & -(A_{3,3})^{-1}A_{3,7} & \\ & & & I & & & -(A_{4,4})^{-1}A_{4,7} & \\ & & & & I & & & \\ & & & & & I & & \\ & & & & & & I & \end{pmatrix},$$

$$R = \begin{pmatrix} I & 0 \\ -A_{cf}(A_{ff})^{-1} & I \end{pmatrix} \tag{11.6}$$

$$= \begin{pmatrix} I & & & & & & \\ & I & & & & & \\ & & I & & & & \\ & & & I & & & \\ -A_{5,1}A_{1,1}^{-1} & -A_{5,2}A_{2,2}^{-1} & & & I & & \\ & -A_{6,2}A_{2,2}^{-1} & -A_{6,3}A_{3,3}^{-1} & & & I & \\ & & -A_{7,3}A_{3,3}^{-1} & -A_{7,4}A_{4,4}^{-1} & & & I \end{pmatrix},$$

and

$$Q = \begin{pmatrix} A_{ff} & 0 \\ 0 & A_{cc} - A_{cf}(A_{ff})^{-1}A_{fc} \end{pmatrix}. \tag{11.7}$$

Note that Q contains in its lower-right block the Schur complement of A with respect to the partitioning $f \cup c$. Therefore,

$$A = R^{-1}QP^{-1}$$

is nothing but the block-LU decomposition of A with respect to this partitioning. As a result, the TL method in Section 6.1 is actually a direct solver that converges in one iteration, even when no relaxations are used ($\nu_1 = \nu_2 = 0$).

The matrices $(A_{1,1})^{-1}$, $(A_{2,2})^{-1}$, $(A_{3,3})^{-1}$, and $(A_{4,4})^{-1}$ that are used in the definitions of P in (11.5) and R in (11.6) are never computed explicitly; all that is needed is to apply them to vectors when P or R is applied to a vector, which can be done by forward elimination and back substitution in their LU decompositions. Thus, the applications of P and R to vectors in the TL method can be done in the individual subdomains simultaneously in parallel. The same is true also for the application of Q_{ff}^{-1} in the middle of the TL method. The application of Q_{cc}^{-1}, on the other hand, is more complicated. In [35], it is done by an inner PCG iteration, with a preconditioner derived from a Fourier analysis of Q_{cc}. This iteration requires only applications of Q_{cc} to vectors, which can be done in parallel as above. Unfortunately, the preconditioner in this inner iteration is limited to problems in which the coefficients are constant within each strip. Therefore, we turn our attention to the line-ILU iteration, and use it in conjunction with the present line reordering to gain more parallelism.

11.6 Reordered Block-ILU Factorization

In this section, we combine the block-ILU method in Section 11.4 with the domain-decomposition approach in Sections 11.3 and 11.5. As above, we assume that the coefficient matrix A is of 9-point stencil on the uniform grid g in (3.6). We assume that each strip in Section 11.5 contains only one horizontal gridline in its interior. As at the end of Section 11.3, we also assume that the strips in Figure 11.6 each contain a single horizontal line of gridpoints, so that the blocks $A_{i,i}$ in (11.4) are tridiagonal matrices of order n. With this assumption, the partitioning $c \cup f$ is actually the "zebra" coloring as in Figure 5.1:

$$c = \{(i,j) \in g \mid i \equiv 0 \bmod 2\}.$$

The line-ILU factorization in Section 11.4 is inherently sequential in the sense that the approximate pivoting of each particular line depends on the pivoting of previous lines. Thus, the approximate pivoting must be done line by line, and cannot be parallelized efficiently. The same is true for the processes of block forward elimination and block back-substitution: they must be done line by line sequentially. It is thus most important to modify the line-ILU method into a more parallelizable version.

In this version, the approximate block pivoting is applied to the reordered matrix in (11.4) rather than the original matrix A in (3.5). In other words, the gridlines are colored in a "zebra" coloring, and the "white" lines (f in Section 11.5) are ordered before the "black" lines (c in Section 11.5). Thus, the approximate pivoting of the white lines can be done simultaneously in parallel. In this process, the matrices F_i are diagonal rather than tridiagonal as in Section 11.4, so the approximate block pivots Z_i are tridiagonal. This way, the low-order system that couples the black lines is of 9-point stencil as well, and can benefit from the same method recursively.

More precisely, for every relevant pair of integers i and j, we use diagonal matrices $P_{i,j}$ of order n to approximate the corresponding blocks in (11.5) spectrally, so they have the same effect on the n-dimensional constant vector:

$$P_{i,j}z = -(A_{i,i})^{-1}A_{i,j}z, \tag{11.8}$$

where

$$z = (1,1,1,1,\ldots) \tag{11.9}$$

is the n-dimensional vector all of whose components are equal to one. The diagonal matrix $F_{i,j}$ contains exactly n nonzero elements, which are determined uniquely by the n equations in (11.8).

Similarly, the diagonal matrices $R_{i,j}$ of order n approximate spectrally the corresponding blocks in (11.6). These diagonal matrices are defined by

$$z^t R_{i,j} = -z^t A_{i,j}(A_{j,j})^{-1}. \tag{11.10}$$

The sparse matrix P that approximates (11.5) is now defined by

$$P = \begin{pmatrix} I & & & P_{1,5} & & & \\ & I & & P_{2,5} & P_{2,6} & & \\ & & I & & P_{3,6} & P_{3,7} & \\ & & & I & & P_{4,7} & \\ & & & & I & & \\ & & & & & I & \\ & & & & & & I \end{pmatrix}. \tag{11.11}$$

Similarly, the sparse matrix R that approximates (11.6) is defined by

$$R = \begin{pmatrix} I & & & & & & \\ & I & & & & & \\ & & I & & & & \\ & & & I & & & \\ R_{5,1} & R_{5,2} & & & I & & \\ & R_{6,2} & R_{6,3} & & & I & \\ & & R_{7,3} & R_{7,4} & & & I \end{pmatrix}. \tag{11.12}$$

Finally, the sparse matrix Q that approximates (11.7) is defined by

$$Q = \begin{pmatrix} A_{ff} & 0 \\ 0 & A_{cc} - R_{cf} A_{ff} P_{fc} \end{pmatrix}. \tag{11.13}$$

Note that R is actually the transpose of a matrix P that would result from applying the above procedure to A^t rather than A. The matrices P, R, and Q are now used in the ML iteration in Section 6.3 (with $\nu_1 = \nu_2 = 0$). In the recursive call to the ML method, Q_{ff}^{-1} is applied in parallel in the individual strips, and Q_{cc}^{-1} is applied approximately using one iteration of the same method itself. Since Q_{cc} is of 9-point stencil, this is indeed possible. Finally, the above ML iteration can be further accelerated by an outer acceleration technique.

Because Q_{cc} above is an approximate Schur complement on the semicoarse grid c, it is scaled differently from an appropriate semicoarse-grid matrix. This is why the convergence rate of the above method is not optimal, and may deteriorate as n increases. This is why we introduce next an additional improvement, which produces a properly scaled semicoarse-grid matrix.

11.7 Matrix-Based Semicoarsening

The matrix-based semicoarsening method uses a modified version of the reordered block-ILU method in (11.11) and (11.12), in which P and R are rectangular rather than square matrices. Furthermore, the semicoarse-grid matrix Q is defined from the Galerkin approach $Q = RAP$ (as in variational multigrid) rather than the approximate Schur complement in (11.13). With this approach, the semicoarse-grid problem is indeed a proper approximation of the original problem. Furthermore, thanks to the fact that Q uses a 9-point stencil, recursion can be used as in the ML algorithm in Section 6.3. When relaxation is also used in the V-cycle ($\nu_1 + \nu_2 > 0$), the Poisson convergence rate is achieved for practical diffusion problems with discontinuous coefficients, even when the discontinuity lines don't align with the coarse grid. Moreover, when line relaxation (either line-GS relaxation or line-ILU) is used in the V-cycle, the Poisson convergence rate is also achieved for highly anisotropic equations.

Let us now describe the matrix-based semicoarsening method in some more detail. Assume that each strip contains only one horizontal gridline in its interior (so the semicoarsening is as in Figure 5.1). Using the notation in Section 11.6, the rectangular matrices P and R are defined by

$$
P = \begin{pmatrix}
P_{1,5} & & & \\
P_{2,5} & P_{2,6} & & \\
& P_{3,6} & P_{3,7} & \\
& & P_{4,7} & \\
I & & & \\
& I & & \\
& & I &
\end{pmatrix},
\tag{11.14}
$$

and

$$
R = \begin{pmatrix}
R_{5,1} & R_{5,2} & & & I & & \\
& R_{6,2} & R_{6,3} & & & I & \\
& & R_{7,3} & R_{7,4} & & & I
\end{pmatrix}.
\tag{11.15}
$$

Thanks to (11.8)–(11.10), P preserves the constant vector, as required in Section 6.2. Note also that R is the transpose of a prolongation matrix defined from A^t rather than A.

Finally, Q is defined by

$$
Q \equiv RAP.
$$

These matrices are now used in the ML algorithm in Section 6.3. Thanks to the fact that Q also uses a 9-point stencil, the recursive call in the ML method can indeed be carried out. The relaxation method used within the V-cycle is usually the "zebra" line relaxation that relaxes the individual horizontal gridlines as a whole. (Actually, this relaxation method is a special case of the Schwarz iteration that uses the domain decomposition in Figure 11.6.) For lines i that are contained in f, the LU decomposition of $A_{i,i}$ is already available from the solution of (11.8). This decomposition can also be used in the line relaxation. The reordered line-ILU method in Section 11.6 can also make a good relaxation method within the V-cycle.

For nonsymmetric problems, one should probably define P by applying the above algorithm not to A but rather to its symmetric part, $(A + A^t)/2$, in the spirit of [40].

The present matrix-based semicoarsening multigrid algorithm was first discovered by Steve Schaffer. It was also rediscovered independently by Joel Dendy and applied successfully to many diffusion problems with anisotropic and discontinuous coefficients [43]. However, Dendy also reports that the method may still diverge for problems like (9.8) and (9.9). This is because the theory developed later in this book is not applicable to the matrix-based semicoarsening method.

Schaffer and Dendy have also applied the present semicoarsening method to 3-D problems. In this case, the blocks $A_{i,i}$ correspond to planes in a cubic grid rather than lines in a square grid. These blocks are inverted implicitly in (11.8)–(11.10) by the 2-D version of the same algorithm itself, and the "zebra" plane relaxation is used in the V-cycle. With this implementation, convergence rates almost as good as the Poisson convergence rate are achieved for 3-D diffusion problems with anisotropic and discontinuous coefficients.

Finally, the matrix-based semicoarsening method was also applied to systems of PDEs in [42].

The main drawback in the matrix-based semicoarsening method is that it is not easily extended to semistructured and unstructured grids. In the following chapters, we turn our attention to a simpler multigrid framework that can be extended to semistructured and unstructured grids and also enjoys a mathematical background. Before going into this, however, we conclude the subject of semicoarsening with an interesting example.

11.8 A Deblurring Problem

A common problem in image processing is the problem of deblurring: a digital grayscale image that had been blurred by a linear low-pass filter is given, and the problem is to produce the original deblurred image. Actually, the low-pass filtering is equivalent to multiplying by a nonnegative matrix. In general, this matrix is not available; here, however, we assume that this matrix is available, so the original image is actually the solution of a system as in (3.5), with A being the low-pass filtering matrix and b being the given blurred image.

Let us now consider a more concrete example. Let A be the symmetric nonnegative matrix with the constant 9-point stencil

$$\begin{bmatrix} 1/4 & 1/2 & 1/4 \\ 1/2 & 1 & 1/2 \\ 1/4 & 1/2 & 1/4 \end{bmatrix} \tag{11.16}$$

on the uniform $n \times n$ grid in (3.6). The equation (3.5) with this matrix is particularly easy to solve because A is separable in the sense that

$$A = \mathcal{XY}. = \mathcal{YX},$$

where $\mathcal{X}$ is a matrix with the 3-point stencil

$$\begin{bmatrix} 0 & 0 & 0 \\ 1/2 & 1 & 1/2 \\ 0 & 0 & 0 \end{bmatrix}$$

(coupling in the x spatial direction), and $\mathcal{Y}$ is a matrix with the 3-point stencil

$$\begin{bmatrix} 0 & 1/2 & 0 \\ 0 & 1 & 0 \\ 0 & 1/2 & 0 \end{bmatrix}$$

(coupling in the y spatial direction). Now, since both $\mathcal{X}$ and $\mathcal{Y}$ are tridiagonal in some order of gridpoints, they have sparse LU decompositions with no fill-in whatsoever:

$$\mathcal{X} = L_\mathcal{X} U_\mathcal{X} \quad \text{and} \quad \mathcal{Y} = L_\mathcal{Y} U_\mathcal{Y},$$

where $L_\mathcal{X}$ and $L_\mathcal{Y}$ are bidiagonal lower triangular matrices and $U_\mathcal{X}$ and $U_\mathcal{Y}$ are bidiagonal upper triangular matrices. Therefore, the LU decomposition of A is

$$A = \mathcal{XY} = L_\mathcal{X}U_\mathcal{X}L_\mathcal{Y}U_\mathcal{Y} = L_\mathcal{X}L_\mathcal{Y}U_\mathcal{X}U_\mathcal{Y}.$$

Thus, A has an LU decomposition with no fill-in at all. Therefore, its ILU factorization coincides with its LU factorization, and actually produces a direct method that converges in one iteration, as is indeed observed numerically.

Furthermore, A remains separable also when multiplied on the left (or right) by a diagonal matrix. However, when even a tiny change is made in (11.16), A is no longer separable, and its ILU factorization is completely different from its LU factorization. In this case, both ILU and GS iterations become extremely inefficient, even when accelerated by PCG or any other acceleration method.

This failure is probably because the algebraic conditions required in the theory in [96] no longer hold. Indeed, in our numerical experiments, the perturbation of (11.16) can be solved by ILU or GS as long as these conditions are met, but not when they are violated. (See also another explanation in [129], pp. 348–349.) This is why we turn our attention here to multigrid algorithms, in the hope to have more robust linear system solvers.

Unfortunately, it is reported in [44] that several versions of BBMG are inefficient for the above deblurring example. Fortunately, one version of the present matrix-based semicoarsening method yields a fairly acceptable convergence rates. This version is based on the observation that the nearly singular eigenvectors of A are no longer nearly constant as in diffusion problems but rather highly oscillating. In fact, if A is applied to the (k, l)th 2-D Fourier (sine) mode as in Section 2.4, then we have [with $h = 1/(n+1)$]

$$\begin{aligned} Av^{(k,l)} &= \mathcal{XY}v^{(k,l)} \\ &= (1 + \cos(\pi kh))(1 + \cos(\pi lh))v^{(k,l)} \\ &= 4\cos^2(\pi kh/2)\cos^2(\pi lh/2)v^{(k,l)}. \end{aligned}$$

This means that A can actually be diagonalized by the 2-D sine transform. As a matter of fact, the sine transform could be used as a solver to the deblurring problem in (11.16). However, this solver would be limited to problems with constant stencil as in (11.16), and would fail to solve more general problems with variable coefficients. In fact, it would fail even for the product of the matrix in (11.16) with some diagonal matrix. This is why the sine transform is used here not as a solver but merely as a mathematical tool to analyze multigrid methods.

From the above spectral analysis, it is evident that small eigenvalues of A are obtained for eigenvectors with $k = n$ or $l = n$, which oscillate frequently in either the x or the y spatial direction. Therefore, the constant vector z in (11.9) should be replaced by the highly oscillating vector

$$z = (1, -1, 1, -1, \ldots) \tag{11.17}$$

to simulate a vector that oscillates frequently in the x spatial direction. Furthermore, the prolongation should be negative in the sense that the prolonged values in f should be multiplied by -1 to simulate a vector that oscillates frequently in the

y spatial direction. Similarly, values from f should be first multiplied by -1 before being used in the restriction. In summary, (11.8)–(11.10) should be modified to read

$$P_{i,j}z = (A_{i,i})^{-1}A_{i,j}z, \qquad (11.18)$$

where

$$z = (1, -1, 1, -1, \ldots), \qquad (11.19)$$

and

$$z^t R_{i,j} = z^t A_{i,j}(A_{j,j})^{-1}. \qquad (11.20)$$

As a matter of fact, in order to simulate well nearly singular vectors of all possible kinds, one should use multiple semicoarse-grid corrections: one to simulate vectors that oscillate in the x direction ($k = n$, $l = 1$), one to simulate vectors that oscillate in the y direction ($k = 1$, $l = n$), one to simulate vectors that oscillate in both the x and y directions ($k = l = n$), and one to simulate well the constant vector ($k = l = 1$). The required four semicoarse-grid corrections are thus as follows.

1. Use (11.8) and (11.10), but with (11.19) rather than (11.9).
2. Use (11.18) and (11.20), but with (11.9) rather than (11.19).
3. Use (11.18)–(11.20).
4. Use (11.8)–(11.10).

This algorithm is introduced in [44] for parallel computers, where the four individual semicoarse-grid corrections could be calculated independently of each other in parallel. Note that here the situation is more optimistic than in Section 10.5, because here the nearly singular error modes that are annihilated in some semicoarse-grid correction are either nearly-constant or highly oscillating in each spatial direction, so they belong (approximately) to the null spaces of the restriction operators of the other semicoarse-grid corrections, and should not be re-enhanced by them. It is indeed reported in [44] that, for a problem as in (11.16), it achieves the convergence factor of 0.8 (with no acceleration).

11.9 Exercises

1. Show that the application of the lower-triangular matrix R in the domain-decomposition solver in Section 11.5 is equivalent to forward elimination of the interior unknowns (the unknowns corresponding to gridpoints that lie in the interior of the strips).
2. Show that it can be done in all the strips simultaneously in parallel.
3. Show that it actually amounts to solving the original PDE in the individual strips, with homogeneous Dirichlet internal boundary conditions on the interfaces between subdomains (i.e., on the interface unknowns that form the semicoarse grid c).
4. Show that the application of the upper-triangular matrix P in the domain-decomposition solver in Section 11.5 is equivalent to back-substitution in the interior unknowns in f (after solving for the interface unknowns in c).

5. Show that it can be done in all the strips simultaneously in parallel.
6. Show that it actually amounts to solving a homogeneous PDE in the individual subdomains, with internal Dirichlet boundary conditions on the interfaces between the subdomains taken from the previous step, in which the Schur-complement system has been solved for the interface unknowns in c.
7. Show that this domain-decomposition solver is analogous to the parallel algorithm in Section 1.18.
8. Show that, when this domain-decomposition solver is applied recursively to solve the Schur-complement system on the interface unknowns in c, it can be viewed as a block version of the cyclic-reduction method in Section 7.2.
9. What is the difference between P in Section 11.7 and P in Section 11.5? What are the advantages and disadvantages of each approach?
10. What is the difference between R in Section 11.7 and R in Section 11.5? What are the advantages and disadvantages of each approach?
11. What is the difference between Q in Section 11.7 and Q in Section 11.5? What are the advantages and disadvantages of each approach?
12. Show that Q in the "zebra" block-ILU method in Section 11.6 has a 9-point stencil (like the original matrix A).
13. Show that the semicoarse-grid matrix Q in Section 11.7 has a 9-point stencil (like the original matrix A).
14. Show that the matrix-based semicoarsening method in Section 11.7 is actually equivalent to the matrix-based semicoarsening method described concisely in Section 6.11.

Part V

Matrix-Based Multigrid
for Semistructured Grids

So far, we have studied multigrid methods in the context of structured (uniform) grids. This model case is particularly suitable to show the power of multigrid methods. Indeed, the spectral analysis available in this case may provide a priori estimates for the convergence rate of multigrid, as either iterative method or preconditioner in the Krylov-subspace acceleration method.

Unfortunately, structured grids are unsuitable for most realistic problems arising in applied science and engineering. Indeed, most applications are defined on a nonrectangular domain with curved boundary, which must be approximated on a highly unstructured finite-element mesh. Even in applications that use rectangular domains, the physical phenomenon may exhibit different behavior in different parts of the domain, which requires nonuniform grids with variable resolution.

Completely unstructured grids may provide the required high resolution wherever necessary. However, they are particularly difficult to implement and manipulate. We therefore seek a grid that is more flexible than structured grids and yet not as expensive in terms of computer resources as general unstructured grids.

A fair compromise between uniform grids and completely unstructured grids can be obtained by local refinement. The required grid is obtained from an initial (coarse) grid by repeated local refinement at those areas where high resolution is needed. The criterion for refinement can be based on the original PDE itself. For example, extra refinement can be used around corners (and other segments of large curvature) in the boundary of the domain and in discontinuity curves in the diffusion coefficients. A more automatic approach, which requires no knowledge of properties of the original PDE, is adaptive refinement. In this approach, extra refinement is used wherever the coarse-grid numerical solution exhibits large variation. In this process, a hierarchy of finer and finer grids is constructed, each of which has extra refinement only in those areas where the previous one is not sufficiently fine to capture subtle variation in the solution. The finest grid is then accepted as the final grid, on which the original PDE should be discretized.

The coarse grids constructed in the above local-refinement process are also used in the multigrid algorithm that solves the linear system produced from the discretization method on the finest grid. Thus, no coarsening is required; the coarse grids that supply correction terms to the finest-grid system are already available from the refinement process that had created this finest grid. The multigrid algorithm that uses these coarse grids to solve the finest-grid system is described and analyzed in Chapter 12 below. This chapter is based on [99] and [100].

A special case of locally refined grids that is particularly convenient to implement is the case of the so-called *semistructured grid*. This grid is constructed from an initial uniform grid by embedding in it a smaller uniform grid with smaller meshsize in the region where higher accuracy is needed. The process is repeated recursively by further embedding smaller and smaller uniform grids with smaller and smaller meshsizes where extra accuracy is required. We refer to the combined grid consisting of the entire hierarchy of uniform grids as the semistructured grid. The PDE is then discretized on the semistructured grid (using finite elements or volumes) to produce the linear system of equations (3.5). The multigrid algorithm to solve (3.5) uses the coarse grids available from the above local-refinement process to supply correction terms in the V-cycle. This multigrid linear system solver is tested in Chapter 13 below, which is based on [99].

Matrix-Based Multigrid for Locally Refined Meshes

In this chapter, we describe a matrix-based multigrid method suitable for complicated nonuniform meshes obtained from local refinement. Under some algebraic assumptions such as diagonal dominance of the coefficient matrices, we derive an aposteriori upper bound for the condition number of the V(0,0)-cycle. This result applies also to diffusion problems with variable and even discontinuous coefficients, even when the discontinuity lines don't align with the coarse mesh. Furthermore, the upper bound is independent of the meshsize and the jump in the diffusion coefficient. Of course, the actual application of the multigrid method uses more efficient cycles such as the V(1,1)-cycle. Still, the theoretical result indicates that the nearly singular eigenvectors of A are indeed well approximated on the coarse grids, so the remaining error modes can be well handled by the relaxation.

12.1 Multigrid and Hierarchical-Basis Linear System Solvers

The subject of this chapter is matrix-based multigrid methods for the solution of elliptic PDEs discretized on locally-refined meshes. The method can be viewed as an extension of the method described in Figure 6.10. The present analysis gives an (a posteriori) upper bound for the condition number of the V(0,0)-cycle. Although this cycle is rarely used in practice, this result can still be interpreted to indicate that the nearly singular eigenvectors of A are well handled by the coarse-grid correction, and that the V(1,1)-cycle that uses relaxation to reduce the rest of the error modes should converge rapidly.

The V(0,0)-cycle can be actually viewed as a *hierarchical-basis method*. In standard hierarchical-basis methods, the problem is rewritten in terms of basis functions supported on linear finite elements of growing size: from small triangles on the fine mesh, to larger and larger triangles on the coarser and coarser meshes. This way, the original linear system obtained from the finite-element discretization is no longer represented in the usual nodal basis but rather in a hierarchical basis formed by a hierarchy of nested subspaces associated with coarser and coarser triangulations.

In matrix-based multigrid, on the other hand, the subspaces in the coarse levels in the hierarchy contain functions that are no longer linear in each coarse element but rather only piecewise linear, in such a way that flux continuity is preserved.

Of course, when the discontinuity lines in the diffusion coefficients align with the coarse meshes this would lead to the same standard hierarchical basis as in geometric and variational multigrid. However, when the discontinuity lines don't align with the coarse mesh, these flux-preserving subspaces are the only ones that can approximate the original problem well and supply good correction terms.

Of course, the transformation to the hierarchical basis is never carried out explicitly. Instead, the restriction operators that restrict the original system to the coarse grids and the prolongation operators that transform the coarse-grid systems back to the original system are carefully designed to be flux-preserving, using the algebraic information stored in the matrix elements.

Hierarchical-basis methods associated with variational multigrid are studied in [7] to [21] and [126] and [127]). The Algebraic Multilevel Iteration (AMLI) method in [4] also uses the hierarchy of nested finite-element spaces that are available in locally (and globally) refined meshes. These methods are not matrix-based methods, because the transforms from the nodal basis to the hierarchical basis and vice versa are defined in terms of the mesh rather than the coefficient matrix A. Although upper bounds for the condition numbers are available for some of these methods, they are derived under the assumption that the discontinuity lines in the diffusion coefficients align with the coarse meshes. This assumption limits considerably the number of coarse levels that can be used in realistic applications, in which the discontinuity lines could be of rather complicated shapes. Furthermore, it prevents one from writing a computer code that is applicable to any kind of discontinuity in the diffusion coefficients. This is why we are careful to avoid this assumption in the present analysis.

12.2 The Two-Level Method

We start with the two-level method, which is later extended into a multilevel method. The coarse grid used in the two-grid linear system solver is already available from the previous refinement level in the local refinement algorithm. More specifically, the coarse grid c consists of the nodes in the mesh S in Section 4.8, from which the fine mesh T has been produced.

The first step is to define a prolongation operator in the spirit of (9.2)–(9.4). For this purpose, consider a coarse-grid function $v \in l_2(c)$ defined on the coarse grid c. In particular, v takes the values $v(n_1)$ and $v(n_2)$ on the two coarse-gridpoints n_1 and n_2 in Figure 4.9. A prolongation operator must define the fine-grid function Pv. Clearly, Pv takes the same values as v at n_1 and n_2:

$$(Pv)(n_1) = v(n_1) \quad \text{and} \quad (Pv)(n_2) = v(n_2).$$

A more challenging task is to define Pv properly also at the midpoint $m = (n_1 + n_2)/2$ that doesn't belong to c. In the present matrix-based approach, this is obtained from a weighted average, with weights taken from the relevant elements in A:

$$(Pv)(m) \equiv \frac{a_{m,n_1}v(n_1) + a_{m,n_2}v(n_2)}{a_{m,n_1} + a_{m,n_2}}.$$

This is in the spirit of the definition in (12.14) below.

Let us now define the two-grid method in some more detail. We assume that A is nonsingular and has nonzero main-diagonal elements. In the sequel, we also use the notation in Section 4.8 above.

As mentioned above, the coarse grid c consists of the nodes in the coarse mesh S in the previous refinement level. The fine-gridpoints that are left outside c are contained in f. In other words, f contains the nodes in the mesh T in the current refinement level that do not serve as nodes in S. (We use this partitioning of nodes also for the partitioning of the corresponding nodal-basis functions.) This notation induces a block partitioning of A:

$$A = \begin{pmatrix} A_{ff} & A_{fc} \\ A_{cf} & A_{cc} \end{pmatrix}$$

and similarly for other matrices of the same order.

For every subgrid $s \subset c \cup f$, let $J_s : l_2(c \cup f) \to l_2(s)$ denote the injection operator defined by

$$(J_s v)_j = v_j, \quad v \in l_2(c \cup f), \ j \in s.$$

Recall also the definition of the row-sum matrix in (2.5) and the absolute value of a matrix in (2.4). This notation is used next to define the transfer operators.

Below we define the matrix $\tilde{A}$, from which the transfer operators R and P are later derived. First, we define the matrix $A^{(0)}$, which contains only the off-diagonal elements $a_{i,j}$ for which the nodes i and j are either completely disconnected in T or at most "poorly connected" in the sense that they are connected by an edge that has been introduced from half-refinement in the previous refinement step or by an edge with a positive weight:

$$A_{i,j}^{(0)} = \begin{cases} 0 & i = j \\ 0 & i \text{ and } j \text{ are vertices of the same triangle in } T \\ & \text{that are connected by an edge that has not been} \\ & \text{created by half refinement of a triangle } s \in S, \\ & \text{and } a_{i,j} a_{i,i} \leq 0 \\ a_{i,j} & \text{otherwise.} \end{cases}$$

The matrix $\tilde{A}$ is now obtained by "lumping" in A those elements that are also in $A^{(0)}$. This way, $\tilde{A}$ contains only those elements $a_{i,j}$ for which i and j are indeed "well-connected" in T:

$$\tilde{A} = A - A^{(0)} - rs(|A^{(0)}|). \tag{12.1}$$

In other words, $\tilde{A}$ is obtained from A by "throwing" onto the main diagonal elements $a_{i,j}$ for which the nodes i and j are not well connected in T in the above sense. The elements that remain in $\tilde{A}$ are later used to define the transfer matrices R and P.

The above method of lumping in (12.1) that "throws" certain off-diagonal matrix elements onto the main diagonal (as in Figure 6.10) has the advantage of being more stable than the method of lumping used in BBMG, which also throws certain off-diagonal matrix elements from A_{ff} onto off-diagonal elements from A_{fc} (as in Figure 6.9). Although the present method is slightly inferior to BBMG in terms of convergence rate for some examples, it is more robust (see Section 9.4) and enjoys the theory in Theorem 12.1 below.

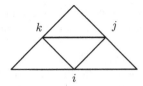

Fig. 12.1. The edge midpoints i, j, and k that are candidates for inclusion in f_a in the matrix-based multigrid method.

A desirable property of a multigrid algorithm is the boundedness of the prolongation operator P. This property is also essential in the analysis in Theorem 12.1 below. In order to improve this boundedness, we split f into two subsets: f_a (nodes that are strongly coupled with c) and f_b (nodes that are only weakly coupled with c). This splitting is done as follows. Let i, j, and k be edge midpoints of a refined triangle $s \in S$ (see Figure 12.1). The splitting is done as follows. Let τ be a parameter defined later. Let $(rs(\tilde{A}_{fc}))_{i,i}$, $(rs(\tilde{A}_{fc}))_{j,j}$, and $(rs(\tilde{A}_{fc}))_{k,k}$ be the diagonal elements of $rs(\tilde{A}_{fc})$ corresponding to i, j, and k, respectively. If

$$\left| \frac{(rs(\tilde{A}_{fc}))_{i,i}}{a_{i,i}} \right| > \tau \max_{l \in \{i,j,k\}} \left| \frac{(rs(\tilde{A}_{fc}))_{l,l}}{a_{l,l}} \right|, \tag{12.2}$$

then include i in f_a. The same procedure is employed for j and k. After completing this procedure for all refined triangles $s \in S$, define $f_b = f \setminus f_a$.

After f_a and f_b have been initialized as above, they are further modified to guarantee stability, as discussed below. First, scan f_b; if a node $i \in f_b$ is found for which

$$\left| \sum_{j \in c \cup f_a} \tilde{A}_{i,j}/a_{i,i} \right| \leq \text{threshold}, \tag{12.3}$$

then drop i from f_b and add it to c.

Then, scan f_a; if a node $i \in f_a$ is found for which

$$\left| \sum_{j \in c} \tilde{A}_{i,j}/a_{i,i} \right| \leq \text{threshold}, \tag{12.4}$$

then drop i from f_a and add it to c.

As we will see below, the above fix guarantees boundedness of the prolongation operator, which is essential in the analysis. The fix also accounts for anisotropy in the diffusion coefficient in the PDE. Indeed, if points in f lie on a line of weak diffusion, then they are weakly coupled with c, and, hence, thrown back to c by the above fix. The result is that the fine grid is locally semicoarsened in the strong-diffusion direction, as is indeed required for highly anisotropic equations (see Section 4.7). (If no multigrid iteration converges, then one may suspect that the boundary-value problem is ill-posed (has no solution or has more than one solution) or the discretization method is inadequate.)

Clearly, c, f_a, and f_b are disjoint. We now redefine

$$f = f_a \cup f_b.$$

Most often, the fix that follows (12.3) and (12.4) produces only a few changes to f_b and f_a, so the number of points in c remains much less than the number of points in the original grid $f \cup c$. In fact, in the numerical experiments in Chapter 13 this fix is not used at all.

The above new definitions induce the following block partitioning of A:

$$A = \begin{pmatrix} A_{ff} & A_{fc} \\ A_{cf} & A_{cc} \end{pmatrix} = \begin{pmatrix} A_{f_b f_b} & A_{f_b f_a} & A_{f_b c} \\ A_{f_a f_b} & A_{f_a f_a} & A_{f_a c} \\ A_{c f_b} & A_{c f_a} & A_{cc} \end{pmatrix} \qquad (12.5)$$

and similarly for other matrices of the same order.

Define the diagonal matrices G_a and G_b by

$$G_a = rs(|\tilde{A}_{f_a c}|) \quad \text{and} \quad G_b = rs(|\tilde{A}_{f_b c}|) + rs(|\tilde{A}_{f_b f_a}|). \qquad (12.6)$$

From the fix that follows (12.3) and (12.4), G_a and G_b are nonsingular. In Section 12.5 below, we show that, for isotropic diffusion problems, [even without using the fix that follows (12.3) and (12.4)] the main-diagonal elements of G_a and G_b have a positive lower bound that is independent of the meshsize and the diffusion coefficient, which implies that the prolongation operator P defined below is bounded independently of the meshsize and the possible jump in the diffusion coefficient. This important property is essential in the condition number estimate in Section 12.8 below.

The matrices R, P, and Q defined below are all square matrices of the same order as A. (This property is helpful in the analysis in Section 12.8 below.) Fortunately, the complexity of inverting Q is much smaller than that of inverting A and is practically the same as in standard multigrid algorithms, which use rectangular matrices R and P.

For simplicity, P and R are first defined implicitly by defining P^{-1} and R^{-1}, respectively. In this way, it is easy to see that the application of P and R to a vector can be done by block back-substitution in P^{-1} and block forward elimination in R^{-1}, respectively.

$$P^{-1} = \begin{pmatrix} G_b & \tilde{A}_{f_b f_a} & \tilde{A}_{f_b c} \\ 0 & G_a & \tilde{A}_{f_a c} \\ 0 & 0 & I \end{pmatrix} = \begin{pmatrix} I & 0 & 0 \\ 0 & G_a & \tilde{A}_{f_a c} \\ 0 & 0 & I \end{pmatrix} \begin{pmatrix} G_b & \tilde{A}_{f_b f_a} & \tilde{A}_{f_b c} \\ 0 & I & 0 \\ 0 & 0 & I \end{pmatrix} \qquad (12.7)$$

and

$$R^{-1} = \begin{pmatrix} G_b & 0 & 0 \\ \tilde{A}_{f_a f_b} & G_a & 0 \\ \tilde{A}_{c f_b} & \tilde{A}_{c f_a} & I \end{pmatrix} = \begin{pmatrix} G_b & 0 & 0 \\ \tilde{A}_{f_a f_b} & I & 0 \\ \tilde{A}_{c f_b} & 0 & I \end{pmatrix} \begin{pmatrix} I & 0 & 0 \\ 0 & G_a & 0 \\ 0 & \tilde{A}_{c f_a} & I \end{pmatrix}, \qquad (12.8)$$

where I denotes the identity matrix of a suitable order.

The operator P^{-1} may be thought of as a transform from the nodal basis to a hierarchical basis that spans a hierarchy of nested finite-element spaces. Unlike standard hierarchical-basis methods, however, the basis functions defined on coarse elements are no longer linear but rather piecewise linear in each individual coarse element, so that flux continuity is preserved [$\mathcal{D}\nabla\tilde{u}$ is continuous, where $\mathcal{D}$ is the diffusion coefficient in (3.2) and $\tilde{u}$ is the approximate solution on a coarse mesh].

[From another point of view, P^{-1} may also be thought of as a two-dimensional extension of the wavelet transform described in Section 1.15. However, unlike the standard wavelet transform that preserves continuity of derivatives, this transform preserves the continuity of the flux $\mathcal{D}\nabla\tilde{u}$, which is a necessary condition for a solution of (3.2). In the particular case where the Poisson equation is considered and the usual 9-point stencil is used, P^{-1} is reduced to the natural 2-D extension of the wavelet transform.]

In the actual algorithm, P^{-1} is never applied, and the above transforms are thus never carried out explicitly. The coarse finite-element spaces that are nested in the hierarchical basis are used only implicitly to provide correction terms for the fine-grid problem.

The operators that are applied in the multigrid algorithm are P and R rather than P^{-1} and R^{-1}. Here are the explicit definitions of these matrices:

$$
P = \begin{pmatrix} (G_b)^{-1} & -(G_b)^{-1}\tilde{A}_{f_b f_a} & -(G_b)^{-1}\tilde{A}_{f_b c} \\ 0 & I & 0 \\ 0 & 0 & I \end{pmatrix} \begin{pmatrix} I & 0 & 0 \\ 0 & (G_a)^{-1} & -(G_a)^{-1}\tilde{A}_{f_a c} \\ 0 & 0 & I \end{pmatrix}
$$
(12.9)

and

$$
R = \begin{pmatrix} I & 0 & 0 \\ 0 & (G_a)^{-1} & 0 \\ 0 & -\tilde{A}_{c f_a}(G_a)^{-1} & I \end{pmatrix} \begin{pmatrix} (G_b)^{-1} & 0 & 0 \\ -\tilde{A}_{f_a f_b}(G_b)^{-1} & I & 0 \\ -\tilde{A}_{c f_b}(G_b)^{-1} & 0 & I \end{pmatrix}.
$$
(12.10)

Finally, we define

$$
Q = \begin{pmatrix} W & 0 \\ 0 & B \end{pmatrix},
$$
(12.11)

where

$$
B = J_c R A P J_c^t,
$$
(12.12)

and

$$
W = R_{ff}\,diag(A_{ff})\,P_{ff}.
$$
(12.13)

Next, we show that the multigrid method that uses the square matrices P, R, and Q in (12.9)–(12.11) can be also formulated with rectangular matrices, yielding a more straightforward implementation. Denote the right block column in P in (12.9) by

$$
\tilde{P} \equiv P J_c^t = \begin{pmatrix} -(G_b)^{-1}\tilde{A}_{f_b f_a} & -(G_b)^{-1}\tilde{A}_{f_b c} \\ I & 0 \\ 0 & I \end{pmatrix} \begin{pmatrix} -(G_a)^{-1}\tilde{A}_{f_a c} \\ I \end{pmatrix}.
$$
(12.14)

Denote also the lower-block row in R in (12.10) by

$$
\tilde{R} \equiv J_c R = \begin{pmatrix} -\tilde{A}_{c f_a}(G_a)^{-1} & I \end{pmatrix} \begin{pmatrix} -\tilde{A}_{f_a f_b}(G_b)^{-1} & I & 0 \\ -\tilde{A}_{c f_b}(G_b)^{-1} & 0 & I \end{pmatrix}.
$$
(12.15)

Finally, denote the coarse-grid matrix by

$$
\tilde{Q} \equiv \tilde{R} A \tilde{P} = B.
$$
(12.16)

With this notation, the ML method in Section 6.3 that uses the square matrices P, R, and Q of (12.9)–(12.11) above is equivalent to the same method with P, R, and Q replaced by $\tilde{P}$, $\tilde{R}$, and $\tilde{Q}$, respectively, provided that the coarse-grid correction (6.3) is also replaced by

$$x_{out} = x_{in} + \tilde{P}e + J_f^t J_f \, diag(A)^{-1}(b - Ax_{in}). \tag{12.17}$$

This equivalent implementation is used in practice in Chapter 13, whereas the original formulation that uses (6.3) with the R, P, and Q defined in (12.9)–(12.11) is useful in the analysis below.

Clearly, (12.17) adds to (6.3) only a single term that can be obtained from a point-Jacobi iteration, restricted to fine-gridpoints that are not coarse-gridpoints. This addition has little effect on the rate of convergence, particularly when post-relaxation is used. In general, the residual $b - Ax_{in} = A(x - x_{in})$ does not contain nearly singular error modes, hence does not contribute much to the convergence. Thus, the coarse-grid correction term $\tilde{P}e$ in (12.17) is much more important than the last term in (12.17). For this reason, in most of the applications in this book the last term in (12.17) is actually dropped and not used in practice. However, the implementation in (12.17) is important from a theoretical point of view. Because it is equivalent to using (6.3) with the square matrices R, P, and Q defined in (12.9)–(12.11), it gives a nonsingular preconditioner, with which upper bounds for the condition number of the preconditioned matrix are available.

In practice, we have found that the definition

$$W = R_{ff} \, diag(A_{ff}) \, diag(P_{ff}) \tag{12.18}$$

[in the spirit of (9.7)] performs slightly better than (12.13). Here, however, we stick to (12.13) because it preserves symmetry, thus allowing the derivation of upper bounds for the condition number of the preconditioned matrix.

The parameter τ in (12.2) can be chosen in several ways, leading to different algorithms. For example, $\tau > 1$ would lead to the trivial partitioning $f_a = \emptyset$ and $f_b = f$. This choice is attractive because it leads to simple definitions of R, P, and Q. Furthermore, the coarse-grid coefficient matrix $B = J_c Q J_c^t$ has the same stencil as the corresponding stiffness matrix on S. This important property is also preserved in further coarsening steps: the coarse-grid coefficient matrices B_i defined in Section 12.7 below have the same stencil as the corresponding stiffness matrices on the corresponding coarse meshes.

A disadvantage of the choice $\tau > 1$, however, is that it produces rather large coarse grids. For example, with Refinement Method 4.2 in a uniform grid as in Figure 4.13, the coarse grids can be as large as in the repeated red-black coarsening in [22]. These coarse grids arise also with Refinement Method 4.1, provided that the fix that follows (12.3) and (12.4) is used to guarantee stability.

A more stable choice seems to be $\tau = 1/2$. With Refinement Method 4.1 and uniform grids as in Figure 4.13, this choice leads to the standard coarse grid (6.1) and the usual 9-point stencil at all coarse grids [see (13.2) below]. For more general triangle meshes, however, the stencil may be much larger than that. Indeed, the stencils of the coarse-grid matrices B_i defined in Section 12.7 below may exceed the stencils of the corresponding stiffness matrices and may contain coupling between nodes that belong to finite elements that do not share any joint node, provided that there exists a third finite element that shares a node with each of them. Fortunately, the coarse-grid stencils cannot be wider than that, so the coarse-grid matrices are still sparse. It is up to the user to choose τ that suits a particular application. In the numerical examples in Chapter 13 we use $\tau = 1/2$; since in these examples semistructured grids are used, the stencil at all the coarse grids is always the 9-point stencil [see (13.2) below].

12.3 Matrix-Induced Inner Products and Norms

Here we define some inner products and norms that are useful in the analysis. Let $(\cdot, \cdot)_2$ denote the usual inner product in $l_2(c \cup f)$ [or $l_2(c)$ or $l_2(f)$, when appropriate]; that is,

$$(u, v)_2 = \sum_i u_i \bar{v}_i,$$

where the sum is over $c \cup f$, c, or f, as appropriate. Unless stated otherwise, the term "symmetry" refers to this inner product. Let $\| \cdot \|_2$ denote the corresponding vector and matrix norms as in Section 2.1. For some SPD matrix D, let

- $(\cdot, \cdot)_D = (D\cdot, \cdot)_2$ denote the inner product induced by D
- $\| \cdot \|_D = \sqrt{(D\cdot, \cdot)_2}$ denote the corresponding vector norm
- $\| \cdot \|_D = \|D^{1/2} \cdot D^{-1/2}\|_2$ denote the matrix norm induced by D.

Note that the above definitions coincide with those in Section 2.1 above (see Lemma 2.17).

We also say that A is symmetric with respect to the inner product $(\cdot, \cdot)_D$ if

$$(Ax, y)_D = (x, Ay)_D$$

for every two vectors x and y in $l_2(c \cup f)$. When A is also positive definite, we say that it is SPD with respect to $(\cdot, \cdot)_D$. In this case, we also refer to $(x, Ax)_D$ as "the energy norm induced by A."

The aim of this notation is to enable the analysis of left-scaled systems of the form

$$D^{-1}Ax = D^{-1}b, \quad (D = diag(A)). \tag{12.19}$$

If A is SPD, then $D^{-1}A$ is symmetric with respect to $(\cdot, \cdot)_D$. Furthermore, both L-matrix property and diagonal dominance are preserved under this scaling. From Corollary 2.3, it follows that the D-induced norm of $D^{-1}A$ is smaller than or equal to two which is particularly helpful in the bounds derived in the present theory.

In Lemma 12.3 below we show that the two-level method applied to the left-scaled linear system (12.19) is mathematically equivalent to that applied to the original system (3.5). Therefore, Theorem 12.1 below may be applied to the left-scaled system (12.19), which yields not only mesh- but also jump-independent upper bound in (12.28) below, whereas the multigrid method is actually applied to the original system (3.5).

The following lemma will be useful in the proof of Lemma 12.5 below.

Lemma 12.1 *Let D be an SPD matrix, and assume that A is symmetric with respect to $(\cdot, \cdot)_D$ and positive semidefinite. Then for every vector x we have*

$$\|Ax\|_D^2 \le \|A\|_D (x, Ax)_D.$$

Proof. From Lemma 2.16, we have

$$
\begin{aligned}
\|Ax\|_D^2 &= \|A^{1/2}(A^{1/2}x)\|_D^2 \\
&\leq \|A^{1/2}\|_D^2 \|A^{1/2}x\|_D^2 \\
&= \|A\|_D (x, (A^{1/2})^2 x)_D \\
&= \|A\|_D (x, Ax)_D,
\end{aligned}
$$

which completes the proof of the lemma.

The purpose of the next lemma is to estimate the energy norm induced by A in terms of the energy norm induced by $blockdiag(A)$. This property will be useful in the proof of Theorem 12.1 below.

Lemma 12.2 *Let*

$$
D = \begin{pmatrix} D_{ff} & 0 \\ 0 & D_{cc} \end{pmatrix}
$$

be a block-diagonal SPD matrix. Assume that A is symmetric with respect to $(\cdot, \cdot)_D$. and positive semidefinite. Then, for every vector $x \in l_2(c \cup f)$,

$$
(x, Ax)_D \leq 2 \left(x, \left(J_f^t J_f A J_f^t J_f + J_c^t J_c A J_c^t J_c \right) x \right)_D .
$$

Proof. Let $\acute{x} = J_f^t J_f x - J_c^t J_c x$. Using Lemma 2.12, we have

$$
\begin{aligned}
0 &\leq (\acute{x}, A\acute{x})_D \\
&= \left(x, \left(J_f^t J_f A J_f^t J_f + J_c^t J_c A J_c^t J_c \right) x \right)_D \\
&\quad - \left(x, \left(J_f^t J_f A J_c^t J_c + J_c^t J_c A J_f^t J_f \right) x \right)_D .
\end{aligned}
$$

The lemma follows from

$$
\begin{aligned}
(x, Ax)_D &= \left(x, \left(J_f^t J_f A J_f^t J_f + J_c^t J_c A J_c^t J_c \right) x \right)_D \\
&\quad + \left(x, \left(J_f^t J_f A J_c^t J_c + J_c^t J_c A J_f^t J_f \right) x \right)_D \\
&\leq 2 \left(x, \left(J_f^t J_f A J_f^t J_f + J_c^t J_c A J_c^t J_c \right) x \right)_D .
\end{aligned}
$$

This completes the proof of the lemma.

12.4 Properties of the Two-Level Method

Here we present some properties of the two-grid method that will be useful in the sequel.

The following lemma shows that left scaling of A has no impact on the two-level method. This property is essential in obtaining a jump-independent upper bound in Theorem 12.1 below.

Lemma 12.3 *Let D be a nonsingular diagonal matrix of the same order as A. Define $\acute{A} = D^{-1}A$, and $\acute{W}$, $\acute{P}$, $\acute{R}$, and $\acute{Q}$ to be the operators obtained by applying the two-level method to the scaled linear system $\acute{A}x = D^{-1}b$. Then*

$$
\acute{P}\acute{Q}^{-1}\acute{R}\acute{A} = PQ^{-1}RA. \tag{12.20}
$$

Proof. Note that

$$\acute{P}^{-1} = \begin{pmatrix} D_{ff}^{-1} & 0 \\ 0 & I \end{pmatrix} P^{-1}$$

and

$$\acute{R}^{-1} = D^{-1}R^{-1} \begin{pmatrix} I & 0 \\ 0 & D_{cc} \end{pmatrix}.$$

Hence,

$$J_c \acute{R} \acute{A} \acute{P} J_c^t = D_{cc}^{-1} J_c RAP J_c^t \text{ and } \acute{W} = W D_{ff}.$$

Consequently,

$$\acute{Q} = \begin{pmatrix} I & 0 \\ 0 & D_{cc}^{-1} \end{pmatrix} Q \begin{pmatrix} D_{ff} & 0 \\ 0 & I \end{pmatrix},$$

and (12.20) follows. This completes the proof of the lemma.

The following two lemmas are necessary in the proof of Theorem 12.1 below. First, we show that the coarse-grid matrix preserves symmetry and positive definiteness of the coefficient matrix.

Lemma 12.4 *Let D be a diagonal SPD matrix of the same order as A. Assume that A is symmetric with respect to $(\cdot, \cdot)_D$. Then R and P are adjoint to each other with respect to $(\cdot, \cdot)_D$, and Q is also symmetric with respect to $(\cdot, \cdot)_D$. If, in addition, A is positive (semi) definite, then so is also Q.*

Proof. From the symmetry of A with respect to $(\cdot, \cdot)_D$ we have that, for every two vectors x and y,

$$(DAx, y)_2 = (Ax, y)_D = (x, Ay)_D = (Dx, Ay)_2 = (x, DAy)_2,$$

implying that

$$DA = (DA)^t = A^t D \tag{12.21}$$

[where 't' denotes the usual adjoint with respect to $(\cdot, \cdot)_2$]. Bearing in mind that D is diagonal and looking at the lower-triangular part of both sides of the equality sign in (12.21), we have from (12.6)–(12.8) that

$$DR^{-1} = P^{-t}D.$$

Therefore, for every two vectors x and y,

$$\left(R^{-1}x, y\right)_D = \left(DR^{-1}x, y\right)_2 = \left(P^{-t}Dx, y\right)_2 = \left(Dx, P^{-1}y\right)_2 = \left(x, P^{-1}y\right)_D.$$

This implies that R^{-1} and P^{-1} are adjoint to each other with respect to $(\cdot, \cdot)_D$, which in turn implies that R and P are also adjoint to each other with respect to $(\cdot, \cdot)_D$. The symmetry of Q with respect to $(\cdot, \cdot)_D$ follows now from the symmetry of A with respect to $(\cdot, \cdot)_D$ and (12.11)–(12.13). Similarly, the assertion about positive (semi) definiteness of Q follows from (12.11)–(12.13) and Lemma 2.12. This completes the proof of the lemma.

The next lemma estimates the norm induced by the "ff" and "fc" blocks in P^{-1} in terms of the energy norm induced by A. This estimate will be useful in the proof of Theorem 12.1 below.

Lemma 12.5 *Let D be a diagonal SPD matrix of the same order as A. Assume that A is symmetric with respect to $(\cdot, \cdot)_D$ and diagonally dominant. Let $x \in l_2(c \cup f)$ be a nonzero vector. Then*

$$\|J_f^t J_f P^{-1} x\|_D^2 \le (\sqrt{2} + 1)^2 \|A\|_D (x, Ax)_D. \tag{12.22}$$

Proof. Because A is symmetric with respect to $(\cdot, \cdot)_D$ and diagonally dominant, it follows from Lemmas 2.1 and 2.10 that it is also positive semidefinite. From Lemma 12.1,

$$\|Ax\|_D^2 \le \|A\|_D (x, Ax)_D.$$

Because D is diagonal, $\tilde{A}$ and $A - \tilde{A}$ are also symmetric with respect to $(\cdot, \cdot)_D$. Because A is diagonally dominant, $\tilde{A}$ and $A - \tilde{A}$ are also diagonally dominant. It follows from Lemmas 2.1 and 2.10 that they are also positive semidefinite. Using also Lemmas 12.1 and 2.15, we have

$$\|\tilde{A}x\|_D^2 \le \|\tilde{A}\|_D (x, \tilde{A}x)_D \le \|A\|_D (x, Ax)_D.$$

Define

$$A_a = \begin{pmatrix} rs(|\tilde{A}_{f_b f_a}|) & \tilde{A}_{f_b f_a} & 0 \\ \tilde{A}_{f_a f_b} & \tilde{A}_{f_a f_a} - G_a & 0 \\ 0 & 0 & 0 \end{pmatrix}$$

and

$$A_b = \begin{pmatrix} \tilde{A}_{f_b f_b} - G_b & 0 & 0 \\ 0 & 0 & 0 \\ 0 & 0 & 0 \end{pmatrix}.$$

Because $\tilde{A}$, A_a, A_b, $\tilde{A} - A_a$, and $\tilde{A} - A_b$ are symmetric with respect to $(\cdot, \cdot)_D$ and diagonally dominant, it follows from Lemmas 2.1 and 2.10 that they are also positive semidefinite. Using also Lemmas 12.1 and 2.15, we have

$$\|A_a x\|_D^2 \le \|A_a\|_D (x, A_a x)_D \le \|\tilde{A}\|_D (x, \tilde{A}x)_D \le \|A\|_D (x, Ax)_D$$
$$\|A_b x\|_D^2 \le \|A_b\|_D (x, A_b x)_D \le \|\tilde{A}\|_D (x, \tilde{A}x)_D \le \|A\|_D (x, Ax)_D.$$

Note that

$$J_{f_a} A_a = J_{f_a} (\tilde{A} - P^{-1})$$
$$J_{f_b} A_b = J_{f_b} (\tilde{A} - P^{-1}).$$

Consequently,

$$| \|J_f^t J_f \tilde{A} x\|_D - \|J_f^t J_f P^{-1} x\|_D |^2 \le \|J_f^t J_f (\tilde{A} - P^{-1}) x\|_D^2$$
$$= \|J_{f_a}^t J_{f_a} A_a x\|_D^2 + \|J_{f_b}^t J_{f_b} A_b x\|_D^2$$
$$\le \|A_a x\|_D^2 + \|A_b x\|_D^2 \le 2\|A\|_D (x, Ax)_D.$$

Therefore,

$$\|J_f^t J_f P^{-1} x\|_D \leq \sqrt{2\|A\|_D (x, Ax)_D} + \sqrt{\|A\|_D (x, Ax)_D}.$$

This completes the proof of the lemma.

12.5 Isotropic Diffusion Problems

A key factor in the success of a multigrid method is the boundedness of the prolongation operator P. Indeed, since P should preserve the nearly singular eigenfunctions of the original differential operator, it must avoid division by small numbers in the weighted averaging used in it. In the present method, this means that the main-diagonal elements in G_a and G_b defined in (12.6) should have a positive lower bound that is independent of the meshsize and the possible jump in the diffusion coefficient. In this section, we show that this requirement is indeed satisfied for a class of isotropic diffusion problems.

Consider the isotropic diffusion problem in Section 4.6. Recall that, under the assumptions made there, the result (4.17) holds. Therefore, with either Refinement Method 4.1 [with $\tau = 1/2$ in (12.2) to guarantee stability] or Refinement Method 4.2, the diagonal elements in G_a and G_b have a positive lower bound that is independent of the meshsize.

Assume further that the two-level method is applied to the left-scaled system (12.19) rather than the original system (3.5). This way, $\tilde{A}$, G_a, and G_b are defined from $D^{-1}A$ (with $D = diag(A)$) rather than A. Since D, G_a, and G_b are diagonal, we have in view of Lemma 2.17 that

$$\|G_a^{-1}\|_{D_{f_a f_a}} = \|G_a^{-1}\|_2 \quad \text{and} \quad \|G_b^{-1}\|_{D_{f_b f_b}} = \|G_b^{-1}\|_2.$$

Furthermore, these quantities are bounded independently of not only the meshsize but also the diffusion coefficient $\tilde{D}$ and the possible jump in it. This property can now be used to bound $\|P_{ff}\|_{D_{ff}}$ as follows. Since $\tilde{A}$ is symmetric, $D^{-1}A$ is symmetric with respect to $(\cdot, \cdot)_D$. In particular, the upper-left block in $\tilde{A}$, $\tilde{A}_{ff}$, is symmetric with respect to $(\cdot, \cdot)_{D_{ff}}$. From Corollary 2.2, we therefore have that $\|\tilde{A}_{ff}\|_{D_{ff}}$ is bounded independently of the meshsize and the possible jump in the diffusion coefficient $\tilde{D}$. From (12.9), we therefore have that

$$
\begin{aligned}
\|P_{ff}\|_{D_{ff}} &= \left\|\begin{pmatrix} G_b^{-1} & -G_b^{-1}\tilde{A}_{f_b f_a}G_a^{-1} \\ 0 & G_a^{-1} \end{pmatrix}\right\|_{D_{ff}} \\
&= \left\|\begin{pmatrix} G_b^{-1} & 0 \\ 0 & G_a^{-1} \end{pmatrix} - \begin{pmatrix} G_b^{-1} & 0 \\ 0 & 0 \end{pmatrix} \tilde{A}_{ff} \begin{pmatrix} 0 & 0 \\ 0 & G_a^{-1} \end{pmatrix}\right\|_{D_{ff}} \\
&\leq \left\|\begin{pmatrix} G_b^{-1} & 0 \\ 0 & G_a^{-1} \end{pmatrix}\right\|_{D_{ff}} + \left\|\begin{pmatrix} G_b^{-1} & 0 \\ 0 & 0 \end{pmatrix}\right\|_{D_{ff}} \|\tilde{A}_{ff}\|_{D_{ff}} \left\|\begin{pmatrix} 0 & 0 \\ 0 & G_a^{-1} \end{pmatrix}\right\|_{D_{ff}} \\
&= \max(\|G_a^{-1}\|_{D_{f_a f_a}}, \|G_b^{-1}\|_{D_{f_b f_b}}) + \|G_a^{-1}\|_{D_{f_a f_a}} \|\tilde{A}_{ff}\|_{D_{ff}} \|G_b^{-1}\|_{D_{f_b f_b}} \\
&= \max(\|G_a^{-1}\|_2, \|G_b^{-1}\|_2) + \|G_a^{-1}\|_2 \|\tilde{A}_{ff}\|_{D_{ff}} \|G_b^{-1}\|_2,
\end{aligned}
$$

so $\|P_{f\!f}\|_{D_{f\!f}}$ is also bounded independently of both the meshsize and the diffusion coefficient $\tilde{D}$ and the possible jump in it. This property will be particularly useful in Theorem 12.1 below.

The above bound for $\|P_{f\!f}\|_{D_{f\!f}}$ can actually improve to be linear in $\|G_a^{-1}\|_2$ and $\|G_b^{-1}\|_2$ rather than quadratic. This is done as follows. Since $\tilde{A}_{f\!f}$ is symmetric with respect to $(\cdot,\cdot)_{D_{f\!f}}$, we have (using the notation introduced in Lemma 2.18) that

$$\begin{pmatrix} 0 & 0 \\ \tilde{A}_{f_a f_b} & 0 \end{pmatrix}^t_{D_{f\!f}} = \begin{pmatrix} 0 & \tilde{A}_{f_b f_a} \\ 0 & 0 \end{pmatrix}.$$

Furthermore, since

$$\begin{pmatrix} G_b^{-1} & 0 \\ 0 & G_a^{-1} \end{pmatrix}$$

is diagonal, it commutes with $D_{f\!f}$, and, hence, is also symmetric with respect to $(\cdot,\cdot)_{D_{f\!f}}$. Therefore,

$$\begin{pmatrix} G_b^{-1} & 0 \\ 0 & G_a^{-1} \end{pmatrix} \tilde{A}_{f\!f} \begin{pmatrix} G_b^{-1} & 0 \\ 0 & G_a^{-1} \end{pmatrix}$$

is also symmetric with respect to $(\cdot,\cdot)_{D_{f\!f}}$, which implies that

$$\begin{pmatrix} 0 & 0 \\ G_a^{-1}\tilde{A}_{f_a f_b}G_b^{-1} & 0 \end{pmatrix}^t_{D_{f\!f}} = \begin{pmatrix} 0 & G_b^{-1}\tilde{A}_{f_b f_a}G_a^{-1} \\ 0 & 0 \end{pmatrix}.$$

Using (12.9), Lemma 2.19, and the definitions of G_a and G_b, we have

$$\|P_{f\!f}\|_{D_{f\!f}} = \left\|\begin{pmatrix} G_b^{-1} & -G_b^{-1}\tilde{A}_{f_b f_a}G_a^{-1} \\ 0 & G_a^{-1} \end{pmatrix}\right\|_{D_{f\!f}}$$

$$\leq \left\|\begin{pmatrix} G_b^{-1} & 0 \\ 0 & G_a^{-1} \end{pmatrix}\right\|_{D_{f\!f}} + \left\|\begin{pmatrix} 0 & G_b^{-1}\tilde{A}_{f_b f_a}G_a^{-1} \\ 0 & 0 \end{pmatrix}\right\|_{D_{f\!f}}$$

$$\leq \max(\|G_a^{-1}\|_{D_{f_a f_a}}, \|G_b^{-1}\|_{D_{f_b f_b}})$$

$$+ \sqrt{\left\|\begin{pmatrix} 0 & 0 \\ G_a^{-1}\tilde{A}_{f_a f_b}G_b^{-1} & 0 \end{pmatrix}\right\|_\infty \left\|\begin{pmatrix} 0 & G_b^{-1}\tilde{A}_{f_b f_a}G_a^{-1} \\ 0 & 0 \end{pmatrix}\right\|_\infty}$$

$$\leq \max(\|G_a^{-1}\|_2, \|G_b^{-1}\|_2)$$

$$+ \sqrt{\|G_a^{-1}\tilde{A}_{f_a f_b}\|_\infty \|G_b^{-1}\|_\infty \|G_b^{-1}\tilde{A}_{f_b f_a}\|_\infty \|G_a^{-1}\|_\infty}$$

$$\leq \max(\|G_a^{-1}\|_2, \|G_b^{-1}\|_2) + \sqrt{\text{threshold}^{-1}\|G_b^{-1}\|_\infty \|G_a^{-1}\|_\infty},$$

where "threshold" is as in (12.4). This improved bound shows again that $\|P_{f\!f}\|_{D_{f\!f}}$ is bounded independently of the meshsize and the possible jump in the diffusion coefficient $\tilde{D}$. This result will be useful in Theorem 12.1 to bound the condition number of the V(0,0)-cycle.

The above assumption that the multigrid method is applied to the left-scaled system (12.19) rather than (3.5) is not really necessary. Indeed, it follows from

Lemma 12.3 that this left-scaling has absolutely no effect on the preconditioned matrix. Thus, in practice one can apply the multigrid method to the original system (3.5), while the upper bound for the condition number is derived for (12.19), to make it independent of not only the meshsize but also the diffusion coefficient $\tilde{D}$ and the possible jump in it.

12.6 Instability and Local Anisotropy

As we've seen above, in isotropic diffusion problems, the prolongation operator is stable in the sense that P_{ff} is moderate in terms of the norm induced by D_{ff}. This property will be used below to obtain an upper bound for the condition number of the linear system preconditioned by multigrid.

This stability of P follows from the property that the main-diagonal elements in P_{ff}^{-1} are bounded away from zero. Otherwise, an application of P to a vector would involve division by (almost) zero, resulting in instability of the multigrid algorithm.

When would such an instability occur? When there exists a fine-gridpoint $i \in f$ that is only weakly coupled with its coarse-grid neighbors in c, but strongly with some other neighbors in f. Indeed, in such a case $(P_{ff}^{-1})_{i,i}$ would be too small in magnitude, because it is the sum of small matrix elements in A. As a result, $\|P_{ff}\|_{D_{ff}}$ would be too large, and the condition number of the multigrid method might be too large as well.

Such instability can indeed follow from local anisotropy. This means that $i \in f$ lies in between two coarse-gridpoints in c in some spatial direction and is only weakly coupled to them, yet is strongly coupled to other neighbors in f in some other spatial direction. We refer to this as local anisotropy because the diffusion is small (weak coupling) in the line connecting i to its coarse-grid neighbors in c, whereas the diffusion is large (strong coupling) along the line connecting i to its fine-grid neighbors in f.

Fortunately, the coarse grid c is designed carefully to avoid such instability. In fact, it is modified in such a way that whenever such a point $i \in f$ is discovered, it is immediately dropped from f and added to c. This stabilizing process guarantees that the upper bound derived below is indeed meaningful, and that the multigrid method should indeed work well.

12.7 The Multilevel Method

So far, we have considered a single refinement step, that is, the refined triangulation T obtained from the original triangulation S as in Section 4.8. Define $T_0 = S$ and $T_1 = T$. Let us now consider the more general case of L refinement steps, where L is a positive integer. In the ith refinement step, $1 \le i \le L$, the refined triangulation T_i is created from the current triangulation T_{i-1} using a refinement step as in Refinement Method 4.1 or 4.2. Let us define a partitioning of the nodes in the final triangulation T_L. Let f_0 be the set of nodes in T_L that are not in T_{L-1}. Similarly, for $i = 1, 2, \ldots, L$, let f_i be the set of nodes in T_{L-i} that are not in T_{L-i-1}. Finally, let f_L be the set of nodes in $T_0 = S$. The reason that the f_i's are indexed in reverse order is that in the analysis of multigrid we are interested in the coarsening process more than in the refinement process.

Furthermore, for $0 \leq k \leq i \leq L$, define $f_i^k = \bigcup_{j=k}^{i} f_j$. With these definitions, the partitioning $c \cup f$ in the two-level method can be written as $f_L^1 \cup f_0$.

Denote $A_0 = A, B_0 = A, G_0 = A, P_1 = P, R_1 = R, B_1 = B, W_1 = W$, and $A_1 = Q$. For $i = 2, 3, \ldots, L$, construct the matrices $\check{R}, \check{P}, B_i$, and W_i from B_{i-1} (with f_L^i serving as the coarse grid) in the same way that R, P, B, and W (respectively) are constructed from A in the two-level method; then, define

$$R_i = \begin{pmatrix} I & 0 \\ 0 & \check{R} \end{pmatrix},$$

$$P_i = \begin{pmatrix} I & 0 \\ 0 & \check{P} \end{pmatrix},$$

and

$$A_i = blockdiag(W_1, W_2, \ldots, W_i, B_i),$$

so that R_i, P_i, and A_i are of the same order as A. The V(0,0)-cycle then takes the form

$$x_{out} = x_{in} + P_{L,1} A_L^{-1} R_{L,1} (b - A x_{in}), \tag{12.23}$$

where for (a) $i > k$, $P_{k,i} = R_{k,i} = I$; for (b) $i = k$, $P_{k,i} = P_i$, and $R_{k,i} = R_i$; and for (c) $i < k$, $P_{k,i} = P_i P_{i+1} \cdots P_k$ is the prolongation from level k to level $i - 1$ and $R_{k,i} = R_k R_{k-1} \cdots R_i$ is the restriction from level $i - 1$ to level k. As discussed in Sections 12.1 and 12.2 above, the V(0,0)-cycle may be viewed as a hierarchical-basis method, and $P_{L,1}^{-1}$ may be viewed as a transform to a hierarchical basis that uses piecewise linear finite elements on the coarse meshes and also as a flux-preserving extension of the wavelet transform.

Using induction on L in Lemma 12.3 above, one can show that it makes no difference if the V(0,0)-cycle in the ML algorithm is applied to the original system (3.5) or the left-scaled system (12.19). Thus, (12.19) can be used for obtaining mesh- and jump-independent bounds for the condition number, while the actual application is done to (3.5).

Lemma 12.6 *Let D be a diagonal SPD matrix of the same order as A. Assume that A is symmetric with respect to $(\cdot, \cdot)_D$. Then, for $1 \leq i \leq L$, R_i and P_i are adjoint with respect to $(\cdot, \cdot)_D$, and A_i is symmetric with respect to $(\cdot, \cdot)_D$. If, in addition, A is positive (semi) definite, then so is also A_i $(1 \leq i \leq L)$.*

Proof. The lemma follows from induction on $i = 1, 2, \ldots, L$ and Lemma 12.4.

Define $G_0 = A$, and, for $1 \leq i \leq L$, define the matrices

$$G_i = J_{f_L^i}^t B_i J_{f_L^i}.$$

Thus, for $0 \leq i \leq L$, we also have

$$G_i = J_{f_L^i}^t J_{f_L^i} A_i J_{f_L^i}^t J_{f_L^i}.$$

Define also the scalars

$$a = \max_{0 \leq k < L} \rho\left(J_{f_k} diag(A_k)^{-1} G_k J_{f_k}^t\right)$$

and, for $0 \leq i < L$,

$$g_i = \|G_i\|_D,$$
$$p_i = \|J_{f_i}^t J_{f_i} P_{i+1} J_{f_i}^t J_{f_i}\|_D, \quad \text{and}$$
$$w_i = \|J_{f_i}^t W_{i+1} J_{f_i}\|_D.$$

Clearly,

$$w_i \leq g_i p_i^2. \tag{12.24}$$

In some cases, a is bounded, and D can be chosen in such a way that the g_is are bounded. For example, assume that A is symmetric and diagonally dominant, and that the matrices B_i $(1 \leq i < L)$ that are constructed during the application of the multigrid method to (3.5) are also diagonally dominant. From Lemma 2.1 we have that, in this case,

$$a \leq 2. \tag{12.25}$$

Define $D_0 = diag(A)$, and, for $i = 1, 2, \ldots, L - 1$, define the diagonal matrices D_i (of the same order as A) by

$$(D_i)_{k,k} = \begin{cases} \min(1, (A_i)_{k,k} \left(\Pi_{j=0}^{i-1}(D_j)_{k,k}^{-1} \right) & k \in f_L^i \\ 1 & \text{otherwise.} \end{cases}$$

Hence, by applying the multigrid method to

$$D^{-1}Ax = D^{-1}b, \quad D = \Pi_{i=0}^{L-1} D_i,$$

one obtains diagonally dominant matrices G_i whose main-diagonal elements are less than or equal to 1 whenever $0 \leq i < L$. Using Lemmas 2.1 and 12.6, one has

$$g_i = \|G_i\|_D = \rho(G_i) \leq 2, \quad 0 \leq i < L.$$

Assume that A is SPD with respect to $(\cdot, \cdot)_D$ for a diagonal SPD matrix D. For $0 \leq i < k \leq L$, let the scalars $C_{k,i}$ denote the squared norms of the matrices $P_{k,i+1} J_{f_L^k}^t J_{f_L^k}$ with respect to the energy norm induced by G_i:

$$C_{k,i} = \max_{x \in l_2(f_L^0), \, J_{f_L^k} x \neq 0} (P_{k,i+1} J_{f_L^k}^t J_{f_L^k} x, G_i P_{k,i+1} J_{f_L^k}^t J_{f_L^k} x)_D / (x, G_i x)_D$$

$$= \max_{x \in l_2(f_L^0), \, J_{f_L^k} x \neq 0} (x, G_k x)_D / (x, G_i x)_D. \tag{12.26}$$

12.8 Upper Bound for the Condition Number

We are now ready to give the upper bound for the condition number of the $V(0,0)$-cycle (12.23). The diagonal dominance assumptions made in Theorem 12.1 are rather strong; however, they hold in the diffusion problems with sharply discontinuous coefficients that are tested in the numerical examples.

In order to estimate the convergence rate of the multigrid method, one needs to estimate the condition number defined in (5.10) and (5.12). The following lemma is a common tool in condition-number estimates.

Lemma 12.7 *Let D be an SPD matrix, and assume that both A and $\mathcal{P}$ are SPD with respect to $(\cdot, \cdot)_D$. Then*

$$\|\mathcal{P}^{-1}A\|_\mathcal{P}\|A^{-1}\mathcal{P}\|_\mathcal{P} = \kappa(\mathcal{P}^{-1}A) = \left(\max_{x\neq 0} \frac{(Ax, x)_D}{(\mathcal{P}x, x)_D}\right)\left(\max_{x\neq 0} \frac{(\mathcal{P}x, x)_D}{(Ax, x)_D}\right). \quad (12.27)$$

Proof. The first equality in (12.27) follows from Lemmas 2.9 and 2.13. For the proof of the second equality in (12.27), note that $\mathcal{P}^{-1}A$ is symmetric with respect to $(\cdot, \cdot)_{D\mathcal{P}}$. From Lemma 2.13, we have

$$\rho(\mathcal{P}^{-1}A) = \max_{x\neq 0} \frac{(\mathcal{P}^{-1}Ax, x)_{D\mathcal{P}}}{(x, x)_{D\mathcal{P}}} = \max_{x\neq 0} \frac{(Ax, x)_D}{(\mathcal{P}x, x)_D}.$$

Similarly, $A^{-1}\mathcal{P}$ is symmetric with respect to $(\cdot, \cdot)_{DA}$. From Lemma 2.13, we have

$$\rho(A^{-1}\mathcal{P}) = \max_{x\neq 0} \frac{(A^{-1}\mathcal{P}x, x)_{DA}}{(x, x)_{DA}} = \max_{x\neq 0} \frac{(\mathcal{P}x, x)_D}{(Ax, x)_D}.$$

This completes the proof of the lemma.

We are now ready for the main theorem in this chapter.

Theorem 12.1 *Let D be a diagonal SPD matrix of the same order as A. Assume that A is diagonally dominant and symmetric with respect to $(\cdot, \cdot)_D$. Assume also that the coarse-grid matrices B_i $(0 \leq i < L)$ are diagonally dominant as well. Then*

$$\kappa\left(P_{L,1}A_L^{-1}R_{L,1}A\right) \leq \left(C_{L,0} + \left(\sqrt{2}+1\right)^2 \sum_{k=0}^{L-1} w_k g_k C_{k,0}\right) 2aL. \quad (12.28)$$

Proof. As in Lemma 12.7, we need first to estimate the energy norm induced by the preconditioner $R_{L,1}^{-1}A_L P_{L,1}^{-1}$ in terms of the energy norm induced by the coefficient matrix A. To this end, we need a few observations.

Note that, for $2 \leq i \leq L$,

$$J_{f_{i-2}^0}P_i J_{f_{i-2}^0}^t = J_{f_{i-2}^0}R_i J_{f_{i-2}^0}^t = I. \quad (12.29)$$

Note also that, for $0 \leq k \leq j \leq i < L$,

$$J_{f_j}^t J_{f_j} P_{j+1}^{-1} = J_{f_j}^t J_{f_j} P_{i+1,k+1}^{-1}. \quad (12.30)$$

From Lemmas 12.5 and 12.6, it follows that for every nonzero vector $x \in l_2(f_L^0)$ and $0 \leq i < L$,

$$\|J_{f_i}^t J_{f_i} P_{i+1}^{-1}x\|_D^2 \leq (\sqrt{2}+1)^2\|G_i\|_D(x, G_i x)_D. \quad (12.31)$$

Using (12.30) and (12.31), we have

$$
\begin{aligned}
(x, R_{L,1}^{-1} A_L P_{L,1}^{-1} x)_D &= \left(P_{L,1}^{-1} x, A_L P_{L,1}^{-1} x\right)_D \\
&= (x, G_L x)_D + \sum_{k=0}^{L-1} \left(J_{f_k}^t J_{f_k} P_{k+1}^{-1} x, J_{f_k}^t W_{k+1} J_{f_k} P_{k+1}^{-1} x\right)_D \\
&\leq (x, G_L x)_D + \sum_{k=0}^{L-1} (\sqrt{2}+1)^2 w_k \|G_k\|_D (x, G_k x)_D \\
&\leq \left(C_{L,0} + (\sqrt{2}+1)^2 \sum_{k=0}^{L-1} w_k g_k C_{k,0} \right) (x, Ax)_D .
\end{aligned}
$$

Next, we need to estimate the energy norm induced by A in terms of the energy norm induced by the preconditioner $R_{L,1}^{-1} A_L P_{L,1}^{-1}$. This is done as follows.

Define the union of sets f_k with even index k by

$$
e_{0,2} = \cup_{k \equiv 0 \mod 2} f_k = f_0 \cup f_2 \cup f_4 \cdots .
$$

Similarly, define the union of sets f_k with odd index k by

$$
e_{1,2} = \cup_{k \equiv 1 \mod 2} f_k = f_1 \cup f_3 \cup f_5 \cdots .
$$

Similarly, define the union of sets f_k with index k that is equivalent to $j \mod 4$ by

$$
e_{j,4} = \cup_{k \equiv j \mod 4} f_k = f_j \cup f_{j+4} \cup f_{j+8} \cdots .
$$

Similarly, define $e_{j,8}$, $e_{j,16}$, and so on.

Consider a nonzero vector $x \in l_2(f_L^0)$, and define $y = P_{L,1}^{-1} x$. Recall (12.29) and the fact that $a \geq 1$. By repeated application of Lemma 12.2, we have

$$
\begin{aligned}
(x, Ax)_D &= (y, R_{L,1} A P_{L,1} y)_D \\
&\leq 2 \left(y, \left(\sum_{j=0}^{1} J_{e_{j,2}}^t J_{e_{j,2}} R_{L,1} A P_{L,1} J_{e_{j,2}}^t J_{e_{j,2}} \right) y \right)_D \\
&\leq 4 \left(y, \left(\sum_{j=0}^{3} J_{e_{j,4}}^t J_{e_{j,4}} R_{L,1} A P_{L,1} J_{e_{j,4}}^t J_{e_{j,4}} \right) y \right)_D \\
&\leq \cdots \leq 2L \left(y, \left(\sum_{k=0}^{L} J_{f_k}^t J_{f_k} R_{L,1} A P_{L,1} J_{f_k}^t J_{f_k} \right) y \right)_D \\
&= 2L \left(y, \left(G_L + \sum_{k=0}^{L-1} J_{f_k}^t J_{f_k} R_{k+1} G_k P_{k+1} J_{f_k}^t J_{f_k} \right) y \right)_D \\
&\leq 2aL(y, A_L y)_D = 2aL(x, R_{L,1}^{-1} A_L P_{L,1}^{-1} x)_D .
\end{aligned}
$$

The theorem follows now from Lemma 12.7. This completes the proof of the theorem.

The bound in (12.28) is an *a posteriori* bound in the sense that the scalars involved in it are evaluated after constructing the coarse-level matrices. For diffusion problems of the form (3.2), however, these scalars are moderate, as is shown in Section 12.5 and in the numerical examples in [100]. In the present numerical examples, the coefficient matrix A is a diagonally dominant L-matrix, and so are the coarse-grid coefficient matrices B_i, $0 \le i < L$.

We try now to bound the scalars $C_{k,0}$ ($0 < k \le L$). We have

$$C_{k,0} \le \Pi_{i=0}^{k-1} C_{i+1,i}. \tag{12.32}$$

$C_{i+1,i}$ can be bounded (using Lemma 12.5) as follows:

$$
\begin{aligned}
(x, G_{i+1}x)_D &= \left(\left(P_{i+1}^{-1} - J_{f_i}^t J_{f_i} P_{i+1}^{-1} \right) x, R_{i+1} G_i P_{i+1} \left(P_{i+1}^{-1} - J_{f_i}^t J_{f_i} P_{i+1}^{-1} \right) x \right)_D \\
&= \left(\left(I - J_{f_i}^t J_{f_i} P_{i+1} J_{f_i}^t J_{f_i} P_{i+1}^{-1} \right) x, G_i \left(I - J_{f_i}^t J_{f_i} P_{i+1} J_{f_i}^t J_{f_i} P_{i+1}^{-1} \right) x \right)_D \\
&\le (x, G_i x)_D + 2p_i \| J_{f_i}^t J_{f_i} P_{i+1}^{-1} x \|_D \| G_i x \|_D + p_i^2 \| G_i \|_D \| J_{f_i}^t J_{f_i} P_{i+1}^{-1} x \|_D^2 \\
&\le (1 + (\sqrt{2}+1) p_i g_i)^2 (x, G_i x)_D. \tag{12.33}
\end{aligned}
$$

However, this would imply that the bound in (12.28) might grow exponentially with L, which is unacceptable.

A better approach is used in [100], where it is indicated that the $C_{k,0}$'s are bounded polynomially in k for a uniform-grid discretization of diffusion problems with variable coefficients. Perhaps it is possible to use the approach introduced in [126] and [127] to bound the $C_{k,0}$'s by a constant times k. Indeed, such a bound is indicated in Section 15.8 below for the simplified implementation that uses $f_a = \emptyset$ and $f = f_b$. [The analysis in Section 15.8 works equally well also here, provided that $\tau > 1$ is used in (12.2)].

Using the estimate in Section 15.8 below, it follows that the condition number of the V(0,0)-cycle grows at most polynomially (rather than exponentially) with L. This property indicates that the nearly singular eigenvectors of A are handled well by the coarse-grid correction. The V(1,1)-cycle that uses relaxation to take care of the other eigenvectors as well should thus have good convergence rates, as is indeed illustrated below.

12.9 Exercises

1. Explain the term "instability" in the prolongation operator P (Section 12.6).
2. Explain the term "local anisotropy."
3. How can instability follow from local anisotropy?
4. How does the careful design of c and f avoid instability?
5. Why is stability important in making the upper bound for the condition number meaningful?
6. Consider the Poisson equation (3.12), discretized by finite differences as in (3.13) on the uniform grid g in (3.6). Consider the multigrid method that uses full coarsening as in (6.1), and the prolongation indicated in Figure 6.10. Show that Theorem 12.1 is relevant in this case as well, and implies that the condition number of the V(0,0)-cycle is moderate. Conclude that the V(1,1)-cycle should converge rapidly.
7. Can the above result be extended also to the highly anisotropic equation (3.27)? Show that, in this case, $\| P_{f_a f_a} \|_2$ increases rapidly as ε approaches zero, so the

bound in Theorem 12.1 is no longer useful. Conclude that full coarsening leads to impractical multigrid iteration for highly anisotropic equations.

8. Consider a diffusion equation with discontinuous coefficients, which are highly anisotropic as in (3.27) in four cells around some gridpoint $(i, j) \in g$ with even i and odd j, and isotropic as in (3.12) elsewhere. Assume that the finite-volume or the bilinear finite-element discretization method is used. Show that the results in the previous exercise still hold. Conclude that even local anisotropy makes full coarsening impractical, and requires local semicoarsening in the strong-diffusion direction only.

9. Use the discussion in Section 12.5 to show that, for isotropic diffusion problems discretized on a finite-element triangulation (with angles that do not exceed $\pi/2$), $\|P_{ff}\|$ is moderate. Use this result in Theorem 12.1 to indicate that the condition number of the V(0,0)-cycle is moderate and, hence, the V(1,1)-cycle should converge rapidly.

10. Write the computer code that implements the above multigrid method for locally refined meshes. (Assume that Refinement Method 4.1 is used.)

11. Assume that the multigrid method is applied to an isotropic diffusion problem on the uniform mesh in Figure 4.13 (refined globally by Refinement Method 4.1). Use the estimates in Section 12.5 to bound $\|P_{ff}\|_{D_{ff}}$.

12. Compute the coefficient matrix on the next (coarser) grid. Is it diagonally dominant as well? Compute the prolongation operator resulting from it. Repeat the above exercise for this prolongation operator as well.

13. Use your multigrid code to solve the individual linear systems in the adaptive-mesh-refinement algorithm in Section 4.10. The solution can be found in Chapter 14 in [103].

Application to Semistructured Grids

Although the diagonal-dominance assumptions used in the analysis in the previous chapter are rather strong, they are nevertheless satisfied in the present numerical examples that use semistructured grids. Below we test the standard V-cycle and more parallelizable cycles such as AFAC and AFACx.

13.1 Semistructured Grids

A special case of locally refined grids that is particularly convenient to implement is the case of the so-called *semistructured grid*. This grid is constructed from an initial uniform grid by embedding in it a smaller uniform grid with smaller meshsize in the region where higher accuracy is needed. The process is repeated recursively by further embedding smaller and smaller uniform grids with smaller and smaller meshsizes where extra accuracy is required. We refer to the combined grid consisting of the entire hierarchy of uniform grids as the semistructured grid. The PDE is then discretized on the semistructured grid (using finite elements or volumes) to produce the linear system of equations (3.5).

The coefficient matrix A contains, thus, not only coupling between neighboring points within uniform grid portions (using standard stencils), but also coupling between regions of high resolution and regions of low resolution at the interface between refined and unrefined regions, using nonstandard stencils.

The discretization methods on semistructured grids are no longer as simple as in the uniform-grid case. Although one can still use the standard 5-point and 9-point stencils in the interiors of the regions where the grid is locally uniform, special stencils are required at the interfaces between regions of different meshsizes. For symmetric differential operators, the coefficient matrix for the discrete system should also be symmetric [118]. One should thus make sure that the dependence of the gridpoints in the refined region on their neighbors in the unrefined region is the same as the dependence of the latter on the former.

The main step in constructing the matrix-based multigrid solver for this linear system is to define the restriction and prolongation operators that transfer information between fine and coarse grids. In the interior of unrefined regions, where the coarse grid coincides with the fine grid, these operators can naturally coincide with the identity operator. In refined regions, however, it is necessary to define these operators in terms of weighted average, in the spirit of the BBMG method.

The final step in the construction of the multigrid solver is the definition of the coarse-grid coefficient matrix. This is done by the Galerkin approximation, as in BBMG.

13.2 The V-Cycle

Here we test the multigrid method for several isotropic diffusion problems with discontinuous coefficients, discretized on semistructured grids that result from local refinement (Refinement Method 4.1). The discrete system of equations is solved iteratively by the matrix-based multigrid algorithm in Chapter 12. No acceleration method is used here.

The computer code is written in $C++$, using the $A++$ class library developed by Dan Quinlan of the Los Alamos National Laboratory, which provides a convenient framework for handling arrays of double-precision numbers. In order to simplify the programming, we use the implementation in (12.14)–(12.17). Since fine-gridpoints that are not coarse-gridpoints are all located in the refined region, and since the prolongation and restriction operators are just the identity at the unrefined region, $\tilde{Q}$ agrees with the coarse-grid operator obtained from the Galerkin approach that uses the transfer operator in Figure 6.10 at the interior of the refined region and agrees with A at the interior of the unrefined region. At the interface between the refined and unrefined regions, $\tilde{R}$ and $\tilde{P}$ are no longer the identity because they involve also coupling with neighboring fine-gridpoints from the refined region; hence, the equations in $\tilde{Q}$ no longer agree with the corresponding equations in A because they also incorporate an additional sum of triple products of elements in $\tilde{R}$, A, and $\tilde{P}$ at neighboring fine-gridpoints. As a result, if $\tilde{Q}$ is stored in an array of double-precision numbers, then only the equations in A that correspond to the refined region and its internal boundary need to be stored. By the *internal boundary* of the refined region we mean the lines of gridpoints in the unrefined region that are near the refined region (complemented by dummy fine points). However, since the equations in A at the interior of the unrefined region are not stored, they cannot be relaxed as required in the multigrid algorithm in Section 6.3. In order to relax these equations as required, we assume without loss of generality that $x_{in} = \mathbf{0}$. (If $x_{in} \neq \mathbf{0}$, then one turns to the residual equation $Ae = b - Ax_{in}$ and uses the V-cycle to solve for $\mathbf{e}$.) Now, since A coincides with $\tilde{Q}$ at rows corresponding to points in the interiors of the unrefined regions, there is no need to relax the fine-level system there, because this can be done equally well in the coarse-level relaxation of the correction term that will be added later to the fine-level approximate solution. For this reason, there is also no need to store these matrix rows in A; it is sufficient to store them in $\tilde{Q}$ only, and store physically in A only those matrix rows that correspond to points in the refined regions and their internal boundaries.

13.3 The AFAC and AFACx Cycles

Besides the standard V(1,1)-cycle, we also test versions of the AFAC [60] and [72] and AFACx [73] cycles. These approaches are particularly attractive for parallel computers because they allow simultaneous processing of levels and, hence, using efficient global load-balancing techniques such as that in [72].

The standard V-cycle described in Section 6.3 is based on an essentially sequential recursion. The processing at a certain level cannot proceed until the recursive call to the ML function is completed and a correction term is supplied from the next coarser level. The AFAC algorithm is based on the observation that, for locally refined meshes, it may be worthwhile to solve a coarse-grid problem that is confined to the refined region in order to obtain a "local" correction term that will enable continuing the process rather than waiting for the global coarse-grid correction term. The AFACx cycle further improves this idea by observing that the above local correction term does not have to be accurate; it is good enough if it is smooth, because then fine-level relaxation would annihilate the high-frequency error modes that are invisible on the next coarser level. The detailed formulation is as follows. Without loss of generality, it is assumed that the initial approximate solution is $x_{in} = \mathbf{0}$. For $1 \leq i \leq L$, define $\tilde{R}_i = J_{f_L^i} R_i$ and $\tilde{P}_i = P_i J_{f_L^i}^t$,

AFACx$(\mathbf{0}, A, b, L, x_{out})$:

1. Initialize x_{out} by $x_{out} = \mathbf{0}$.
2. Initialize r_0 by $r_0 = b$.
3. For $i = 1, 2, \ldots, L$, define $r_i = \tilde{R}_i r_{i-1}$.
4. For $0 \leq i < L$, relax the equations in B_{i+1} that correspond to coarse-gridpoints in the refined region of the next coarser level, that is, points in f^{i+1} and points in f_L^{i+2} that are surrounded by points from f^{i+1}. This relaxation uses zero initial guess and right-hand side values that are taken from the corresponding components in r_{i+1}. The result of this relaxation is prolonged using $\tilde{P}_{i+1}$ to a vector q_i that is supported only in the refined region and its internal boundary, namely, f^i and the points in f_L^{i+1} that are near points from f^i. Define $z_i \equiv q_i$. Relax the equations in B_i that correspond to points in the refined region (namely, f_i and the points in f_L^{i+1} that are surrounded by points from f_i) using the initial guess q_i and the corresponding right-hand side values taken from r_i. Put the difference between the result of this relaxation and z_i in the vector e_i that is supported only in the refined region.
5. Define $e_L \equiv J_{f_L}^t B_L^{-1} r_L$.
6. For $i = L, L - 1, \ldots, 1$, do: update x_{out} by

$$x_{out} \leftarrow \tilde{P}_i(x_{out} + e_i) + J_{f_{i-1}}^t J_{f_{i-1}} J_{f_L^{i-1}}^t diag(B_{i-1})^{-1} r_{i-1}.$$

7. Update x_{out} by

$$x_{out} \leftarrow x_{out} + e_0.$$

The main advantage of this cycle in comparison to the usual V-cycle is that the computation of the correction terms e_i above can be done simultaneously in parallel.

The only difference between AFAC and AFACx is that with AFACx the relaxation is a point relaxation (in our examples on semistructured grids we use the 4-color point-GS relaxation), whereas in AFAC both point and block relaxation methods are used: the latter relaxation that uses B_i is point relaxation, but the former relaxation that uses the coarser matrix B_{i+1} is a block relaxation that solves for all the coarse gridpoints in the refined region of the next coarser grid simultaneously. In other words, the relaxation that uses B_{i+1} is replaced by an exact solve for the unknowns in the refined region of the next coarser grid using homogeneous Dirichlet internal boundary conditions. In the numerical examples, however, we will see that this exact solve is not very helpful, which means that AFACx is more efficient than AFAC.

13.4 The Numerical Examples

We test the isotropic diffusion problem

$$-(\tilde{D}u_x)_x - (\tilde{D}u_y)_y = \mathcal{F} \qquad (13.1)$$

in a square with Neumann mixed boundary conditions. The scalar function $\tilde{D}$ is equal to $\tilde{D}_{subsquare}$ in a small subsquare at the upper-right corner of the domain (see Figure 13.1) and to 1 in the rest of the domain. The mixed boundary conditions

$$\tilde{D}u_n + 0.5u = \mathcal{G}_2$$

(where **n** is the outer normal vector and $\mathcal{G}_2$ is a given function) are imposed on the left and bottom edges, and Neumann boundary conditions are imposed on the right and upper edges. The coarsest grid is a uniform 5×5 grid of cell size 1. The refinement is done in the upper-right quarter of the domain as in Figure 13.2. This procedure is repeated recursively, so that only the upper-right quarter of a refined region is refined again. A linear finite-element discretization method is used on the locally refined mesh as in Figure 13.3. The number of levels is determined in such a way that the discontinuity lines of $\tilde{D}$ lie within the region of highest resolution in the mesh and align with the finest mesh (but not with the coarse grids). The coefficient matrix for this discretization method is an irreducibly diagonally dominant L-matrix (in the sense explained in the exercises at the end of Chapter 3), as are also the coarse-grid coefficient matrices.

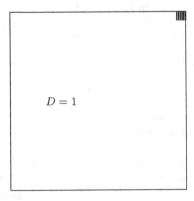

Fig. 13.1. The diffusion coefficient $\tilde{D}$. $\tilde{D} = \tilde{D}_{subsquare}$ in the small black subsquare.

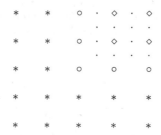

Fig. 13.2. The semistructured grid created by one step of local refinement. The coarse grid f_1 contains the points denoted by $*$, $\circ$, and $\diamond$. The set of points added in the localrefinement step, f_0, contains the points denoted by $\cdot$. The refined region contains the points denoted by $\cdot$ and $\diamond$. Its internal boundary contains the points denoted by $\circ$.

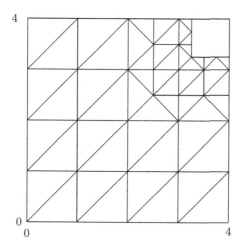

Fig. 13.3. The locally refined finite-element mesh. The mesh in the upper-right blank subsquare is further refined recursively in the same manner until the finest level is reached, where the usual right-angled triangulation is used.

From (12.2) and (4.17), it follows that fine-gridpoints that lie in between two coarse-gridpoints belong to f_a, and the rest of the fine-gridpoints that are not coarse-gridpoints belong to f_b. This yields the following grid partitioning:

$$
\begin{array}{ccccc}
f_b \cdot & f_a \cdot & f_b \cdot & f_a \cdot & f_b \cdot \\
f_a \cdot & c * & f_a \cdot & c * & f_a \cdot \\
f_b \cdot & f_a \cdot & f_b \cdot & f_a \cdot & f_b \cdot, \\
f_a \cdot & c * & f_a \cdot & c * & f_a \cdot \\
f_b \cdot & f_a \cdot & f_b \cdot & f_a \cdot & f_b \cdot
\end{array}
\tag{13.2}
$$

where points in c, f_a, and f_b are denoted by '$c *$', '$f_a \cdot$', and '$f_b \cdot$', respectively.

The number of levels in the multigrid linear system solver is the same as the number of levels of refinement used for creating the semistructured grid. The 4 color point Gauss–Seidel relaxation is used in the V-cycle. Only 16 unknowns are relaxed

Table 13.1. Convergence factors (cf) for V(1,1), AFAC, and AFACx cycles with the present matrix-based multigrid method for diffusion problems with discontinuous coefficients discretized on the locally refined finite-element mesh.

Description of Examples

Example	Size of Domain	Size of Subsquare	$\tilde{D}_{subsquare}$	Levels	Points per Level
(1)	4×4	$2^{-8} \times 2^{-8}$	10^3	10	4×4
(2)	16×16	$2^{-16} \times 2^{-16}$	10^6	20	16×16
(3)	6×6	$(65/32) \times (65/32)$	10^3	6	finest: 68×68

Numerical Results

Example	V(1,1)	AFAC	AFACx
(1)	.250	.263	.25
(2)	.249	.253	.33
(3)	.249	.257	.306

at each level, that is, only the unknowns corresponding to gridpoints in the refined region (see Figure 13.2). We refer to this example as Example (1) in Table 13.1. The convergence factors (cf) defined in (10.12) are reported in this table.

The convergence factors in Table 13.1 indicate that the V(1,1)-cycle converges rapidly. Furthermore, the convergence factors for AFAC and AFACx are also good. In fact, the V(1,1)-cycle converges at most twice as rapidly as AFAC and AFACx, in agreement with results in [60] and [72].

Example (2) is different from Example (1) in that the size of the subsquare is smaller, the diffusion coefficient there is larger, and more levels of refinement and more gridpoints per level are used. The convergence factors are about the same as those for Example (1). In Example (3), the subsquare is much larger than in the previous examples: its area is about 1/9th of the area of the entire domain. All levels, including the finest one, still cover the subsquare. Each coarse grid contains the even-numbered points of the next finer grid. The index of the point of discontinuity is deliberately taken to be odd because we are interested in the most difficult case for which the discontinuities are invisible on the coarse grids.

Note that, unlike in the previous examples, the number of points per grid is not constant, but rather decreases from one grid to a coarser grid. The unrefined region at each grid is still two gridpoints wide as in Figure 13.3, but the numbers of points per level and the numbers of gridpoints along the interfaces between refined and unrefined regions are much larger than those in Figure 13.3. The convergence factors for this example are also about the same as those for the previous examples.

Similar convergence rates are also obtained for a finite-volume discretization, provided that it is defined at the interfaces between refined and unrefined regions in such a way that the coefficient matrix A is symmetric. This is a desirable and important property for discretization methods for symmetric PDEs [118]. Indeed, the discretization error (i.e., the difference between the solution to the boundary-value problem and the solution to the discrete system on the grid) is A^{-1} times the truncation error [see (3.10)]. Therefore, an accurate discretization method should have not only small truncation error but also small $\|A^{-1}\|_2$. For symmetric matrices A, we have $\|A^{-1}\|_2 = \rho(A^{-1})$ (Lemmas 2.9 and 2.14); but for nonsymmetric matrices, $\|A^{-1}\|_2$ could be much larger than that, resulting in poor approximation even when the truncation error is small.

The convergence of the present multigrid method deteriorates whenever discretization methods with nonsymmetric coefficient matrices are used for symmetric PDEs; even the cure in [40] does not help in this case. Careful definition of the equations at points at the interfaces between refined and unrefined regions is thus necessary in the finite-volume discretization to make sure that A is symmetric. The success or failure of multigrid linear system solvers can thus serve also as an indicator about the adequacy of the discretization method used. When good multigrid solvers fail to converge, the user may suspect that the PDE is ill-posed or the discretization method is inadequate.

Finally, we consider a problem that is the symmetric extension across the upper-right corner of Example (1) in Table 13.1 (see Figure 13.4). This problem is equivalent to two problems as in Example (1) in Table 13.1 with continuity enforced at their joint corner. (Although the boundary-value problem is ill-posed, the algebraic problem may well serve as a test problem.) The refinement and discretization methods are as before, except that they are also extended symmetrically across the

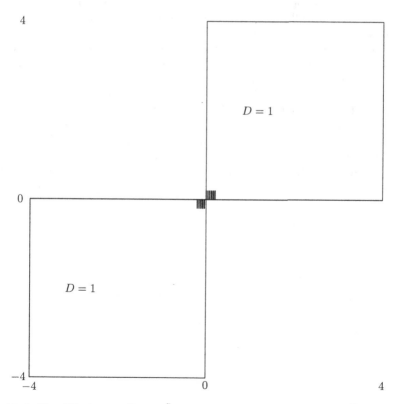

Fig. 13.4. The diffusion coefficient $\tilde{D}$ in the irregular domain example. $\tilde{D} = 1000$ in the small black subsquares. The size of each of these subsquares is $10^{-8} \times 10^{-8}$.

corner, and the multigrid method is also implemented as before. The convergence factor for the V(1,1)-cycle in this case is cf = 0.45. Similar results were also obtained for the Poisson equation on this domain, implying that the difficulty in this example lies in the irregular shape of the domain.

We suspect that the reason for this somewhat slow convergence is in the ill-posedness of the PDE. Indeed, the same convergence rate is also obtained also for the ill-posed example in (9.10) using global refinement and BBMG. The conclusion is, thus, that multigrid methods may be helpful not only in solving large linear systems of equations but also in detecting ill-posedness of PDEs. By converging slowly, the multigrid method indicates that something is wrong in the linear system, and the user should check the well-posedness of the PDE and the adequacy of the discretization method.

13.5 Scaling the Coefficient Matrix

As expected from Lemma 12.3, the above convergence rates are not influenced by scaling of the linear system from the left in advance. However, there is an example for which the multigrid method applied to the original system (3.5) stagnates, but it does converge when applied to the left-scaled system (12.19). The reason for this lies in numerical roundoff errors that appear when the diffusion coefficient takes very small values, as discussed below.

The example in which scaling has an important effect is similar to Example (3) in Table 13.1; in the square the only changes are that the domain is of size 10×10, and $D = 1$ in all of it except in a 2×2 subsquare in the middle of it, where $\tilde{D} = \tilde{D}_{subsquare} = 10^{-4}$ (see Figure 13.5). The refinement is done towards the middle of the domain as follows. The finest level is a 71×71 grid covering the subsquare. (The subsquare is of size 64×64 fine-gridpoints.) Each coarse level uses the even-numbered points of the next finer level plus a margin of another extra three gridpoints in the unrefined region (see Figure 13.6). As in Section 13.4, the discontinuities align with points in the region of highest resolution only, and not with points in the coarser

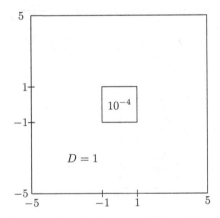

Fig. 13.5. The diffusion coefficient $\tilde{D}$ for the finite-volume local refinement discretization. $\tilde{D} = 10^{-4}$ in the subsquare.

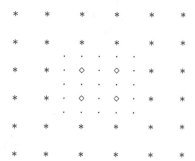

Fig. 13.6. The semistructured grid created from one local refinement step. (The finite-volume discretization method is applied to such a grid.) The coarse grid f_1 contains the points denoted by $*$ and $\diamond$. The set of points added in the local-refinement step, f_0, contains the points denoted by $\cdot$. The refined region contains the points denoted by $\cdot$ and $\diamond$. Only these points are relaxed at the fine level.

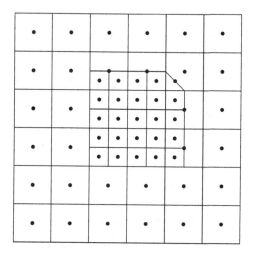

Fig. 13.7. The locally refined finite-volume mesh.

grids; the fine-gridpoints where the discontinuity is present are odd-numbered, so they do not appear on the coarse grids. This approach is taken in order to make the problem more difficult and challenging; indeed, using even-numbered discontinuity points improves the convergence factors a little, as expected. The finite-volume discretization method (Section 3.12) is used as follows. The integral of the left-hand side in (13.1) over a volume in Figure 13.7 is replaced by a line integral over the boundary of this volume using Green's theorem. Then, the normal derivatives in this line integral are replaced by finite differences using the points in Figure 13.7. These approximations are of second-order accuracy in the interior of refined regions and of first-order accuracy at the interfaces between refined and unrefined regions. The discretization is designed so that the coefficient matrix is symmetric also at these interfaces; otherwise, the rate of convergence of the multigrid linear system solver deteriorates.)

Table 13.2. Convergence factors (cf) for the matrix-based multigrid method applied to the finite-volume discretization (with local refinement) of the diffusion problem with $D_{subsquare} = 10^{-4}$. The linear system is scaled in advance from the left to avoid numerical roundoff errors in the coarse-grid matrices.

Levels	V(1,1)	AFAC	AFACx
6	.250	.416	.421

Note that a small oblique shift of the refined region in the grid in Figure 13.7 in the southwest direction yields the grid of Figure 13.6. Therefore, the unknowns may be stored as in Figure 13.6, and the coarse grid is logically rectangular as well as the fine grid. With this implementation, the stencil contains at most nine coefficients, and the present coarsening method is applicable. The full coarsening used in the refined regions as in (13.2) guarantees that the coarse-grid matrices are also of 9-point stencil at most. Six levels of refinement are used to produce the semistructured grid for the discretization, and six-level multigrid cycles are used to solve the resulting linear system of equations. The rest of the details of implementation are as in Section 13.4.

It turns out that the multigrid method stagnates for the above example when applied to (3.5), but converges nicely when applied to the left-scaled system (12.19) (see Table 13.2). We have observed that, because of roundoff errors, the coarse-grid matrices for the original system (3.5) have positive row-sums for equations corresponding to the subsquare, implying that the coarse-grid approximation is inappropriate. In fact, the coarse-grid equations at the subsquare tend towards trivial equations as the grid becomes coarser and coarser. These equations cannot possibly yield suitable correction terms for the original problem. Fortunately, this difficulty disappears when the linear system is left-scaled before constructing the transfer and coarse-grid operators.

To guarantee numerical stability, the coarse-grid matrices are constructed in a way that is both stable and efficient; that is, first $J_c RA$ is calculated, and then, using it, $(J_c RA)PJ_c^t$ is calculated. The results in Table 13.2 show that left-scaling can indeed help in avoiding numerical instability. A similar observation is made in the examples in Section 13.4 when $\tilde{D}_{subsquare}$ is made very small rather than very large: left-scaling is useful in preventing stagnation of the multigrid linear-system solver due to numerical roundoff errors.

13.6 A Black-Box Multigrid Version

In the spirit of the above implementation, we have also extended BBMG for problems on semistructured grids as follows. The prolongation operator

$$\tilde{P}_i : l_2(f_L^i) \rightarrow l_2(f_L^{i-1})$$

is constructed as in Section 9.1 in the refined region, and is the identity in the unrefined region, where all the gridpoints remain in the coarse grid as well. (This approach is in the spirit in [45], where the unknowns that correspond to no gridpoint remain in all the coarse levels.) The restriction operator

$$\tilde{R}_i : l_2(f_L^{i-1}) \to l_2(f_L^i)$$

is the transpose of $\tilde{P}_i$ with respect to the inner product for which the coefficient matrix at the $(i-1)$st level, B_{i-1}, is symmetric. Then, we define the next coarse-grid coefficient matrix B_i by

$$B_i \equiv \tilde{R}_i B_{i-1} \tilde{P}_i.$$

As before, the 4-color point Gauss–Seidel relaxation is used in the refined region, and the relaxation in the unrefined region is deferred to the next coarser level. This version yields practically the same convergence rates as those reported in Sections 13.4 and 13.5. For the left-scaled problem in Section 13.5, it is absolutely essential to use the above-mentioned inner product (with respect to which the B_{i-1}'s are symmetric) to have good convergence. Indeed, with the standard definition $\tilde{R} \equiv \tilde{P}^t$, BBMG exhibits slow convergence (even with the fix in [40]).

There is only one example for which the extension of BBMG converges more rapidly than our main multigrid version in Section 12.7. This example is like that in Section 13.5, but with $\tilde{D}_{subsquare} = 10^3$ rather than 10^{-4}. With the V(1,1)-cycle, no matter whether scaling is used, the extension of BBMG yields a convergence factor cf $= 0.19$, whereas the main method in Section 12.7 yields cf $= 0.35$.

So, although the upper bound for the condition number in Theorem 12.1 above applies to the multigrid method in Section 12.7 only, one can still learn from it how to extend BBMG to semistructured grids as well. Still, the method in Section 12.7 is more general than the extension of BBMG, because it applies not only to semistructured grids but also to more general locally refined meshes.

13.7 Exercises

1. Assume that A has a 9-point stencil on the grid in (13.2). Show that Q has a 9-point stencil as well.
2. Consider the serial computer in the exercises at the end of Chapter 1. Compute the sequential time of the V(1,1)-cycle in Section 13.2 applied to a uniform mesh as in Figure 4.13, refined globally by Refinement Method 4.1.
3. Consider the parallel computer in the exercises at the end of Chapter 1. Compute the parallel time for AFACx cycle in Section 13.3 for the above linear system. (Assume that AFACx requires twice as many iterations as the V(1,1)-cycle.)
4. Compute the speedup and average speedup as functions of the number of processors P and the number of unknowns N.
5. As explained in [72], the main advantage of AFAC and AFACx is the opportunity to use an efficient load-balancing scheme to distribute the entire workload in advance rather evenly among the processors. Modify your above speedup estimates by incorporating this point as well.
6. Apply your code from the exercises at the end of Chapter 12 to problems on semistructured grids as in Section 13.4.
7. Are the coarse-grid matrices diagonally dominant as well?
8. Is $\|P_{ff}\|_{D_{ff}}$ reasonably bounded?
9. Does AFACx indeed require twice as many iterations as the V(1,1)-cycle?

Part VI

Matrix-Based Multigrid
for Unstructured Grids

Locally refined meshes and, in particular, semistructured grids provide accurate approximation to irregular solution functions. The process of repeated local refinement yields a grid that is fine wherever the solution has large variation (large gradient), and coarser wherever the solution is relatively smooth. However, locally refined meshes are not always suitable for complicated domains, for which the initial coarse mesh may be inappropriate. To approximate well such domains, one must often use completely unstructured grids that cannot be obtained from local refinement but rather by a careful study of the properties of the particular domain under consideration. Unlike locally refined grids, such grids have no natural hierarchy of coarser and coarser grids that may be used in a multigrid linear system solver. Furthermore, completely unstructured finite-element meshes have no hierarchy of nested finite-element function spaces to approximate the original problem on coarse levels. The coarse grids in the multigrid linear system solver must therefore be defined algebraically, using the information in the original algebraic system. In fact, the coarse grids are just nested subsets of unknowns, each of which is referred to as "level." Because these subsets no longer relate to any physical grids, it is more appropriate to refer to the linear system solver that uses them as a multilevel rather than multigrid method.

The family of algebraic multilevel methods is, thus, yet more "algebraic" than the family of matrix-based multigrid methods studied in the previous parts in this book. Indeed, in algebraic multilevel methods the original coefficient matrix is used to form not only the transfer and coarse-level operators but also the coarse levels themselves, namely, the subsets of unknowns chosen to be used to supply correction terms. Because the method is independent of any property of the particular application, it has to be implemented once and for all, yielding a computer program that can in principle be used for general linear systems.

In this part of the book, we describe two multigrid approaches to solve large sparse linear systems arising from the discretization of elliptic PDEs on completely unstructured grids. The first approach uses domain decomposition to design the coarse grid and the transfer operators to and from it. This approach is not entirely algebraic, because the definition of coarse grid comes from the geometric properties of the domain and its decomposition. This is why it is still referred to as "multigrid" rather than "multilevel." The second approach, on the other hand, introduces an entirely algebraic multilevel method that uses only information from the coefficient matrix to design both the coarse level (the subset of unknowns with no geometric interpretation whatsoever, on which a low-order system is solved to supply a correction term) and the transfer operators to and from it. Because no geometry is used in its definition, this method is called *algebraic multilevel* rather than *algebraic multigrid*. (The name "algebraic multigrid" is preserved for the original AMG method, see Section 6.10.) This way, the method can be implemented once and for all, yielding a computer code ready to be used for different kinds of applications, including the locally refined meshes discussed in the previous part and completely unstructured meshes. Indeed, the method is tested here numerically for many difficult examples, including highly anisotropic problems, complicated domains, highly nonsymmetric problems, and systems of coupled PDEs.

The Domain-Decomposition Multigrid Method

In this chapter, we introduce a domain decomposition two-grid iterative method for solving large sparse linear systems that arise from the discretization of elliptic PDEs on general unstructured grids that do not necessarily arise from local refinement. In this method, the coarse grid consists of the vertices of the subdomains in the domain decomposition. Assuming that the coefficient matrix is SPD and diagonally dominant, we supply an upper bound for the condition number of the $V(0,0)$ cycle.

We call the present method the Domain-Decomposition Multigrid (DDMG) method [98]. This method can be viewed as an extension of BBMG to the most complicated case of unstructured grids. Indeed, it extends the approach illustrated in Figure 6.8 to nonuniform grids.

14.1 The Domain-Decomposition Approach

In this chapter, we use domain decomposition to design a two-grid iterative linear system solver. We assume that a general (unstructured) finite-element mesh is given; it is not assumed that the mesh has been obtained from local refinement or any other inductive process that can guide the definition of coarse grids. Instead, we assume that a domain decomposition is available, which can be used in the construction of the coarse grid. The transfer and coarse-grid matrices are then derived from the coefficient matrix alone.

Domain decomposition is a well-known mathematical methodology for the numerical solution of elliptic boundary-value problems. It can be used (a) in the original mathematical model to reformulate the original PDE, (b) in the discretization stage to form parallelizable numerical schemes or (c) in the discrete system of algebraic equations to introduce iterative linear system solvers. Here we use it for purpose (c) above; in fact, we use it to produce a multigrid algorithm.

Domain decomposition is particularly attractive in parallel computing, because the individual subtasks in the individual subdomains can be carried out simultaneously in parallel. It is also attractive in the context of cache-oriented programming, provided that the data required to carry out a particular subtask in a particular subdomain fit in the cache (see Section 5.10). Another advantage of domain decomposition is the opportunity to use available algorithms (such as multigrid) and software (such as finite-element packages) to solve the individual subproblems in the individual subdomains.

Here we use domain decomposition not as a discretization method but rather as a multigrid algorithm to solve the discrete system obtained from a given finite-element discretization of an elliptic boundary-value problem such as the diffusion equation (3.2). The diffusion coefficients may be variable and even discontinuous, and the discontinuity lines don't have to align with the edges of the subdomains, neither in the analysis below nor in the actual application of the method.

14.2 The Domain-Decomposition Multigrid Method

Consider the finite-element mesh for the solution of an elliptic boundary-value problem as in (3.2) and the domain decomposition in Figure 14.1. Assume that the linear finite-element discretization is used to produce the linear system (3.5) with the coefficient (stiffness) matrix A. We define the Domain-Decomposition Multigrid (DDMG) method for solving this system as follows. Let the coarse grid c be the set of corners or vertices of subdomains (interior nodes that are shared by at least three subdomains, boundary points that are shared by at least two subdomains, corner boundary points, and possibly some other suitable nodes on the edges of subdomains); (see Figure 14.2).

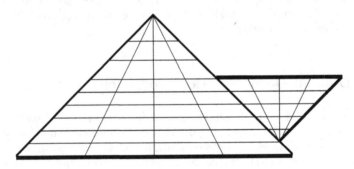

Fig. 14.1. The unstructured grid and the domain decomposition.

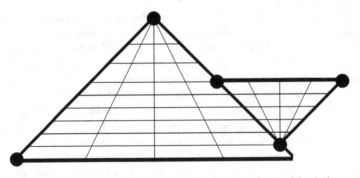

Fig. 14.2. The coarse-gridpoints are denoted by '•.'

In the following, by "edge of a subdomain" we refer to the set of nodes in a segment of the (internal) boundary of the subdomain that connects one point from c (at one end of the segment) to the next point from c (at the other end of the segment). (The endpoints of the segment are included in the edge as well.)

The key factor in the DDMG method is the prolongation operator P, which prolongs a function v defined only at nodes in c into a function Pv defined in every node in the mesh. This is done in two steps: first Pv is defined in each edge of each subdomain by solving a tridiagonal subsystem; then, the resulting values of Pv in the edges of each subdomain are used to define Pv also in its interior by solving a a local (Dirichlet) subproblem in it. The complete definition is detailed next.

Define the matrix $A^{(0)}$ by:

$$A^{(0)}_{i,j} = \begin{cases} a_{i,j} & \text{if } i \text{ and } j \text{ do not lie on the same edge} \\ & \text{but do lie on two different edges} \\ 0 & \text{otherwise} \end{cases}$$

and the matrix $\tilde{A}$ by

$$\tilde{A} = A - A^{(0)} + rs(A^{(0)}).$$

In other words, $\tilde{A}$ is obtained from A by "throwing" onto the main diagonal those elements $a_{i,j}$ that couple nodes from different edges.

We are now ready to define two disjoint subsets of nodes, on which the two prolongation steps will be defined. We start with f_a, the subset of nodes where the first prolongation step is defined. Initialize f_a to be the set of nodes outside c that lie on some edge of some subdomain. Once f_a has been initialized, it is further modified to guarantee stability as follows. For every $i \in f_a$, if

$$- \sum_{j \in c \cup f_a, \ j \neq i} \tilde{A}_{i,j}/a_{i,i} \leq \text{threshold}, \tag{14.1}$$

then i is dropped from f_a and added to c.

Once this procedure has been completed, we are ready to define f_b, the subset of nodes on which the second prolongation step will be defined. This subset contains the nodes that are neither in c nor in f_a. In other words, f_b is the set of nodes in the interiors of the subdomains.

Clearly, c, f_a, and f_b are disjoint. Define $f = f_a \cup f_b$. Define the matrix G_a as follows.

1. Initialize G_a by

$$G_a = \tilde{A}_{f_a f_a} + rs(\tilde{A}_{f_a f_b}).$$

2. If, for some $i \in f_a$,

$$(G_a)_{i,i}/a_{i,i} < \text{threshold},$$

then update $(G_a)_{i,i}$ by

$$(G_a)_{i,i} \leftarrow \tilde{A}_{i,i} - rs(\tilde{A}_{f_a f_a})_{i,i} - rs(\tilde{A}_{f_a c})_{i,i}.$$

From the fix that follows (14.1), the new value of $(G_a)_{i,i}$ is bounded away from zero. This property guarantees stability of the prolongation operator defined later.

Note also that G_a is a block-diagonal matrix, with tridiagonal blocks that correspond to the different edges. Thus, its inversion requires the solution of independent tridiagonal systems, which can be done efficiently in parallel.

Define G_b by

$$G_b = \tilde{A}_{f_b f_b}.$$

Note that G_b is block-diagonal, with blocks that correspond to the subdomain interiors. Thus, its inversion requires the solution of independent problems in the individual subdomains, which can be done efficiently in parallel.

The restriction and prolongation matrices are analogous to those in Section 12.2. Note that these are square matrices of the same order as A, thus they are slightly different from the rectangular restriction and prolongation matrices used traditionally in multigrid algorithms. Still, this difference has little practical effect, because the submatrices that are added to make the restriction and prolongation matrices square are easily invertible, thanks to their block-diagonal structure. The main reason for using square restriction and prolongation matrices here is to have the upper bound for the condition number derived below. (In practice, one could use the rectangular prolongation and restriction matrices obtained by omitting the upper-left blocks R_{ff} and P_{ff} from the present matrices.)

Here are the definitions of the prolongation matrix P and the restriction matrix R:

$$P = \begin{pmatrix} (G_b)^{-1} & -(G_b)^{-1}\tilde{A}_{f_b f_a} & -(G_b)^{-1}\tilde{A}_{f_b c} \\ 0 & I & 0 \\ 0 & 0 & I \end{pmatrix} \begin{pmatrix} I & 0 & 0 \\ 0 & (G_a)^{-1} & -(G_a)^{-1}\tilde{A}_{f_a c} \\ 0 & 0 & I \end{pmatrix} \tag{14.2}$$

$$R = \begin{pmatrix} I & 0 & 0 \\ 0 & (G_a)^{-1} & 0 \\ 0 & -\tilde{A}_{c f_a}(G_a)^{-1} & I \end{pmatrix} \begin{pmatrix} (G_b)^{-1} & 0 & 0 \\ -\tilde{A}_{f_a f_b}(G_b)^{-1} & I & 0 \\ -\tilde{A}_{c f_b}(G_b)^{-1} & 0 & I \end{pmatrix}. \tag{14.3}$$

Finally, the coarse-grid matrix Q is defined by

$$Q = \begin{pmatrix} W & 0 \\ 0 & B \end{pmatrix}, \tag{14.4}$$

where

$$B = J_c R A P J_c^t, \tag{14.5}$$

and

$$W = R_{ff} \, diag(A_{ff}) P_{ff}. \tag{14.6}$$

The coarse-grid matrix Q is different from standard coarse-grid matrices by the submatrix W, which is added at the upper-left block to make sure that Q is of the same order as A. This definition is used here for the sake of the analysis below. The submatrix W is easily invertible, and adds little to the cost of the multigrid algorithm.

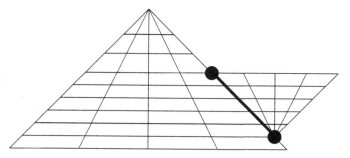

Fig. 14.3. First prolongation step: from the endpoints of the edge (denoted by '•') to the entire edge.

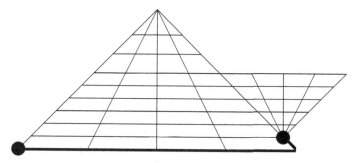

Fig. 14.4. First prolongation step: from the coarse grid (c) to the edges (f_a).

The application of the prolongation operator is actually done in two steps:

1. First, determine the values of the prolonged function in the edges by solving a homogeneous 1-D problem (tridiagonal system) on each edge with the Dirichlet boundary conditions at the endpoints of this edge taken from the known values in c (Figures 14.3 and 14.4).
2. Next, determine the values of the prolonged function in the subdomain interiors as well by solving homogeneous subproblems in the individual subdomains, with the Dirichlet boundary conditions on the edges taken from the first prolongation step (Figure 14.5).

This is in the spirit of the algorithm in Section 6.6.

Thus, applications of the operators R or P require the solution of the local 1-D problems in the tridiagonal blocks in G_a in the individual edges and the local subproblems in the blocks in G_b in the individual subdomains. Because these local problems are independent of each other, they can be solved simultaneously in parallel. Similarly, the definition of Q requires applications of $J_c RAP J_c^t$ to the standard unit vectors in $l_2(c)$ (which have the value 1 at one of the nodes in c and 0 elsewhere). This can also be done in parallel, as discussed above. Furthermore, when the data required to solve each local subproblem fit in the cache, one could use cache-oriented programming to minimize access to the secondary memory (see Section 5.10).

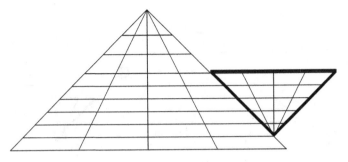

Fig. 14.5. Second prolongation step: from the edges $(c \cup f_a)$ to the interior of each sub-domain (f_b).

The above local subproblems (corresponding to the blocks in G_a and G_b) can be solved by an inner multigrid iteration. This iteration, however, could still be rather time consuming. To avoid it, one could approximate the blocks in P_{fc} by suitable row-sum blocks, in the spirit of Section 6.11 above. This approach has never been tested, and is left to future research. Here we prefer to stick to the original approach, which enjoys the theory below.

The relaxation method used in the DDMG method can be a version of the alternating Schwarz method. In this relaxation, the original equations are relaxed in c in a first stage, then relaxed simultaneously in f_a in a block-Jacobi stage, and finally relaxed simultaneously in f_b in another block-Jacobi stage. (See also Section 16.4, where a special case of this approach is formulated algebraically.) In the analysis below, however, we assume that no relaxation is used, and derive an upper bound for the condition number of the V(0,0)-cycle.

14.3 Upper-Bound for the Condition Number

The following theorem gives an upper-bound for the condition number of DDMG, implemented in a V(0,0)-cycle (with no relaxations). This implies that the nearly singular eigenvectors of A are handled well by the coarse-grid correction. This indicates that the V(1,1)-cycle, in which relaxation is used to handle the rest of the error modes as well, should converge rapidly, at least for diagonally dominant SPD problems.

Theorem 14.1 *Let D be a diagonal SPD matrix of the same order as A. Assume that A is a diagonally dominant L-matrix that is symmetric with respect to $(\cdot, \cdot)_D$. Then, we have for DDMG*

$$\kappa(PQ^{-1}RA) \le 4\left((1 + 2\|P_{ff}\|_{D_{ff}}\|A\|_D)^2 + 4\|P_{ff}\|_{D_{ff}}^2\|A\|_D^2\right).$$

Proof. The theorem is actually a special case of Theorem 12.1 with $L = 1$. [Note that Lemma 12.5 applies also for DDMG in a stronger version, in which the factor $\sqrt{2}$ in (12.22) is replaced by 1. Indeed, the proof of this lemma is valid also for DDMG with $A_b = (0)$. Thus, we have

$$\|J_f^t J_f P^{-1} x\|_D^2 \le 4\|A\|_D(x, Ax)_D.$$

The theorem then follows from (12.28), (12.24), (12.25), and (12.33), with $L = 1$ and $\sqrt{2}$ replaced by 1.] This completes the proof of the theorem.

Note that, when D is as in (12.19) and A is left-scaled in advance as in (12.19), we have from Corollary 2.3 that $\|A\|_D \le 2$. Furthermore, Theorem 14.1 takes in this case a more concrete interpretation, as stated in the following corollary.

Corollary 14.1 *Assume that the isotropic diffusion equation (13.1) is discretized on a linear finite-element mesh using triangles with angles of at most $\pi/2$. Then, for DDMG, $\kappa(PQ^{-1}RA)$ is bounded independently of the meshsize and the possible jump in the diffusion coefficient $\tilde{D}$, provided that the number of elements per subdomain is bounded.*

Proof. From (4.17), A is a diagonally dominant L-matrix. The boundedness of the number of elements per subdomain and the fix that follows (14.1) guarantee that $\|P_{ff}\|_{D_{ff}}$ is bounded independently of the meshsize. In view of (12.19) and the discussion in Section 12.5, $\|P_{ff}\|_{D_{ff}}$ is bounded also independently of the possible jump in the diffusion coefficient $\tilde{D}$. The corollary follows from Theorem 14.1 and Lemma 12.3.

Note that there is no need to assume that the possible discontinuities in the diffusion coefficient $\tilde{D}$ align with the edges of the subdomains. Thus, once DDMG has been implemented for a particular domain and a particular domain decomposition, the computer code can be used for different kinds of diffusion coefficients.

The above corollary indicates that, when the meshsize tends to zero, the number of subdomains should also increase to guarantee the boundedness of $\|P\|_2$ and the condition number. When the number of subdomains is very large, c is large as well, and the coarse-grid problem is hard to solve directly. Instead, it can be solved approximately recursively by one DDMG iteration, using a secondary domain decomposition with larger subdomains. In this case, DDMG is implemented as a multigrid rather than two-grid method.

14.4 High-Order Finite Elements and Spectral Elements

In high-order finite-element and spectral-element discretization methods [80], each element in the mesh may contain a large number of nodes to increase the accuracy of the numerical approximation. Because all the nodes in a particular finite or spectral element are coupled to each other, the coefficient matrix A is less sparse than when finite differences or linear finite elements are used. The definition of a multigrid linear system solver is therefore more difficult. Indeed, even for structured grids as in Figure 14.6, it is not clear how to define the coarse grids and the transfer and coarse-grid matrices.

One possible approach is to use a low-order scheme that uses the same nodes as the original high-order scheme. Once the resulting matrix is scaled as the original one (say, left-scaled), it can be used as a preconditioner in a PCG or any other Krylov-subspace acceleration method [83].

In [106], this approach is used for structured grids as in Figure 14.6. Two preconditioners are tested: one uses finite differences on the nodes in Figure 14.6, and

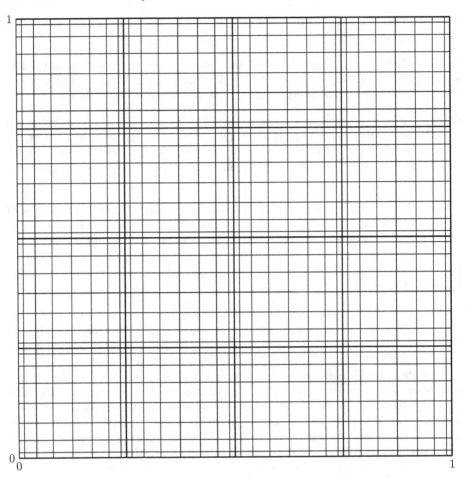

Fig. 14.6. A mesh of 16 spectral elements (defined by the thick lines), each of which uses polynomials of degree at most 8.

the other uses bilinear finite elements on the cells in Figure 14.6. Both the original matrix and the preconditioning matrices are left-scaled in advance, so they have the same scaling. The preconditioning step is done approximately using a single Auto-MUG or BBMG V(1,1)-cycle. (Because of the left-scaling, the transpose in BBMG should be interpreted as in Sections 13.5 and 13.6.)

Another possible approach tested in [106] is to use the above low-order scheme only to construct the multigrid hierarchy, then drop it and replace it by the original high-order scheme on the original fine grid. The resulting V(1,1)-cycle that uses the original high-order scheme on the finest grid in Figure 14.6 (with two-step Jacobi relaxation based on the red-black coloring, see Section 16.2 below) and the transfer and coarse-grid operators derived from the low-order scheme yields a more stable iterative solver than in the previous approach.

Here we propose a third approach in the spirit of domain decomposition. In this approach, each finite or spectral element is viewed as a subdomain. The DDMG method may now be applied to the original high-order scheme, with no need to use any low-order scheme.

As a matter of fact, the multigrid algorithm in Chapter 12 can also be thought of as a DDMG algorithm, with the finite elements in the coarse mesh viewed as small subdomains. This observation establishes a close relation between multigrid and domain-decomposition methods, as discussed in Chapter 6.

14.5 Exercises

1. Use the discussion in Section 12.5 to show that, for isotropic diffusion problems discretized on a finite-element triangulation (with angles that do not exceed $\pi/2$), $\|P_{ff}\|$ is moderate, provided that the number of nodes per subdomain is moderate.

2. Use the above result in Theorem 14.1 to show that the condition number of the DDMG V(0,0)-cycle is moderate.

3. Use the above result to show that the DDMG V(1,1)-cycle should converge rapidly.

4. Explain the advantage of DDMG in terms of cache access.

5. Show that DDMG can be viewed as a version of the multigrid method whose prolongation is indicated in Figure 6.8.

6. Show that the ith column in Q can be calculated by

$$Qe^{(i)} = R(A(Pe^{(i)})),$$

where $e^{(i)}$ is the ith column of the identity matrix of the same order as Q.

7. Show that all the columns of Q can be calculated simultaneously in parallel.

8. Show that Q is sparse in the sense that two coarse-gridpoints are coupled in Q if and only if they are corners of the same subdomain. Conclude that the dimension of $e^{(i)}$ above could actually be much smaller than the order of Q. In fact, $e^{(i)}$ should be defined only at the coarse-gridpoint corresponding to i and at the coarse-gridpoints surrounding it, in which $Qe^{(i)}$ is supported.

9. Write the computer code that implements the DDMG method for diffusion problems in the L-shaped domain in Figure 6.4. Use the cyclic-reduction method in Section 7.2 to solve the individual tridiagonal systems in the individual edges between subdomains and your BBMG code from the exercises at the end of Chapter 9 to solve the individual subsystems in the individual subdomains in the applications of R and P and the construction of Q.

10. Test your code for the Poisson equation with Dirichlet boundary conditions. Do you obtain the Poisson convergence factor of 0.1?

11. Test your code for problems with various kinds of boundary conditions. Is the convergence rate still good?

12. Test your code for isotropic diffusion problems with discontinuous coefficient with discontinuity lines that don't align with the edges of the subdomain. Is the convergence rate still good?

13. Test your code for highly anisotropic equations as in Section 3.9. Does the convergence rate deteriorate?

14. Modify your code to use the block-GS relaxation, with the blocks being the subdomains themselves. Does the convergence rate improve?

15. Modify your code to use the alternating "zebra" line-GS relaxation. Does the convergence rate improve?

15

The Algebraic Multilevel Method

.

The multilevel method introduced here is algebraic in the sense that it is defined in terms of the coefficient matrix only. In particular, the coarse level is just a subset of unknowns with no geometric interpretation, hence the name "multilevel" rather than "multigrid." For diagonally dominant SPD problems, we derive an a posteriori upper-bound for the condition number of the V(0,0)-cycle. For diffusion problems, this upper-bound indicates that the condition number grows only polynomially with the number of levels, independent of the meshsize and the possible discontinuities in the diffusion coefficient. This indicates that the nearly singular eigenvectors of A are handled well by the coarse-level correction, so the V(1,1)-cycle that uses relaxation to handle the rest of the error modes as well should converge rapidly.

15.1 The Need for Algebraic Multilevel Methods

An *algebraic multilevel method* is a multilevel method that is defined in terms of the coefficient matrix A only, with no mention of the underlying PDE or discretization method from which the linear system has been produced. Because the coarse levels are just nested subsets of unknowns with no geometric meaning, the name "multilevel" is more suitable than "multigrid." (The name "algebraic multigrid" (AMG) is preserved for the original algebraic multilevel method in [25] and [86]; see Section 6.10.)

Algebraic multilevel methods are particularly attractive for unstructured-grid problems, where no geometric method is available to construct the coarse grids. The elements in the coefficient matrix A are used to define the coarse level, namely, the subset of unknowns on which the original system can be approximated to supply a correction term to the original error.

The present algebraic multilevel method is carefully designed to suit also problems with highly anisotropic variable coefficients on unstructured meshes as in Figure 4.7. For this purpose, the present analysis proves most useful. Indeed, in order to have a good upper-bound in Theorem 12.1, the prolongation matrix P must be reasonably bounded. To guarantee this, the main-diagonal elements in G_a and G_b should be not too small in magnitude. In other words, there should be no point in f_b that is only weakly coupled to $c \cup f_a$ in A. If the algebraic multilevel method detects such a point, then it immediately drops it from f_b and adds it to c, avoiding any instability in the prolongation operator P. This can actually be viewed as local semi-coarsening in the (local) strong-diffusion direction only, in the spirit of [70] and [86].

Thanks to its algebraic formulation, the algebraic multilevel method can be implemented in a computer code that uses matrix manipulations only, and doesn't have to be modified for each particular application. On the other hand, algebraic multilevel methods cannot exploit the information contained in the original PDE or discretization method, which may lead to nonoptimal convergence rates. This, however, is a price worth paying for having a purely algebraic linear system solver.

15.2 The Algebraic Multilevel Method

In this section, we describe the algebraic multilevel method for the solution of large sparse linear systems. The method is defined in terms of the coefficient matrix only; the underlying PDE and discretization method are never used. In this sense, the method is indeed algebraic: it is an iterative method for the solution of the algebraic system (3.5).

In the multigrid algorithms that have been used so far in this book, the coarse grids are constructed from the original, fine grid, using its geometric properties. In structured grids, the coarse grid is obtained from the fine grid by dropping every other gridline in both the x and y spatial directions (6.1) or only in the y spatial direction (6.15). In locally refined grids, the coarse grid is obtained from the coarse mesh in the previous refinement level in the local-refinement algorithm that has been originally used to create the fine grid. Finally, in unstructured grids on which a domain decomposition is imposed, the coarse grid is obtained from the vertices (corners) of the subdomains. For this reason, these three types of algorithms are not strictly algebraic, as they are not defined in terms of the algebraic system (3.5) alone. In the algebraic multilevel method, on the other hand, the coarse "grid" (or, more precisely, the coarse level) is defined purely algebraically from (3.5) as a subset of unknowns with no apparent geometric meaning whatsoever. This is why it is well defined not only for the above grids but also for completely unstructured grids.

In the algebraic multilevel method, the coarse level is a maximal subset of unknowns that are at most weakly coupled in A. Here is how it is defined. Let N be the order of the coefficient matrix A [the number of unknowns in (3.5)]. Let α, γ, τ, δ, ζ, and ϵ be real nonnegative parameters that will be used below as thresholds (actually, τ, δ, and ζ are positive). The smaller these parameters are, the smaller is the coarse level defined below.

We first describe the two-level implementation, in which only one coarse level is used. The multilevel version that uses more levels is implemented recursively in the same way. One can use the above parameters also in this recursion, or use slightly smaller parameters to further reduce the number of unknowns in the coarser levels.

The matrix $A^{(\gamma)}$ defined below contains off-diagonal elements in A that are small in magnitude or have a real part with the same sign as the corresponding main-diagonal element in A:

$$
(A^{(\gamma)})_{i,j} = \begin{cases} a_{i,j} & \text{if } i \neq j \text{ and } |a_{i,j}| < \gamma \cdot \min(|a_{i,i}|, |a_{j,j}|) \\ a_{i,j} & \text{if } i \neq j \text{ and } \Re(a_{i,j}))\Re(a_{i,i}) > 0 \\ a_{i,j} & \text{if } i \neq j, \ A = A^*, \text{ and } \Re(a_{i,j}))\Re(a_{j,j}) > 0 \\ 0 & \text{otherwise.} \end{cases} \tag{15.1}
$$

The above matrix $A^{(\gamma)}$ is now used to define $\tilde{A}^{(\gamma)}$, which is obtained from the original matrix A by "throwing" the elements in $A^{(\gamma)}$ onto the main diagonal: Define also the matrix

$$\tilde{A}^{(\gamma)} = A - A^{(\gamma)} + rs(A^{(\gamma)}). \tag{15.2}$$

In other words, $\tilde{A}^{(\gamma)}$ is obtained from A by "throwing" onto the main diagonal those off-diagonal elements that are small in magnitude in comparison with the corresponding main-diagonal element or have the same sign as its real part. The matrices $A^{(\alpha)}$ and $\tilde{A}^{(\alpha)}$ are defined in the same way, except that γ in (15.1), (15.2) is replaced by α [$\tilde{A}^{(\alpha)}$ will be used in the coarsening procedure below]. Note that these matrices are symmetric (or hermitian) whenever A is. Furthermore, if A is an L-matrix with positive main-diagonal elements and $\gamma = 0$, then $\tilde{A}^{(\gamma)} = A$.

15.3 The Coarsening Procedure

We are now ready to define the coarsening procedure (the definition of the coarse level), which is just a variant of Algorithm 6.3.

1. First, the coarse level c is initialized to contain all the unknowns:

$$c = \{1, 2, 3, \ldots, N\}.$$

2. Then, for $i = 1, 2, 3, \ldots, N$ do the following:
   - If $i \in c$, then, for every $1 \le j \le N$ for which

$$j \ne i \quad \text{and} \quad |a_{i,j}| \ge \alpha \cdot \min(|a_{i,i}|, |a_{j,j}|),$$

   drop j from c; that is, update c by

$$c \leftarrow c \setminus \{j\}. \tag{15.3}$$

At this point, c is a maximal set of unknowns that are at most weakly coupled in A. We now add some more unknowns to c as follows. Let f_a be the set of indices $1 \le i \le N$ for which $i \notin c$ and

$$\left| \sum_{j \in c} \tilde{A}^{(\alpha)}_{i,j} \right| \ge \tau |a_{i,i}|. \tag{15.4}$$

That is, f_a is the set of unknowns that are strongly coupled to c through elements in A with real-part sign that is different from that of the corresponding main-diagonal element. (Note that when A is an L-matrix with positive main-diagonal elements, f_a contains all the unknowns that are strongly coupled to c.) We are now ready to define f_b, the set of indices of unknowns that are only weakly coupled to c (through elements of A with the appropriate sign) but strongly coupled to $c \cup f_a$ (through elements of A with the appropriate sign).

1. Initialize f_b to be the empty set: $f_b = \emptyset$.
2. For $i = 1, 2, 3, \ldots, N$, do the following.
   - If $i \notin c \cup f_a$, then do one of the following.

a) If i is strongly coupled to $c \cup f_a$ in the sense that

$$\left| \sum_{j \in c \cup f_a} \tilde{A}_{i,j}^{(\alpha)} \right| \geq \delta |a_{i,i}|, \tag{15.5}$$

then add i to f_b by

$$f_b \leftarrow f_b \cup \{i\},$$

b) And if, on the other hand, i is only weakly coupled to $c \cup f_a$ in the sense that

$$\left| \sum_{j \in c \cup f_a} \tilde{A}_{i,j}^{(\alpha)} \right| < \delta |a_{i,i}|, \tag{15.6}$$

then add i to c by

$$c \leftarrow c \cup \{i\}.$$

Thus, f_b is the set of unknowns that are weakly coupled to c in the above sense, but strongly coupled to $c \cup f_a$. Finally, define $f = f_a \cup f_b$. This completes the coarsening procedure.

Most often, (15.6) holds only for a few unknowns, so the number of unknowns in c is much smaller than the original number of unknowns in $f \cup c$. Indeed, in the numerical experiments in [100], (15.6) never holds in the first three levels, and holds only for six to eight unknowns in the fourth level. In the numerical experiments in the next chapter, the number of unknowns in a coarse level is at most half the number of unknowns in the previous finer level.

In its final form, c is a maximal set of unknowns that are only weakly coupled to each other through elements in A with real parts of the appropriate sign. As we will see below, this property guarantees that the prolongation operator P is well bounded, which is most important in bounding the condition number of the $V(0,0)$-cycle.

When parallel implementation is considered, one can modify the above coarsening procedure into a parallelizable procedure. This can be done by using a domain decomposition and applying the above coarsening procedure in each subdomain separately. The coarse level is then defined as the union of the local coarse levels in the individual subdomains. This approach may lead to a slightly larger coarse levels than in the main version introduced above.

15.4 The Transfer and Coarse-Level Matrices

The matrix $\tilde{A}^{(\gamma)}$ is now used to define the prolongation and restriction operators P and R. The above definition of $\tilde{A}^{(\gamma)}$ guarantees that the main-diagonal elements in P^{-1} and R^{-1} are not too small in magnitude, which guarantees that P and R are well bounded (see Section 12.5).

For simplicity, we omit the superscript and use hereafter the notation $\tilde{A} \equiv \tilde{A}^{(\gamma)}$.

We start by defining the diagonal matrices that will eventually be used to form the main blocks in P^{-1} and R^{-1}. The diagonal matrix G_a is defined as follows.

1. Initialize G_a by

$$G_a = rs(\tilde{A}_{f_a f_a}) + rs(\tilde{A}_{f_a f_b}). \tag{15.7}$$

2. If, for some $i \in f_a$,

$$\Re((G_a)_{i,i})\Re(a_{i,i}) \leq 0 \quad \text{or} \quad |(G_a)_{i,i}| < \zeta|a_{i,i}|,$$

then update $(G_a)_{i,i}$ by

$$(G_a)_{i,i} \leftarrow -rs(\tilde{A}_{f_a c})_{i,i}.$$

From (15.4), the diagonal elements in G_a are bounded away from zero. The diagonal matrix G_b is defined as follows.

1. Initialize G_b by

$$G_b = rs(\tilde{A}_{f_b f_b}). \tag{15.8}$$

2. If, for some $i \in f_b$,

$$\Re((G_b)_{i,i})\Re(a_{i,i}) \leq 0 \quad \text{or} \quad |(G_b)_{i,i}| < \zeta|a_{i,i}|,$$

then update $(G_b)_{i,i}$ by

$$(G_b)_{i,i} \leftarrow -rs(\tilde{A}_{f_b f_a})_{i,i} - rs(\tilde{A}_{f_b c})_{i,i}.$$

From (15.5) and (15.6), the diagonal elements in G_b are bounded away from zero. The matrices R and P are now defined in the spirit of Section 12.2:

$$P = \begin{pmatrix} (G_b)^{-1} & -(G_b)^{-1}\tilde{A}_{f_b f_a} & -(G_b)^{-1}\tilde{A}_{f_b c} \\ 0 & I & 0 \\ 0 & 0 & I \end{pmatrix}$$
$$\cdot \begin{pmatrix} I & 0 & 0 \\ 0 & (G_a)^{-1} & -(G_a)^{-1}\tilde{A}_{f_a c} \\ 0 & 0 & I \end{pmatrix}, \tag{15.9}$$

$$R = \begin{pmatrix} I & 0 & 0 \\ 0 & (G_a)^{-1} & 0 \\ 0 & -\tilde{A}_{c f_a}(G_a)^{-1} & I \end{pmatrix} \begin{pmatrix} (G_b)^{-1} & 0 & 0 \\ -\tilde{A}_{f_a f_b}(G_b)^{-1} & I & 0 \\ -\tilde{A}_{c f_b}(G_b)^{-1} & 0 & I \end{pmatrix}, \tag{15.10}$$

These matrices will be used in the analysis below. In practice, we often use the rectangular prolongation and restriction matrices $\tilde{P}$ and $\tilde{R}$, defined by

$$\tilde{P} \equiv PJ_c^t = \begin{pmatrix} -(G_b)^{-1}\tilde{A}_{f_b f_a} & -(G_b)^{-1}\tilde{A}_{f_b c} \\ I & 0 \\ 0 & I \end{pmatrix} \begin{pmatrix} -(G_a)^{-1}\tilde{A}_{f_a c} \\ I \end{pmatrix}, \tag{15.11}$$

and

$$\tilde{R} \equiv J_c R = \begin{pmatrix} -\tilde{A}_{c f_a}(G_a)^{-1} & I \end{pmatrix} \begin{pmatrix} -\tilde{A}_{f_a f_b}(G_b)^{-1} & I & 0 \\ -\tilde{A}_{c f_b}(G_b)^{-1} & 0 & I \end{pmatrix}. \tag{15.12}$$

In fact, $\tilde{P}$ in (15.11) is closely related to the prolongation matrix defined in Algorithm 6.4. Finally, the coarse-level matrix $\tilde{Q}$ is defined as follows.

1. First, $\tilde{Q}$ is initialized by

$$\tilde{Q} = \tilde{R}A\tilde{P} = B. \tag{15.13}$$

2. Then, elements in $\tilde{Q}$ that are small in magnitude in comparison with the corresponding main-diagonal element are "thrown" onto the main diagonal as follows.
   a) Define the matrix $\tilde{Q}^{(\epsilon)}$ (which has the same order as $\tilde{Q}$) by

$$\tilde{Q}_{i,j}^{(\epsilon)} = \begin{cases} \tilde{Q}_{i,j} & \text{if } i \neq j \text{ and } |\tilde{Q}_{i,j}| < \epsilon \cdot \min(|\tilde{Q}_{i,i}|, |\tilde{Q}_{j,j}|) \\ 0 & \text{otherwise.} \end{cases} \tag{15.14}$$

   b) Update $\tilde{Q}$ by

$$\tilde{Q} \leftarrow \tilde{Q} - \tilde{Q}^{(\epsilon)} + rs(\tilde{Q}^{(\epsilon)}). \tag{15.15}$$

This modification of $\tilde{Q}$ guarantees that it is sufficiently sparse, so the coarse-level system is not too hard to solve.

The coarse-level matrix $\tilde{Q}$ is often used in practice. In the analysis, though, we use the coarse-level matrix Q, which is of the same order as the original matrix A:

$$Q = \begin{pmatrix} W & 0 \\ 0 & B \end{pmatrix}, \tag{15.16}$$

where

$$B = \tilde{Q} \tag{15.17}$$

and

$$W = R_{ff}\,diag(A_{ff})P_{ff}. \tag{15.18}$$

15.5 The Relaxation Method

In order to complete the definition of the V-cycle, one must also specify the relaxation method. On serial computers, usually the (symmetric) point Gauss-Seidel relaxation method or ILU (with no fill-in) is used for this purpose. On parallel computers, on the other hand, it is advisable to turn to more parallelizable relaxation methods such as colored relaxation (Section 5.8), parallelizable ILU (Section 5.14), two-step Jacobi relaxation (Section 16.2), block-GS relaxation [70], or the approximate-inverse iteration [48].

As a matter of fact, the coarsening procedure in Section 15.3 can be also used to define a colored relaxation as follows. Let c serve as the first color. Then, apply the coarsening procedure once again to f and let the resulting coarse level serve as the second color. By repeating this process, one obtains an algebraic coloring of the original set of unknowns. Although the unknowns in each color may be weakly coupled to each other, they may still be relaxed simultaneously in a Jacobi sweep on this particular color. This is the two-step Jacobi relaxation, defined in more detail in Section 16.2 below.

15.6 Properties of the Two-Level Method

In this section, the coarse-level matrix $\tilde{Q}$ defined above is represented in a way that will be useful in the analysis below. In the sequel, we assume $\tau = \infty$, so $f_a = \emptyset$, $f_b = f$, and G_a disappears. This choice simplifies both the analysis and implementation. Indeed, the square matrices R and P take the form

$$R = \begin{pmatrix} G_b^{-1} & 0 \\ -\tilde{A}_{cf} G_b^{-1} & I \end{pmatrix} \qquad (15.19)$$

and

$$P = \begin{pmatrix} G_b^{-1} & -G_b^{-1} \tilde{A}_{fc} \\ 0 & I \end{pmatrix}. \qquad (15.20)$$

Furthermore, the rectangular matrices $\tilde{R}$ and $\tilde{P}$ take the form

$$\tilde{R} = J_c R = \begin{pmatrix} -\tilde{A}_{cf} G_b^{-1} & I \end{pmatrix} \qquad (15.21)$$

and

$$\tilde{P} = P J_c^t = \begin{pmatrix} -G_b^{-1} \tilde{A}_{fc} \\ I \end{pmatrix}. \qquad (15.22)$$

Denote the Schur complement of A (with respect to the partitioning $\{1, 2, \ldots, N\} = f \cup c$) by

$$S(A; c) = A_{cc} - A_{cf}(A_{ff})^{-1} A_{fc}, \qquad (15.23)$$

and similarly for other matrices of the same order.
Define also the matrix

$$X(A) = A_{cf} \left(G_b^{-1} - (A_{ff})^{-1} \right) A_{ff} \left(G_b^{-1} - (A_{ff})^{-1} \right) A_{fc}. \qquad (15.24)$$

Lemma 15.1 *Assume that $\tau = \infty$ and $\epsilon = \gamma = 0$ are used in the algebraic multilevel method. Assume also that A is an L-matrix with positive main-diagonal elements. Then*

$$\tilde{Q} = S(A; c) + X(A).$$

Proof. From the assumptions in the lemma, it follows that $\tilde{R}$ is as in (15.21), $\tilde{P}$ is as in (15.22), $\tilde{Q}$ is as in (15.13), $\tilde{A}_{cf} = A_{cf}$, and $\tilde{A}_{fc} = A_{fc}$. Thus, we have

$$\begin{aligned}
\tilde{Q} &= A_{cc} - 2 A_{cf} G_b^{-1} A_{fc} + A_{cf} G_b^{-1} A_{ff} G_b^{-1} A_{fc} \\
&= S(A; c) - (S(A; c) - S(\tilde{A}; c)) - A_{cf} G_b^{-1} A_{fc} + A_{cf} G_b^{-1} A_{ff} G_b^{-1} A_{fc} \\
&= S(A; c) - A_{cf} \left(G_b^{-1} - (A_{ff})^{-1} \right) A_{ff} (A_{ff})^{-1} A_{fc} \\
&\quad - A_{cf}(A_{ff})^{-1} A_{ff} G_b^{-1} A_{fc} + A_{cf} G_b^{-1} A_{ff} G_b^{-1} A_{fc} \\
&= S(A; c) + X(A).
\end{aligned}$$

This completes the proof of the lemma.

15.7 Properties of the Multilevel Method

The multilevel method is implemented in the same spirit as in Section 12.7 above, except that here the operators are defined as in Section 15.4. We therefore use the same notation as in Section 12.7, except that here f_0 contains the unknowns that are excluded from the coarse level c, f_1 contains the unknowns in c that are excluded from the next coarser level, and so on, until f_L, which contains the unknowns in the coarsest level. The $f_{L,i}$s in Section 12.7 are also interpreted here according to these definitions.

The following lemma is helpful in estimating the scalars $C_{k,0}$ used in the upper bound for the condition number of the V(0,0)-cycle. For $0 \leq i < L$, define the Schur complement of G_i (with respect to the partitioning $f_L^i = f_L^{i+1} \cup f_i$) by

$$S_i = J^t_{f_L^{i+1}} S\left(B_i; f_L^{i+1}\right) J_{f_L^{i+1}},$$

For $0 \leq i < L$, also define the matrices

$$X_i = J^t_{f_i} \left(J_{f_i} P_{i+1} J^t_{f_i} - \left(J_{f_i} G_i J^t_{f_i}\right)^{-1}\right) J_{f_i} G_i J^t_{f_L^{i+1}} J_{f_L^{i+1}}.$$

Lemma 15.2 *Assume that $\tau = \infty$ and $\epsilon = \gamma = 0$ are used in the algebraic multilevel method. Let D be a diagonal SPD matrix of the same order as A. Assume that A is symmetric with respect to $(\cdot, \cdot)_D$ and positive definite. Let $0 \leq i < k \leq L$ be fixed. Assume that the matrices B_j ($i \leq j < k$) are L-matrices. Let $x \in l_2(f_L^0)$. Then,*

$$(x, G_k x)_D \leq (x, G_i x)_D + \sum_{j=i}^{k-1} \|X_j x\|^2_{DA_j}. \tag{15.25}$$

Proof. From Lemma 12.6, it follows that the matrices DA_j ($j \geq 0$) are SPD, so the right-hand side in (15.25) is well defined. By repeated application of Lemma 15.1, we have

$$(x, G_k x)_D = (x, S_{k-1} x)_D + \|X_{k-1} x\|^2_{DA_{k-1}}$$
$$\leq (x, G_{k-1} x)_D + \|X_{k-1} x\|^2_{DA_{k-1}}$$
$$= (x, S_{k-2} x)_D + \|X_{k-2} x\|^2_{DA_{k-2}} + \|X_{k-1} x\|^2_{DA_{k-1}}$$
$$\leq (x, G_{k-2} x)_D + \|X_{k-2} x\|^2_{DA_{k-2}} + \|X_{k-1} x\|^2_{DA_{k-1}}$$
$$\leq \cdots \leq (x, G_i x)_D + \sum_{j=i}^{k-1} \|X_j x\|^2_{DA_j}.$$

This completes the proof of the lemma.

15.8 Upper-Bound for the Condition Number

Here we derive an (a posteriori) upper-bound for the condition number of the V(0,0)-cycle in much the same way as in Theorem 12.1 above. Then, we use Lemma 15.2 to interpret this upper-bound in 2-D diffusion problems. We keep using the notation in Section 12.7 (in its present algebraic interpretation) and the notation in Section 15.7 above.

Theorem 15.1 *Assume that the algebraic multilevel method is implemented with* $\epsilon = \gamma = 0$. *Let D be a diagonal SPD matrix of the same order as A. Assume that A is a diagonally dominant L-matrix that is symmetric with respect to $(\cdot,\cdot)_D$. Assume also that the matrices B_i $(0 \le i < L)$ are diagonally dominant L-matrices. Then*

$$\kappa\left(P_{L,1} A_L^{-1} R_{L,1} A\right) \le \left(C_{L,0} + \left(\sqrt{2}+1\right)^2 \sum_{k=0}^{L-1} w_k g_k C_{k,0}\right) 2aL. \qquad (15.26)$$

Proof. The proof is the same as the proof of Theorem 12.1.

The scalars w_k, g_k, and a can be estimated as in Section 12.7. In fact, (15.5) and (15.6) imply that the p_ks are bounded, which, using (12.24), implies that the w_ks are bounded as well. For discretizations of diffusion problems like (3.2), by assuming that the V$(0,0)$-cycle is applied to the left-scaled system (12.19) rather than the original system (3.5), one obtains a not only mesh-independent but also jump-independent upper-bound in (15.26). An induction on L in Lemma 12.3 shows that this assumption should be made only in the theory; the actual implementation can use the original system (3.5), since the preconditioned matrix is the same in both cases.

Using Lemma 15.2, the scalars $C_{k,0}$ can also be estimated for diffusion problems as in (3.2) as follows. The asymptotic estimate is derived under the assumptions that the meshsize $h \to 0$ and $L \to \infty$. Note that $C_{k,0}$ is obtained at the solution $x \in l_2(f_L^0)$ of the problem

$$\text{minimize} \quad (x, Ax)_D \quad \text{subject to} \quad (x, G_k x)_D = 1.$$

From the Lagrange theory, this minimum is obtained at a vector $x \in l_2(f_L^0)$ for which $2DAx$ (the gradient of $(x, Ax)_D$) is a scalar times $2DG_k x$ (the gradient of $(x, G_k x)_D$). This implies that $J_{f_{k-1}^0} DAx = 0$, and $J_{f_L^k} x$ is the solution of the generalized eigenproblem

$$S\left(DA; f_L^k\right) v = \lambda D_{f_L^k f_L^k} B_k v$$

[where $S()$ is defined in (15.23)] with minimal generalized eigenvalue λ. Since D is diagonal, this problem is actually equivalent to the eigenproblem

$$S\left(A; f_L^k\right) v = \lambda B_k v$$

with minimal generalized eigenvalue λ. To estimate $C_{k,0}$, we just need to estimate how small λ can be. Since both $S\left(DA; f_L^k\right)$ and $D_{f_L^k f_L^k} B_k$ represent variational discretizations of (3.2), small λ implies small values of the components of DAx also at the points in f_L^k.

Furthermore, since x has sharp variation at points in f_L^k, the components of DAx are at least of order h_k^2 at points in f_L^k (where h_i is the typical meshsize in f_L^i). Since the number of points in f_L^k is at least of order h_k^{-2}, we have that $(x, Ax)_D$ is at least of order 1. On the other hand, since the eigenfunctions of (3.2) are continuous, the factor $\left(G_b^{-1} - (A_{ff})^{-1}\right) A_{fc} x$ in (15.24) is small. More precisely, for $0 \le i < k$, since the $D_{f_L^i f_L^i} B_i$s represent variational discretizations of (3.2) in f_L^i, the components of $X_i x$ are of $O(h_k)$ at the $O(h_k^{-2})$ points in f_L^i that lie near points from f_L^k and $O(h_i)$

at the rest of the $O(h_i^{-2})$ points in f_L^i. Assuming that $\|G_i\|_D = O(1)$, we have that

$$\|X_i x\|_{DA_i} = O((x, Ax)_D) \quad (0 \le i < k).$$

Using the above arguments and Lemma 15.2, we have that

$$C_{k,0} = O(k).$$

The bound in (15.26) is thus of order L^3 as $L \to \infty$. This moderate growth indicates that the nearly singular eigenvectors of A are approximated well on the coarse levels. Hence, it is expected that the V(1,1)-cycle that also uses relaxation to take care of the high-frequency error modes should converge rapidly. Indeed, the numerical examples in [100] show that the algebraic multilevel method achieves the Poisson convergence rate not only for the Poisson equation but also for diffusion problems with discontinuous coefficients, regardless of whether the discontinuity lines align with the coarse grids. It is also verified numerically in [100] that the matrices B_i $(0 \le i < L)$ are diagonally dominant L-matrices and that the g_is, p_is, and w_is $(0 \le i < L)$ are bounded by moderate bounds.

15.9 The Approximate Schur Complement Method

Instead of the Galerkin approach that initializes $\tilde{Q}$ as in (15.13), one could use a Schur-complement approach as follows:

$$\tilde{Q} = A_{cc} - R_{cf} \begin{pmatrix} G_b & 0 \\ 0 & G_a \end{pmatrix} P_{fc}. \tag{15.27}$$

[Compare this definition with (11.13).]

With this approach, $\tilde{Q}$ may be sparser than with the main version in (15.13). However, it is found in [100] that the coarse-level matrix in (15.27) does not scale right (at least for diffusion equations), and hence is not as appropriate and stable as the original Galerkin matrix in (15.13). Thus, the Schur-complement approach converges only when supplemented with outer acceleration. Therefore, in the present applications, we stick to the main version in (15.13).

15.10 Exercises

1. Show that the scalars $C_{k,0}$ can be interpreted as the squared DA-induced norm (the so-called energy norm) of the operator that is composed of two steps: first, injecting from the finest level onto the kth level; then, prolonging from the kth level back to the finest level.
2. Use the discussion at the end of Section 15.8 to indicate that these scalars are indeed moderate, at least for 2-D diffusion problems. Use Theorem 15.1 to indicate that the V(1,1)-cycle of the algebraic multilevel method should converge rapidly.
3. Use the first exercise above to show that the boundedness of the scalars $C_{k,0}$ means that the prolongation operators preserve the nearly singular eigenfunctions of the original differential operator, as is indeed required in Section 6.2.

4. Show that the boundedness of the scalars $C_{k,0}$ also means that the prolongation operators preserve the continuity of the flux vector

$$(D_1 u_x, D_2 u_y)$$

of the original differential operator.

5. Write the computer code that implements the algebraic multilevel method. The solution can be found in Section 17.10 in [103].

6. Apply your code to the Poisson equation with Dirichlet boundary conditions on the uniform mesh in Figure 4.13. Do you obtain the Poisson convergence factor of 0.1?

7. Is $\|P_{ff}\|_{D_{ff}}$ reasonably bounded?

8. Use an IMSL routine to solve the generalized eigenproblem

$$S(A; c)v = \lambda Q_{cc} v$$

with $v \in l_2(c)$ and minimal λ.

9. Repeat the previous two exercises for the next (coarser) level as well.

10. Is the coarse-level matrix diagonally dominant as well?

11. Modify your code for problems with various kinds of boundary conditions. Is the convergence rate still good?

12. Apply your code to isotropic diffusion problems with discontinuous coefficients. Are the convergence rates still good?

13. Is $\|P_{ff}\|_{D_{ff}}$ reasonably bounded?

14. Use an IMSL routine to solve the generalized eigenproblem

$$S(A; c)v = \lambda Q_{cc} v$$

with $v \in l_2(c)$ and minimal λ.

15. Repeat the previous two exercises for the next (coarser) level as well.

16. Is the coarse-level matrix diagonally dominant as well?

17. Apply your code to highly anisotropic equations as in Section 3.9. Is the convergence rate still good? If not, then also use a PCG outer acceleration to accelerate the basic multilevel iteration. (Use symmetric GS relaxation in the V-cycle to guarantee that the multilevel preconditioner is indeed SPD.)

18. Is $\|P_{ff}\|_{D_{ff}}$ reasonably bounded?

19. Use an IMSL routine to solve the generalized eigenproblem

$$S(A; c)v = \lambda Q_{cc} v$$

with $v \in l_2(c)$ and minimal λ.

20. Repeat the previous two exercises for the next (coarser) level as well.

21. Is the coarse-level matrix diagonally dominant as well?

Applications

Although the above analysis of the algebraic multilevel method is limited to the diagonally dominant SPD case, the method is actually applicable to a much wider class of problems. Indeed, it is applied below to off-diagonally dominant problems such as diffusion problems with oblique anisotropy, the Maxwell equations, and the highly nonsymmetric convection diffusion equation.

16.1 Highly Anisotropic Equations

In this section, we apply the algebraic multilevel method to highly anisotropic equations with oblique anisotropy, discretized by finite differences as in Section 3.9. Although the grid is uniform, so standard multigrid methods (with semicoarsening or line relaxation) could also be used, these methods cannot be extended to problems with discontinuous coefficients, complicated domains, or unstructured grids. Thus, it is particularly interesting to test the algebraic multilevel method for this kind of problem.

As argued above, the multigrid approach should work well for well-posed elliptic PDEs discretized by adequate discretization methods. As we will see below, the present algebraic multilevel method indeed converges rapidly (with the Poisson convergence rate) for highly anisotropic problems of the form in (3.28). These good results, however, are limited to the case in which the grid aligns with the diffusion directions. When it does not, the discretization may become inadequate, and the performance of the algebraic multilevel method may indeed deteriorate.

Consider the highly anisotropic diffusion equation (3.33) in the unit square $0 < x, y < 1$ with Dirichlet boundary conditions. In this equation, the diffusion directions are oblique: $\xi = 2^{-1/2}(x - y)$ is the weak-diffusion direction, and $\eta = 2^{-1/2}(x + y)$ is the strong-diffusion direction. Unfortunately, these directions do not align with the grid in (3.6), which is used in the discretization as in Section 3.10. This makes the problem far more difficult; indeed, the resulting coefficient matrix is no longer diagonally dominant, which poses a particularly interesting challenge for the algebraic multilevel solver.

The adequacy condition in (3.35) implies that ε should be much larger than the meshsize h_x and h_y. This is also why the performance of the algebraic multilevel linear system solver deteriorates as ε decreases.

Note that the coefficient matrix for the finite-difference discretization method in Section 3.10, although SPD, is neither diagonally dominant nor an L-matrix. This is in agreement with our aim here: to test the algebraic multilevel method for difficult examples that do not have these good properties.

16.2 Two-Step Jacobi Relaxation

For relaxation, we use the two-step Jacobi method indicated in Section 15.5. This relaxation method consists of two stages. First, the unknowns in c are relaxed simultaneously in a "half Jacobi relaxation." The inverse of the preconditioner for this step is

$$\mathcal{P}^{-1} = \begin{pmatrix} 0 & 0 \\ 0 & diag(A_{cc})^{-1} \end{pmatrix}.$$

In the second step, the unknowns in f are relaxed simultaneously in another "half" Jacobi relaxation. The inverse of the preconditioner for this step is

$$\mathcal{P}^{-1} = \begin{pmatrix} diag(A_{ff})^{-1} & 0 \\ 0 & 0 \end{pmatrix}.$$

(See also the end of Section 6.8, where such a step is used to improve the prolongation matrix P.) The entire relaxation sweep consists of the first step followed by the second step. This relaxation sweep is referred to as the two-step Jacobi relaxation.

Actually, we use here a symmetric version of the two-step Jacobi relaxation, which may be considered as a three-step Jacobi relaxation. In this version, the unknowns in c are relaxed simultaneously once again in a third "half" relaxation. In other words, the third step is the same as the first step. With this version, the multilevel preconditioner is SPD whenever A is, which allows the use of the Preconditioned Conjugate Gradient (PCG) method to accelerate the convergence of the basic multilevel iteration. (It is well known that PCG is about twice as efficient as CGS for SPD systems with an SPD preconditioner, as is indeed verified numerically here.)

The algebraic multilevel iteration is implemented as follows. Eight levels are used, from 512^2 unknowns in the finest level to about 200 unknowns in the coarsest level. The coarse-level matrices B_i are almost as sparse as A; in fact, they contain at most nine nonzero elements per row. The V(1,1)-cycle is used ($\nu_1 = \nu_2 = \nu_c = 1$ in the ML algorithm in Section 6.3). The matrices R, P, and Q in the ML algorithm are replaced by the present matrices $\tilde{R}$, $\tilde{P}$, and $\tilde{Q}$, respectively.

The parameters used in the algebraic multilevel method are $\alpha = \gamma = \epsilon = 0.02$, $\delta = \zeta = 0.1$, and $\tau = \infty$. The initial guess is random, and its preconditioned residual serves as the first direction vector in PCG.

The preconditioned convergence factor [see (10.13)] for the above iteration is reported in Table 16.1. It can be seen that, especially for very small ε, the performance of the algebraic multilevel method is worse than in the problem in (3.28), where the Poisson convergence rate is achieved even with no acceleration. The reason for this lies probably in the inadequacy of the discretization method, which produces large discretization errors.

Table 16.1. Preconditioned convergence factors (pcf) for the algebraic multilevel method applied to the highly anisotropic equation with oblique diffusion directions. The meshsize is $1/512$ in both the x and y spatial directions. Outer PCG acceleration is also used.

ε	pcf
10^{-1}	0.3
10^{-2}	0.4
10^{-3}	0.5
10^{-4}	0.53
10^{-5}	0.55

16.3 The Maxwell Equations

Although the algebraic multilevel method can be analyzed in the SPD diagonally dominant case only, it can still be applied to much more general cases. Here we show its power for a most difficult system of PDEs: the Maxwell equations.

We consider the Maxwell equations discretized as in [3] on a three-dimensional $30 \times 30 \times 30$ staggered grid. The parameters of the equations used in [3] are $\omega = 1000$ Hz and $\sigma_c = 100$ S/m.

Again, we find it sufficient to use $\tilde{R}$, $\tilde{P}$, and $\tilde{Q}$ rather than R, P, and Q (respectively) in the ML algorithm in Section 6.3. Only three levels are used, with $103,500$ unknowns in the first level, $52,640$ unknowns in the second level, and $7,379$ unknowns in the third level. We use the parameter $\epsilon = 0.02$ in (15.14) to guarantee that $\tilde{Q}$ is sparse. Indeed, the maximal number of nonzero elements per row in the coarse-level matrices is only eighteen (compared to nine in A). The rest of the parameters used in the method are $\gamma = \alpha = \zeta = \delta = .02$ and $\tau = \infty$ [which means that (15.21) and (15.22) are used]. We use the V(1,5)-cycle with point Gauss–Seidel relaxation; that is, $\nu_1 = 1$, $\nu_2 = 5$, and $\nu_c = 20$ in the ML algorithm in Section 6.3. Because of the near singularity of A, five postrelaxations are needed to annihilate the unstable high-energy error modes resulting from the nearly-singular coarse-level problems.

A zero initial guess is used, and its preconditioned residual serves as the initial direction vector in CGS. Twenty-three algebraic multilevel V-cycles are required within the outer CGS iteration to reduce both the l_2-norm of the residual and l_2-norm of the preconditioned residual by ten orders of magnitude. The preconditioned convergence factor is thus pcf $= 0.4$.

This example is not highly indefinite: there are about thirty gridpoints per wave length (see Section 10.6). Thus, an approximate Poisson solver may also serve as a good preconditioner. Indeed, a single BBMG V-cycle with alternating plane relaxation is used in [3] as an approximate Poisson solving preconditioner for the present Maxwell equations, and yields a convergence factor of about 0.3. However, this nonalgebraic preconditioner is limited to structured grids only. Furthermore, when problems with highly oscillating solutions are considered, a Poisson solver can no longer capture the singularity in the original system, and is thus unlikely to serve as a good preconditioner.

16.4 The Convection-Diffusion Equation

In this section, we apply the algebraic multilevel method to the convection-diffusion equation, discretized by the upwind scheme as in Section 3.11. Although the grid is uniform, it is still interesting to test the algebraic multilevel method for these highly nonsymmetric problems.

the analysis in Section 15.8 above doesn't apply to the present nonsymmetric problem. Therefore, it makes sense to modify the algebraic multilevel method in the spirit of the method in [40]. (This approach is also tested in Section 7.3 in [97] and yields acceptable convergence rates for a most difficult convection-diffusion example with closed characteristics.)

The modified algorithm is as follows. First, R and $\tilde{R}$ are defined as in Section 15.4. Then, the $\tilde{A}$ defined in (15.2) is replaced by its symmetric part:

$$\tilde{A} \leftarrow (\tilde{A} + \tilde{A}^t)/2.$$

Then, G_a, G_b, P, and $\tilde{P}$ are redefined using this new $\tilde{A}$. The coarse-level matrix is then defined as in Section 15.4.

In the beginning, we had considered using a convection-diffusion example with closed characteristics. This kind of problem is particularly challenging, because no downstream relaxation is available to eliminate convection errors. However, it turns out [27] [102] that first-order discretization methods such as the upwind scheme in Section 3.11 are inadequate in the sense that, when both the meshsize h and the diffusion coefficient ε approach zero at the same time, the numerical solution may exhibit $O(1)$ errors that are bounded away from zero. In [29], the first-order upwind scheme is modified by introducing an isotropic artificial diffusion term; however, it is shown in [102] that this produces an adequate scheme only for the particular example studied in [27], but not in general.

Inadequacy is also the reason why the present algebraic multilevel method, as well as other multigrid methods, stagnates for this example. Indeed, since the coefficient matrix A is no longer related to an underlying PDE, its nearly singular eigenvectors may have large variation that cannot be captured on the coarse levels.

Although it is reported in [124] that BBMG works well for convection-diffusion equations (even with closed characteristics) discretized by the upwind scheme, this convergence is limited to special cases. Indeed, we have applied BBMG to the model convection equation

$$2u_x + u_y = \mathcal{F}$$

in the unit square with periodic boundary conditions (or in a torus), where $\mathcal{F}$ has zero integral over each individual closed characteristic that spirals around the torus. As shown in [102], first-order schemes are accurate for this particular example (and adequate for its singular perturbation) only when $\mathcal{F} \equiv 0$, but not in general. The discretization uses the first-order upwind scheme (Section 3.11) on the uniform $n \times n$ grid in (3.6). We chose n to be a power of 2, so that periodic boundary conditions can also be used on the coarse grids. We have used BBMG with point-GS relaxation in the downstream direction, which is considered the strongest possible relaxation method for this kind of problem. Indeed, after one sweep of this relaxation, the residual vanishes throughout the grid, except at the left and bottom gridlines. The same is true also in the coarse grids.

As in [124], we have observed that the BBMG V(1,1)-cycle converges rapidly. However, this is limited to an implementation with coarse grids that consist of the even-numbered points of the previous finer grid, as in (6.1). When the coarse grid consists of the odd-numbered points, that is,

$$c = \{1, 3, 5, \ldots, n-1\} \times \{1, 3, 5, \ldots, n-1\},$$

the BBMG method actually stagnates. The reason for this may lie in the ill-posedness of the original problem: it should have been posed as an initial boundary-value problem rather than a boundary-value problem.

In view of the above discussion, we chose to test the algebraic multilevel method for a more conventional problem with no closed characteristics (as in [128]). We use the first-order upwind scheme to discretize the following convection-diffusion equation,

$$-\varepsilon(u_{xx} + u_{yy}) + \sin(\pi(y - 0.5))\cos(\pi x/2)u_x - \cos(\pi(y - 0.5))\sin(\pi x/2)u_y = \mathcal{F}$$
(16.1)

in the unit square, with Dirichlet boundary conditions and diffusion parameter $\varepsilon = 10^{-3}$. In this example, the characteristics are no longer closed; in fact, the convection enters the square through the upper part of its left edge, reverses direction as in a horseshoe, and leaves the square through the lower part of its left edge. (The field of characteristic directions is displayed in Figure 16.1.)

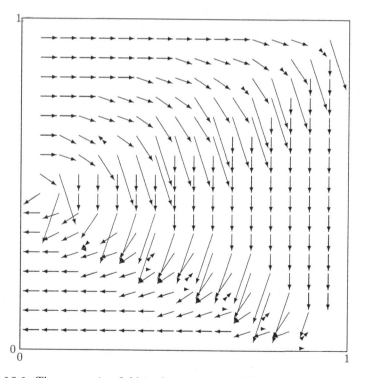

Fig. 16.1. The convection field in the convection-diffusion example (the field of characteristic directions).

The first-order upwind scheme is used on a 512×512 (and also 1000×1000) uniform grid. The resulting linear system is solved by the present algebraic multilevel method with the parameters $\tau = \infty$, $\epsilon = \gamma = \alpha = 0.02$, and $\delta = \zeta = 0.1$. The V(1,1)-cycle is implemented as in Section 6.3, but with R, P, and Q replaced by $\tilde{R}$, $\tilde{P}$, and $\tilde{Q}$, respectively.

The V(1,1)-cycle is used ($\nu_1 = \nu_2 = \nu_c = 1$). No acceleration is used in this section. In order to show the power of the coarse levels, we choose a nonoptimal GS relaxation, in which the gridpoints are ordered in the upstream order in most of the domain.

It turns out that the coarse-level matrices are almost as sparse as A. Unfortunately, it also turns out that, in order to have good convergence for large problems as well, the number of levels must be limited. In particular, when the original grid is of size 512×512, at most five levels can be used. On the other hand, when the original grid is of size 1000×1000, up to six levels can be used. In both cases, the number of unknowns in the coarsest level is between 4000 and 6000, and the convergence factor is about 0.3 (with no acceleration).

The limit on the number of levels is in line with the results in [102], where the adequacy of the first-order upwind scheme is shown as long as no closed characteristics are present and $\varepsilon^{-1}h^2 \ll 1$ as both ε and h approach zero at the same time. In geometric multigrid, where the PDE is rediscretized also on the coarse grids, adequacy must be observed in the coarse-grid problems as well, which requires that the coarsest grid is not too coarse. In the algebraic multilevel method, on the other hand, the situation is somewhat better: when the original discretization on the fine grid is adequate, and the transfer operators preserve the nearly singular eigenfunctions of the original differential operator (which are smooth along characteristics), the Galerkin approximation used on the coarse levels should automatically preserve adequacy. Unfortunately, the present implementation that uses $\tau = \infty$ produces rather unstable transfer operators that do not necessarily agree with most of the nearly singular eigenfunctions. This may slow down the convergence of the algebraic multilevel method as the number of levels increases.

A more stable method is obtained when a moderate parameter τ is used, for which both f_a and f_b are nontrivial. With this implementation, the rectangular matrices $\tilde{P}$ and $\tilde{R}$ are constructed from matrix products as in (15.11) and (15.12), respectively. With $\tau = 0.1$, for example, one can use two more levels, at the expense of somewhat worse convergence factors of $0.5 < \mathrm{cf} < 0.6$. Although this implementation is inferior to the previous one that uses $\tau = \infty$, it may be useful in more singular problems with smaller diffusion coefficient ε. Indeed, with the diffusion coefficient set to $\varepsilon = 10^{-4}$, the implementation that uses $\tau = \infty$ diverges, whereas the implementation that uses $\tau = 0.1$ converges with convergence factor $\mathrm{cf} = 0.8$. (five levels are used, with 512^2 unknowns in the finest level and about 8000 unknowns in the coarsest level; the rest of the details of implementation are as before.)

When a yet more singular equation with $\varepsilon = 10^{-5}$ is encountered, one must use a yet more stable implementation with $\zeta = \delta = \tau = 0.25$ to have convergence factor $\mathrm{cf} = 0.92$. (Six levels are used, with 512^2 unknowns in the finest level and about $10{,}000$ unknowns in the coarsest level.) This is a rather slow convergence.

The convergence factor in the last example can be improved by using the two-step Jacobi relaxation method in Section 16.2. With this relaxation method, the convergence factor improves to $\mathrm{cf} = 0.85$ even for a most singular problem with $\varepsilon = 10^{-5}$.

The two-step Jacobi relaxation method has the advantage of being highly parallelizable. Unfortunately, the algebraic multilevel V-cycle that uses it as a relaxation method converges rather slowly. One should thus consider using other relaxation methods within the multilevel algorithm. Possible candidates for this task are the approximate inverse method (used in [48]), the alternating Schwarz iteration, and ILU versions ([111] and Sections 5.13 and 5.14).

16.5 ILU Relaxation

It seems that the difficulty in supplying a good coarse-grid correction to convection-diffusion equations lies in the boundary layers. Indeed, it is explained in [102] that, when both the diffusion coefficient ε and meshsize h approach zero at the same time, the upwind scheme is accurate outside the boundary layers, but not necessarily inside them. As a result, the coarse-grid correction term may be insufficient inside boundary layers. Furthermore, the sharp variation observed in boundary layers can be completely invisible on the coarse grids.

In order to reduce error modes inside boundary layers as well, one should rely on relaxation methods like ILU (and its parallelizable version), which march (at least locally) along gridpoints to reduce errors due to the convection terms. This approach is indeed used next.

When the ILU method of Section 5.13 (with no fill-in) is used as a relaxation method in the V(1,1)-cycle applied to the convection-diffusion equation (16.1) discretized by the upwind scheme on a uniform 512×512 grid, the convergence factor improves to cf ≤ 0.4, independent of the diffusion coefficient $\varepsilon > 0$. (No acceleration is used in this section.)

We have also used the parallelizable ILU version in Section 5.14 (with $p = 100$ and $O = 50$) as a relaxation method in the algebraic multilevel V(1,1)-cycle. The convergence factor improves to cf ≤ 0.5, independent of the diffusion coefficient $\varepsilon > 0$.

For comparison, we have also tested the ILU and parallelizable ILU iterative methods on their own, with no multilevel setting at all. It turns out that, for arbitrarily small $\varepsilon > 0$, the convergence factors are cf $= 0.85$ for ILU and cf $= 0.9$ for the parallelizable ILU, These are acceptable convergence rates for such a difficult example; however, both ILU and the parallelizable ILU (with or without acceleration) deteriorate considerably for $\varepsilon \gg 1$, which is just a slight perturbation of the Poisson equation. Thus, the ILU versions are less robust than the multilevel method that uses them as relaxation.

In Section 5.14, we have also applied ILU and parallelizable ILU to several sparse linear systems from the Harwell–Boeing collection (Table 5.1). It seems, though, that these examples are too small to benefit from the algebraic multilevel method. Indeed, it is only for the relatively large "sherman3" example that the algebraic multilevel method may be of some use. For this example, if the parallelizable ILU method in Table 5.1 is replaced by a V(1,1)-cycle that uses it as a relaxation method, then the number of CGS iterations is reduced by about 70% (with two levels, the order of the coarse-level matrix being about four times as small as the order of the original matrix, and $\nu_1 = \nu_2 = \nu_c = 1$). This reduction in iteration number may be of use on message-passing parallel architectures, where the inner products required in CGS form the bottleneck in the process.

We have also tried to write f as the union of three subsets and decompose $\tilde{R}$ and $\tilde{P}$ as products of three rectangular matrices rather than two as in (15.11) and (15.12). Unfortunately, this change does not improve the convergence for these difficult problems. Perhaps stronger relaxation methods such as the algebraic line relaxation used in [70] could help here. It seems that one must have some information about the nature of the underlying PDE to have a good multigrid algorithm. This thought leads to the semialgebraic approach in Chapter 17 below.

16.6 Towards Algebraic Semicoarsening

The singularly perturbed convection-diffusion equation (3.36) may also benefit from "algebraic" semicoarsening. Since the nearly-singular eigenfunctions of the original differential operator are smooth in the characteristic direction, the semicoarsening could be done in this direction only. In other words, the coarse grid could consist of every other point in the characteristic direction.

Unfortunately, it is not easy to follow the characteristic direction, which does not necessarily align with the grid. (An attempt to do this algebraically, using the anti-symmetric part of A is proposed in [100].)

With the above semicoarsening, the prolongation is done in the characteristic direction only. Therefore, the prolongation is stable and agrees with the nearly singular eigenfunctions of the original differential operator. (Such an approach is proved useful in the context of highly anisotropic equations in [70], where semi-coarsening in the strong-diffusion direction is used.) This approach, however, is not tested here, and left to future research.

16.7 A Diffusion Problem in a Complicated Domain

Here we use the algebraic multilevel method as a preconditioner for the individual linear systems in the adaptive-refinement algorithm in Section 4.10 for a difficult diffusion problem with anisotropic and discontinuous coefficients in a complicated domain.

We consider the diffusion equation (3.2) in the domain in Figure 16.2, with the indicated boundary conditions: Dirichlet conditions at the boundary segment near the asterisk in the figure, homogeneous Neumann conditions on the horizontal boundary segments on the left, and mixed conditions on the circular part of the boundary. The diffusion coefficients are also as indicated in the figure: they are large and anisotropic in the lower-left part of the domain.

The boundary-value problem is discretized by linear finite elements. The initial (coarsest) mesh contains nine triangles and eleven nodes only (Figure 16.3). This mesh is refined further by the adaptive-refinement algorithm in Section 4.10. For example, the mesh resulting from the fifth level of refinement in this algorithm is displayed in Figure 16.4.

In Table 16.2, we report the number of PCG iterations required to solve (to sixth-order accuracy) the individual linear systems in the various levels of refinement in the adaptive-refinement algorithm. These results provide a comparison between different possible preconditioners: symmetric point-GS (denoted by PCG-SGS), ILU

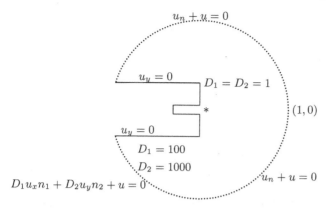

Fig. 16.2. The domain for the diffusion example. The diffusion coefficients D_1 and D_2 are discontinuous and anisotropic. The boundary conditions are of Dirichlet type near the asterisk, homogeneous Neumann type at the horizontal boundary segments, and mixed elsewhere (with $\mathbf{n} = (n_1, n_2)$ being the outer normal vector).

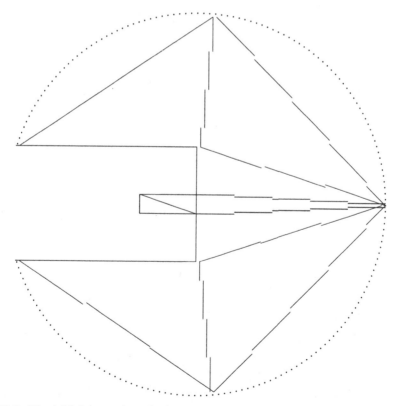

Fig. 16.3. The initial (coarse) mesh that contains eleven nodes and nine triangles only, and is refined further in the adaptive-refinement algorithm both in the interior of the domain and at the circular boundary.

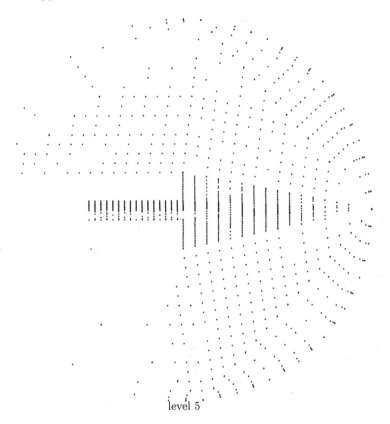

level 5

Fig. 16.4. Distribution of nodes in the fifth level of refinement in the diffusion example. The strong diffusion in the lower-left quarter of the domain prevents large variation, so no extra refinement is needed there.

Table 16.2. Number of PCG iterations used in each refinement level in the adaptive-refinement algorithm applied to the diffusion example. The computation time of a multi-level iteration is like that of three symmetric Gauss–Seidel iterations. (The set-up time is negligible.)

Level	Nodes	PCG-AML	PCG-AMG	PCG-ILU	PCG-SGS
1	11	5	5	3	5
2	34	5	5	6	11
3	106	8	9	9	23
4	340	12	12	17	41
5	1069	18	18	33	71
6	3050	22	22	63	121
7	8242	36	35	121	217
8	18337	54	51	340	382

(denoted by PCG-ILU), the algebraic multilevel method in Section 6.9 (denoted by PCG-AML), and the algebraic multigrid method in Section 6.10 (denoted by PCG-AMG).

The ILU preconditioner uses partial fill-in: only matrix elements that are as large (in magnitude) as 0.05 times the corresponding main-diagonal element are allowed in the incomplete triangular factors L and U, whereas smaller ones are dropped. The algebraic multilevel and algebraic multigrid preconditioners use the symmetric point-GS relaxation within the V(1,1)-cycle (with threshold = 0.01 in Section 6.9).

It is apparent from the table that the multilevel preconditioners are more efficient than the ILU and symmetric point-GS preconditioners. Although a single iteration of the multilevel method costs the same as three symmetric point-GS iterations, this is worthwhile to reduce the overall number of PCG iterations required for convergence. In fact, the advantage of the multilevel methods is even greater when yet larger problems are considered.

16.8 Exercises

1. Use your code from the exercises at the end of Chapter 15 as a preconditioner in the PCG iteration to solve the individual linear systems arising in the adaptive-refinement algorithm in Section 4.10 for the diffusion problem in Section 16.7. The solution can be found in Chapter 19 in [103].

2. Consider the linear system in the fifth level of refinement in the adaptive-refinement algorithm in the previous exercise. In the multilevel preconditioner for this system, is $\|P_{ff}\|_{D_{ff}}$ reasonably bounded?

3. For the linear system considered in the previous exercise, use an IMSL routine to solve the generalized eigenproblem

$$S(A; c)v = \lambda Q_{cc} v$$

with $v \in l_2(c)$ and minimal λ.

4. Repeat the previous two exercises also for the next (coarser) level in the algebraic multilevel method.

Semialgebraic Multilevel Method
for Systems of Partial Differential Equations

In this chapter, we introduce a semialgebraic multilevel method for systems of PDEs. In this approach, the information in the original PDE is used to distinguish between different unknown functions, and transfer them separately to and from the coarse levels. Apart from this modification, the algorithm is as in the algebraic multilevel method. The advantage of the semialgebraic approach is illustrated for the linear elasticity and Stokes systems of PDEs.

17.1 Semialgebraic Multilevel Methods

Algebraic multilevel methods use information exclusively from the coefficient matrix A to form the coarse "grid" c and the transfer operators R and P. In some cases, however, it makes sense to use also information from the original PDE. For example, the original PDE could help to distinguish between different kinds of unknown functions and avoid mixing them with each other when information is transferred from fine to coarse grid and vice versa. Still, the definition of the multilevel algorithm is mostly based on the linear system (3.5) rather than the geometric properties of the underlying grid or mesh. This is why we refer to this family of methods as semialgebraic multilevel methods.

Consider, for example, the case of a vector PDE, or system of PDEs that couple several unknown functions to each other. In this case, it makes sense to let $A^{(\gamma)}$ in (15.1) also contain all the elements in A that couple different unknown functions with each other. This way, $\tilde{A}$ no longer couples different unknown functions with each other. As a result, each unknown function is transferred separately between the coarse and fine grids, independent of the other unknown functions. (This approach is proposed in [42] in the context of uniform grids and black-box multigrid.) With the above approach, the diagonal matrices G_a and G_b are defined by

$$G_a = rs(|\tilde{A}_{f_a c}|) \quad \text{and} \quad G_b = rs(|\tilde{A}_{f_b c}|) + rs(|\tilde{A}_{f_b f_a}|) \qquad (17.1)$$

rather than (15.7) and (15.8). The precise definition is given below.

In the sequel, we show how this approach is actually used in two important systems of PDEs: the linear elasticity equations and the Stokes equations.

17.2 Standard Differential Operators

In order to formulate systems of PDEs, we need some common notation for differential operators that operate upon scalar and (two-dimensional) vector functions. The divergence operator that operates upon vector functions is denoted by

$$\nabla \cdot = (\partial/\partial x, \partial/\partial y) \, .$$

The gradient operator that operates upon scalar functions is denoted by

$$\nabla = \begin{pmatrix} \partial/\partial x \\ \partial/\partial y \end{pmatrix} \, .$$

The Laplacian operator that operates upon scalar functions is denoted by

$$\triangle = \nabla \cdot \nabla .$$

The following differential operators are actually 2×2 matrices of scalar differential operators. The gradient of the divergence is denoted by

$$\nabla \nabla \cdot = \begin{pmatrix} \partial/\partial x \\ \partial/\partial y \end{pmatrix} (\partial/\partial x, \partial/\partial y) = \begin{pmatrix} \dfrac{\partial^2}{\partial x^2} & \dfrac{\partial^2}{\partial x \partial y} \\ \dfrac{\partial^2}{\partial y \partial x} & \dfrac{\partial^2}{\partial y^2} \end{pmatrix} \, .$$

For example, for the vector function $u(x, y) = (u^{(1)}(x, y), u^{(2)}(x, y))$,

$$\nabla \nabla \cdot u = \begin{pmatrix} u^{(1)}_{xx} + u^{(2)}_{yx} \\ u^{(1)}_{xy} + u^{(2)}_{yy} \end{pmatrix} \, .$$

For functions with discontinuous derivatives, this operator is not necessarily the same as its transpose:

$$(\nabla \nabla \cdot)^t u = \begin{pmatrix} u^{(1)}_{xx} + u^{(2)}_{xy} \\ u^{(1)}_{yx} + u^{(2)}_{yy} \end{pmatrix} \, .$$

This can also be written more compactly as

$$(\nabla \nabla \cdot)^t u = \begin{pmatrix} \nabla \cdot u_x \\ \nabla \cdot u_y \end{pmatrix} \, ,$$

where

$$u_x = \begin{pmatrix} u^{(1)}_x \\ u^{(2)}_x \end{pmatrix} \quad \text{and} \quad u_y = \begin{pmatrix} u^{(1)}_y \\ u^{(2)}_y \end{pmatrix} \, .$$

The vector Laplacian operator that operates upon vector functions is denoted by

$$\triangle = \begin{pmatrix} \triangle & \\ & \triangle \end{pmatrix}$$

(where '$\triangle$' on the right-hand side is interpreted as the scalar Laplacian, which operates in this case on the individual components of the vector function). For example,

$$\triangle u = \begin{pmatrix} \triangle u^{(1)} \\ \triangle u^{(2)} \end{pmatrix}.$$

The vector gradient operator that operates upon vector functions is denoted by

$$\nabla = (\nabla \mid \nabla)$$

(where the '∇' in the right-hand side is interpreted as the standard gradient operator, which operates in this case on the individual components of the vector function). For example,

$$\nabla u = \left(\nabla u^{(1)} \mid \nabla u^{(2)} \right).$$

17.3 The Linear Elasticity Equations

We start with the system of PDEs known as the linear elasticity equations. In this system, there are three scalar unknown functions: $p(x, y)$ (the scalar pressure function) and the two-dimensional vector $u(x, y) \equiv (u^{(1)}(x, y), u^{(2)}(x, y))$. These unknown functions are coupled in the following system of PDEs:

$$\begin{pmatrix} -\lambda^{-1} & -\nabla \cdot \\ \nabla & -\mu \left(\triangle + (\nabla \nabla \cdot)^t \right) \end{pmatrix} \begin{pmatrix} p \\ u \end{pmatrix} = \begin{pmatrix} \mathcal{G} \\ \mathcal{F} \end{pmatrix} \tag{17.2}$$

in the given domain $\Omega \subset R^2$, where $0 < \lambda \leq \infty$ and $0 < \mu < \infty$ are given parameters and $\mathcal{G}$ and $\mathcal{F} = (\mathcal{F}^{(1)}, \mathcal{F}^{(2)})$ are given functions in Ω.

In order to have a well-posed boundary-value problem, the above system of PDEs must be accompanied with boundary conditions. Let $\Gamma_D \subset \partial\Omega$ be the subset of the boundary of Ω on which Dirichlet boundary conditions are imposed:

$$u(x, y) = \alpha(x, y) \quad (x, y) \in \Gamma_D \tag{17.3}$$

where $\alpha = (\alpha^{(1)}, \alpha^{(2)})$ is a given vector function in Γ_D. Let $\Gamma = \partial\Omega \setminus \Gamma_D$ be the subset on which mixed boundary conditions are imposed:

$$-p\mathbf{n} + \mu \left((\nabla u)^t + \nabla u \right) \mathbf{n} + \hat{\beta} u = \gamma, \quad (x, y) \in \Gamma, \tag{17.4}$$

where $\hat{\beta}$ is a given 2×2 symmetric and positive semidefinite matrix function in Γ, γ is a given vector function in Γ, and $\mathbf{n}$ is the outer normal vector in Γ.

In the above, the boundary-value problem is given in its strong formulation. In the next section, we also derive the weak formulation, which is the basis for the finite-element discretization method.

17.4 The Weak Formulation

The *weak formulation* is obtained by multiplying the three equations in (17.2) by some three functions: the first equation is multiplied by a function $q(x, y) \in L_2(\Omega)$, and the second and third equations are multiplied (respectively) by the functions $v^{(1)}(x, y)$, and $v^{(2)}(x, y)$ that belong to the Sobolev space of order 1 [their derivatives are in $L_2(\Omega)$] and vanish on Γ_D. The three equations are then integrated over Ω, using Green's formula and the boundary conditions in (17.4) in the second and third equations. This gives

$$
-\int_\Omega \left(\lambda^{-1} p + \nabla \cdot u\right) q \, dx dy
$$

$$
= \int_\Omega \mathcal{G} q \, dx dy \int_\Omega \left(-p v_x^{(1)} + \mu \left(\nabla u^{(1)} + u_x\right) \cdot \nabla v^{(1)}\right) dx dy + \int_\Gamma (\hat{\beta} u)^{(1)} v^{(1)} d\Gamma
$$

$$
= \int_\Omega \mathcal{F}^{(1)} v^{(1)} dx dy + \int_\Gamma \gamma^{(1)} v^{(1)} \int_\Omega \left(-p v_y^{(2)} + \mu \left(\nabla u^{(2)} + u_y\right) \cdot \nabla v^{(2)}\right) dx dy
$$

$$
+ \int_\Gamma (\hat{\beta} u)^{(2)} v^{(2)} d\Gamma
$$

$$
= \int_\Omega \mathcal{F}^{(2)} v^{(2)} dx dy + \int_\Gamma \gamma^{(2)} v^{(2)} d\Gamma.
$$

In the next section, the weak formulation is used to obtain the finite-element discretization.

17.5 The Finite-Element Discretization

The *finite-element discretization* uses a triangulation of Ω with t triangles and n nodes in $\Omega \setminus \Gamma_D$ (where t and n are positive integers). This discretization is obtained from the above weak formulation by assuming that:

1. The numerical solution functions $u^{(1)}$ and $u^{(2)}$ are linear in each triangle, continuous in the entire mesh, and agree with $\alpha^{(1)}$ and $\alpha^{(2)}$ (respectively) in the nodes that lie in Γ_D.
2. The test functions $v^{(1)}$ and $v^{(2)}$ are linear in each triangle, continuous in the entire mesh, and vanish in the nodes that lie in Γ_D
3. The numerical solution function $p(x, y)$ and the test function $q(x, y)$ are constant in each individual triangle in the mesh.

Clearly, $u^{(1)}$, $u^{(2)}$, $v^{(1)}$, and $v^{(2)}$ have n degrees of freedom each, whereas p and q have t degrees of freedom each. More specifically, $u^{(1)}$, $u^{(2)}$, $v^{(1)}$, and $v^{(2)}$ can be written uniquely as linear combinations of n nodal basis functions each, whereas p and q can be written uniquely as linear combinations of t "elemental" basis functions each, where an elemental basis function has the value 1 in some triangle and 0 elsewhere. This leads to a linear system of the form (3.5), with coefficient (stiffness) matrix of the form

$$
A = \begin{pmatrix} A_{pp} & A_{pu} \\ A_{up} & A_{uu} \end{pmatrix}, \tag{17.5}
$$

where A_{pp} is the submatrix of order t that corresponds to the unknown values of the numerical solution p in the individual triangles, and A_{uu} is the submatrix of order $2n$ that corresponds to the unknown values of the numerical solution functions $u^{(1)}$ and $u^{(2)}$ at the n nodes in $\Omega \setminus \Gamma_D$.

Clearly, A_{pp} is diagonal; this property will be used below. Furthermore, A is symmetric; however, as we will see below, it is neither positive definite nor negative definite. In fact, it is indefinite: it has both positive and negative eigenvalues.

In the next section, we define the semi-algebraic multilevel preconditioner for A.

17.6 The Semialgebraic Multilevel Preconditioner

Let us first define the *semialgebraic multilevel method* for the solution of a linear system of the form

$$Se = r, \tag{17.6}$$

where S is a matrix of order $2n$, r is a given $2n$-dimensional vector, and e is the $2n$-dimensional vector of unknowns. Let us write S in the block form

$$S = \begin{pmatrix} S_{1,1} & S_{1,2} \\ S_{2,1} & S_{2,2} \end{pmatrix},$$

where $S_{1,1}$ and $S_{2,2}$ are submatrices of order n. The semialgebraic multilevel method for (17.6) is almost the same as the algebraic multilevel method in Section 6.9; the only difference is that Algorithm 6.4 to derive P is applied to the block-diagonal part of S,

$$\begin{pmatrix} S_{1,1} & \\ & S_{2,2} \end{pmatrix},$$

rather than S. This way, P prolongs the unknowns corresponding to $u^{(1)}$ separately from the unknowns corresponding to $u^{(2)}$, along the guidelines in Section 17.1. Then, R and Q are defined as usual by $R = P^t$ and $Q = RSP$ (since in the present application S is symmetric), as usual.

In our applications, we use the semialgebraic multilevel method in a V(1,1)-cycle with the symmetric point-GS relaxation. This way, whenever S is SPD, an outer PCG acceleration can be used. This completes the definition of the semialgebraic multilevel preconditioner for the solution of (17.6).

The same idea as above can also be applied to the algebraic multigrid method in Section 6.10, to produce the semialgebraic multigrid method. More specifically, the semi-algebraic multigrid method is obtained by applying Algorithm 6.5 to the block-diagonal part of S,

$$\begin{pmatrix} S_{1,1} & \\ & S_{2,2} \end{pmatrix}$$

rather than S, and then defining $R = P^t$ (at least for symmetric S) and $Q = RSP$.

Let us now use the above semialgebraic multilevel method to solve a larger linear system as in (3.5) with the coefficient matrix A (of order $t + 2n$) constructed in Section 17.5. For this purpose, let us write A in the block-LU decomposition induced by the partitioning in (17.5):

$$A = \begin{pmatrix} I & \\ A_{up} A_{pp}^{-1} & I \end{pmatrix} \begin{pmatrix} A_{pp} & \\ & S(A; u) \end{pmatrix} \begin{pmatrix} I & A_{pp}^{-1} A_{pu} \\ & I \end{pmatrix} \tag{17.7}$$

where

$$S(A; u) = A_{uu} - A_{up} A_{pp}^{-1} A_{pu} \tag{17.8}$$

is the Schur complement of A with respect to the partitioning in (17.5).

Clearly, A_{pp} is diagonal with negative main-diagonal elements. As a consequence, $S(A; u)$ is SPD. It therefore follows from (17.7) that A is indefinite.

From (17.7), we have

$$A^{-1} = \begin{pmatrix} I & -A_{pp}^{-1} A_{pu} \\ & I \end{pmatrix} \begin{pmatrix} A_{pp}^{-1} & \\ & S(A; u)^{-1} \end{pmatrix} \begin{pmatrix} I & \\ -A_{up} A_{pp}^{-1} & I \end{pmatrix}. \tag{17.9}$$

In the following, we use the above A as a preconditioner in a GMRES iteration for the solution of the Stokes equations. For this purpose, one needs frequently to apply A^{-1} to a $(t + 2n)$-dimensional vector. The main part in this task is the application of $S(A; u)^{-1}$ to a $2n$-dimensional subvector. This is done by an inner PCG iteration (with initial guess $\mathbf{0}$) with the semialgebraic multilevel preconditioner. Actually, it is usually not necessary to solve the Schur-complement subsystem exactly. In the numerical application below, it is solved to second-order accuracy only, using 20–50 PCG iterations.

17.7 Preconditioner for the Stokes Equations

The Stokes equations are obtained from (17.2) by setting $\lambda = \infty$, so the upper-left term in the 2×2 coefficient matrix in (17.2) vanishes. The boundary conditions on Γ guarantee that the problem is still well-posed. If $\Gamma = \emptyset$, then p is determined only up to an additive constant; an extra condition of the form

$$\int_\Omega p \, dx dy = C \tag{17.10}$$

(for some known constant C) must be imposed to determine p uniquely and guarantee well-posedness.

The weak formulation is as in Section 17.4, and the stiffness matrix A is constructed as in Section 17.5 (with $\lambda^{-1} = 0$).

With sufficiently large λ, the stiffness matrix for the linear elasticity equations in (17.7) can serve as a preconditioner for the Stokes equations. Indeed, we have considered the Stokes equations with $\mu = 1$ in a domain as in Figure 17.1, discretized on a triangulation of 2713 nodes resulting from adaptive refinement (as in Figure 17.2). As a preconditioner for this problem, we have used the corresponding matrix in (17.7) with $\mu = 1$ and $\lambda = 20$. Because λ is not too large, the application of the inverse of the preconditioner is not too hard: indeed, the solution of the Schur-complement subsystem in each application of (17.9) can be done approximately (to second-order accuracy) by 20–50 PCG iterations (with the semialgebraic multilevel preconditioner in Section 17.6) only.

PCG is applicable to the SPD Schur-complement subsystem with the SPD semialgebraic multilevel preconditioner, but not to the original (indefinite) stiffness matrix of the Stokes equations. For this original system, one must use a more general acceleration method such as GMRES. In our numerical experiments, we have used the more stable GMRES(20,10) cycle in [107]. Ten such cycles (a total of

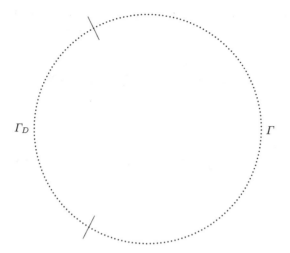

Fig. 17.1. The circular domain in which the linear elasticity and Stokes equations are solved in the present examples.

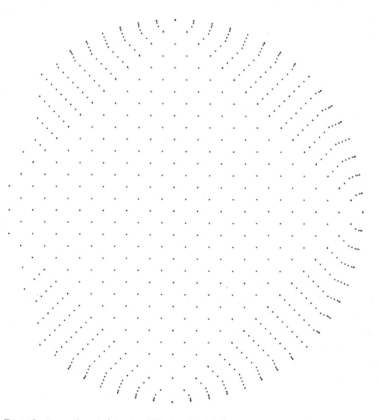

Fig. 17.2. Distribution of nodes in the fifth level of refinement in the linear elasticity and Stokes equations.

310 iterations) with the linear elasticity preconditioner described above are required
to reduce the l_2 norm of the preconditioned residual by six orders of magnitude.

17.8 The Reduced Linear Elasticity Equations

The first equation in (17.2) can be written as

$$p = -\lambda(\nabla \cdot u + \mathcal{G}). \tag{17.11}$$

When this equation is substituted in the second equation in (17.2), we obtain

$$-\nu\nabla\nabla \cdot u - \frac{1-\nu}{2}\left(\Delta + (\nabla\nabla\cdot)^t\right)u = \frac{\nu}{\lambda}\mathcal{F} + \nu\nabla\mathcal{G},$$

where

$$\nu = \frac{\lambda}{2\mu + \lambda}$$

is the so-called Poisson ratio. For most materials, $0.3 \le \nu \le 0.35$; the only exception
is rubber, for which $\nu = 0.5$.

The above is the strong formulation of the reduced linear elasticity equations.
The weak formulation of these equations can be obtained in a similar way by sub-
stituting (17.11) in the weak formulation in Section 17.4:

$$\int_\Omega \left(\nu v_x^{(1)}\nabla \cdot u + \frac{1-\nu}{2}\left(\nabla u^{(1)} + u_x\right) \cdot \nabla v^{(1)}\right)dxdy + \frac{\nu}{\lambda}\int_\Gamma (\hat{\beta}u)^{(1)}v^{(1)}d\Gamma$$

$$= \int_\Omega \left(\frac{\nu}{\lambda}\mathcal{F}^{(1)} + \nu\mathcal{G}\right)v^{(1)}dxdy + \frac{\nu}{\lambda}\int_\Gamma \gamma^{(1)}v^{(1)}d\Gamma$$

$$\int_\Omega \left(\nu v_y^{(2)}\nabla \cdot u + \frac{1-\nu}{2}\left(\nabla u^{(2)} + u_y\right) \cdot \nabla v^{(2)}\right)dxdy + \frac{\nu}{\lambda}\int_\Gamma (\hat{\beta}u)^{(2)}v^{(2)}d\Gamma$$

$$= \int_\Omega \left(\frac{\nu}{\lambda}\mathcal{F}^{(2)} + \nu\mathcal{G}\right)v^{(2)}dxdy + \frac{\nu}{\lambda}\int_\Gamma \gamma^{(2)}v^{(2)}d\Gamma.$$

In its weak formulation, the problem is to find functions $u^{(1)}$ and $u^{(2)}$ [with derivatives in $L_2(\Omega)$] that agree with $\alpha^{(1)}$ and $\alpha^{(2)}$ (respectively) in Γ_D and satisfy the
above equations for any two functions $v^{(1)}$ and $v^{(2)}$ [with derivatives in $L_2(\Omega)$] that
vanish in Γ_D. It is well known that, in this weak formulation, the problem is indeed
well-posed in the sense that it indeed has a unique solution (see, e.g., Chapter 20
in [103]).

The above weak formulation is now discretized on a finite-element triangulation
by assuming that the functions $u^{(1)}$, $u^{(2)}$, $v^{(1)}$, and $v^{(2)}$ are continuous in the mesh
and linear in each triangle in it.

The adaptive-refinement algorithm in Section 4.10 can also be used to construct a suitable triangulation. In this algorithm, however, (4.18) should be modified
to read

$$\min\left(|\tilde{u}^{(1)}(i) - \tilde{u}^{(1)}(j)|, |\tilde{u}^{(2)}(i) - \tilde{u}^{(2)}(j)|\right) \ge \text{threshold}, \tag{17.12}$$

Table 17.1. Number of PCG iterations required in each linear system solve in the adaptive-refinement algorithm applied to the linear elasticity equations in the circle (with Poisson ratio $\nu = 1/3$). The semialgebraic multilevel preconditioner (and the semialgebraic multigrid preconditioner) costs the same as three symmetric Gauss–Seidel iterations. (The set-up time is negligible.)

Level	Nodes	PCG-AML	PCG-AMG	PCG-SGS
1	4	13	13	54
2	13	9	9	13
3	45	12	14	32
4	173	17	16	73
5	679	33	32	177
6	2690	57	52	398
7	10329	93	85	895

where i and j are nodes in the coarse mesh, and $\tilde{u}^{(1)}$ and $\tilde{u}^{(2)}$ are numerical approximations to $u^{(1)}$ and $u^{(2)}$ (respectively) on the coarse mesh. This way, the midpoint $(i + j)/2$ is added to the fine mesh constructed in this refinement level if (and only if) either $u^{(1)}$ or $u^{(2)}$ changes considerably from i to j.

The numerical approximation $\tilde{u}^{(1)}$, $\tilde{u}^{(2)}$ must be calculated on the coarse mesh of each refinement level, and also on the final (finest) mesh obtained from the last refinement level. This is done by solving the linear system (3.5), with the coefficient matrix A being the stiffness matrix on the relevant triangulation. This is done by the semialgebraic multilevel method in Section 17.6 above (with S replaced by A). The V(1,1)-cycle uses the symmetric point-GS relaxation. Since this preconditioner (as well as the stiffness matrix A, see Chapter 20 in [103]) is SPD, the basic semialgebraic multilevel iteration can be further accelerated by PCG.

We apply the above adaptive-refinement algorithm to a problem with $\nu = 1/3$ in the circular domain in Figure 17.1. The mesh obtained from this algorithm at the fifth level of refinement is illustrated in Figure 17.2.

The numbers of iterations required to converge (to reduce the l_2 norm of the preconditioned residual by six orders of magnitude) are reported in Table 17.1. We test three preconditioners: symmetric point-GS (denoted by PCG-SGS), the semialgebraic multilevel preconditioner (denoted by PCG-AML), and the semialgebraic multigrid preconditioner (denoted by PCG-AMG). (Results with the ILU preconditioner are not reported, because it is much inferior even to SGS.) The semialgebraic multilevel and semialgebraic multigrid preconditioners use V(1,1)-cycles with symmetric point-GS relaxation.

It is apparent from the table that the semialgebraic multilevel preconditioner (and the semialgebraic multigrid preconditioner) is superior to the symmetric point-GS preconditioner. This advantage becomes more and more apparent as the number of nodes grows.

17.9 Towards Problems with Constraints

As we see above, the algorithms to construct the prolongation and restriction matrices (Algorithms 6.4 and 6.5) should be modified in problems arising from systems of PDEs in such a way that different unknown functions are prolonged and restricted separately from each other. Here we discuss problems in which it makes sense to modify the coarsening procedure in Algorithm 6.3 and Section 15.3 as well, using preliminary information about the original PDE. This is just another example of using the algebraic approach not religiously but rather with sense and with open mindness towards using extra information from the original PDE whenever available.

Consider, for example, the case of an optimization problem, formulated as a system of PDEs with constraints. Because the constraints usually involve global coupling of the entire set of unknowns, they are unsuitable for the coarsening procedure; they should probably be dropped from the matrix to which this procedure is applied. In other words, the coarsening procedure should probably be applied not to the original matrix but rather to the submatrix obtained from the discretization of the PDEs alone, not the constraints. If Lagrange-multiplier functions are used, they should probably be redefined and reformulated on the coarse mesh, in the spirit of geometric multigrid. Similarly, the constraint equations should also be dropped from the matrix from which the prolongation and restriction matrices are derived, in the spirit of the semialgebraic multilevel method.

17.10 Towards Semialgebraic Block Lumping

In the Schur complement $S(A; u)$ of the linear elasticity equations and the linear system obtained from the discretization of the reduced linear elasticity equations, we assumed that the unknowns are ordered function by function: first unknowns corresponding to $u^{(1)}$ and then unknowns corresponding to $u^{(2)}$. This, however, is not the only way. We could also order the unknowns node by node: first the unknowns representing the values of $u^{(1)}$ and $u^{(2)}$ at the first node, then the unknowns representing the values of $u^{(1)}$ and $u^{(2)}$ at the second node, and so on, until the last node is reached. In other words, the vector of unknowns is divided into n two-dimensional subvectors (where n is the number of nodes in $\Omega \setminus \Gamma_D$), each of which corresponds to a particular node in the mesh. This ordering induces a block form for the coefficient matrix A, in which each block is a submatrix of order 2, which couples a particular node to some other node in terms of both $u^{(1)}$ and $u^{(2)}$.

This form can be used in a block version of the algebraic multilevel method, in which 2×2 blocks and their arithmetics are used instead of the standard matrix elements and their arithmetics. For example, the matrices G_a and G_b are now block-diagonal rather than diagonal, and their inversion requires the inversion of the individual 2×2 blocks on their main block-diagonals. In the following, we explain why this approach is indeed appropriate.

The lumping used in matrix-based multigrid [see (9.2) and (9.3)] is motivated by the assumption that the prolongation operator should be rather accurate for the constant function. This is a fair assumption for scalar PDEs, where the solution is continuous and, therefore, can be approximated locally by the constant function, but not necessarily for systems of PDEs, where there is no continuity relation between different unknown functions, not even at the same point. Thus, it makes no sense to lump matrix elements that couple different unknown functions onto the main diagonal. It makes much more sense to lump (or add) them to the corresponding matrix element in the block main-diagonal, so they still continue to couple the same functions as before, although no longer at different points. In other words, the row-sum operation in (17.1) is interpreted no longer elementwise but rather blockwise, where a block is a 2×2 submatrix that couples two (distinct or not) nodes in terms of both $u^{(1)}$ and $u^{(2)}$. The resulting algorithm is semialgebraic in the sense that it uses not only the original matrix but also the fact that it arises from a system of two coupled PDEs.

The above approach is also semialgebraic in the sense that it uses not only the elements in the original matrix A but also the underlying mesh. Indeed, each two-dimensional subvector in the above splitting corresponds to a particular node in the mesh. (On further coarser levels, the nodes are no longer physical but rather algebraic and virtual.) This approach can also combine with the approach in Section 16.6 to produce a semialgebraic (block) coarsening procedure that may be suitable for nonsymmetric systems of PDEs.

17.11 A Domain-Decomposition Two-Level Method

The semialgebraic multilevel method in Section 17.6 above uses mostly information from the coefficient matrix A. It uses the original system of PDEs only to distinguish between different kinds of functions, such as $u^{(1)}$, $u^{(2)}$, and p. Here we propose a more geometric approach, based on domain decomposition and variational multigrid.

The present approach for solving the Stokes (and linear elasticity) equations uses domain decomposition in a two-level setting, as in Section 6.6 above. The key to this method is to define the prolongation operator. For scalar PDEs such as the diffusion problem, this is illustrated in Figures 6.6 and 6.7. For systems of PDEs such as the Stokes and linear elasticity equations, the definition of the prolongation operator is slightly more complicated.

Consider a coarse-grid vector function $u = (u^{(1)}, u^{(2)})$ to be prolonged to the entire fine grid. The values of the vector function are specified at the corners of the subdomains (the junction points in Figure 17.3). The p values, on the other hand, are specified to have some constant value in each subdomain (see Figure 17.3). Let us first design the first prolongation step, in which u values are defined also in the line that connects two corners of a subdomain. This is done by solving a homogeneous Dirichlet–Neumann subproblem in the strip that contains this line (Figure 17.4). In this subproblem, Dirichlet conditions (as in Γ_D) are used on the left and right edges of the strip, and Neumann conditions (as in Γ) are used on the top and bottom edges of the strip. The u Dirichlet values at the left and right edges are taken from the original values at the subdomain corners, and are constant in each edge.

On the top and bottom edges of the strip, Neumann conditions as in (17.4) (with $\hat{\beta} = 0$) are used. The right-hand sides in these conditions are calculated by using the corresponding constant p values in the top and bottom subdomains in Figure 17.3 and assuming the homogeneous values $\nabla u = (0)$ at the top and bottom edges of the strip. For example, at the bottom edge, the boundary conditions are

$$-p\mathbf{n} + \mu \left((\nabla u)^t + \nabla u \right) \mathbf{n} = -p^{(b)}\mathbf{n},$$

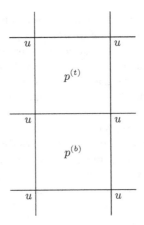

Fig. 17.3. The coarse-grid function to be prolonged. The values of $u = (u^{(1)}, u^{(2)})$ are specified at the subdomain corners. The values of p are specified to be constant in each subdomain; for example, the constant $p^{(t)}$ in the top subdomain, and the constant $p^{(b)}$ in the bottom subdomain.

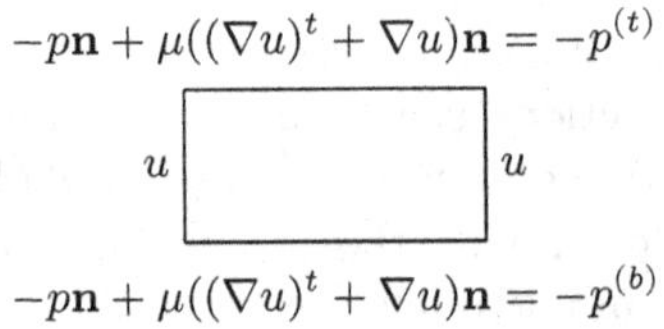

$$-p\mathbf{n} + \mu((\nabla u)^t + \nabla u)\mathbf{n} = -p^{(t)}$$

u ⬚ u

$$-p\mathbf{n} + \mu((\nabla u)^t + \nabla u)\mathbf{n} = -p^{(b)}$$

Fig. 17.4. The first prolongation step, designed to determine the prolonged u values in the subdomain edges, using the u values at the subdomain corners and the constant p values at the subdomains, available from the given coarse-grid function. Each subdomain edge is surrounded by a thin strip, as illustrated in the figure. A Dirichlet–Neumann subproblem is then solved in the strip, as indicated in the figure. This determines uniquely the u values in the subdomain edge, as required.

where $\mathbf{n} = (0, -1)^t$ is the outer normal vector, $p^{(b)}$ at the right-hand side is the constant value available from the coarse-grid function at the bottom subdomain, and u and p at the left-hand side are the unknown functions inside the strip. This procedure determines uniquely the u values in the line contained in the strip.

The same procedure is used for each line that connects two corners of a subdomain. The entire procedure determines the prolonged u values in the entire internal boundary. These values are now used to calculate the prolonged u and p values throughout the fine grid.

(The first prolongation step can be improved further by "lumping" u as in Figure 6.8. This produces a 1-D problem in each subdomain edge to solve for the prolonged u values.)

In the second prolongation step, the prolonged u and p values are calculated also in the interior of each subdomain. This is done by solving a homogeneous Dirichlet subproblem, using the u values calculated in the first prolongation step as Dirichlet data. Consider, for instance, the subdomain in Figure 17.5. The u values

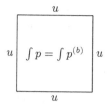

Fig. 17.5. The second prolongation step, designed to determine the prolonged u and p values in the subdomain interiors, using both the original coarse-grid function and the u values calculated in the first prolongation step. In each subdomain, a Dirichlet subproblem is solved, using the u values calculated in the edges in the first prolongation step. The constant $p^{(b)}$ available from the coarse-grid function is used to determine the additive constant for p.

in its internal boundary are known: at the corners, they are available from the original coarse-grid function; and at the edges of the subdomain, they have just been calculated in the first prolongation step. Thus, they can be used as Dirichlet data to solve a homogeneous subproblem in the subdomain, and obtain the required u and p values throughout it.

The p values, however, are determined only up to an additive constant. Still, this constant is easily obtained, because the integral of the prolonged p values over the subdomain must be the same as the integral of the original (constant) p value in the coarse-grid function [see (17.10)]. More specifically, the prolonged p values in the subdomain must satisfy

$$\int pdxdy = \int p^{(b)} dxdy,$$

where the integral is over the subdomain, $p^{(b)}$ in the right-hand side is the constant value available from the coarse-grid function, and p in the left-hand side is the (yet unknown) prolonged function in the subdomain.

Thus, the prolonged u and p values have been calculated in the entire fine grid, as required. This completes the definition of the prolongation operator P.

As usually done in symmetric problems, the restriction operator is now defined by $R = P^t$. As in variational multigrid, the coarse-grid matrix is now defined by $Q = RAP$. This completes the definition of the domain-decomposition two-level method.

17.12 Exercises

1. Show that the submatrix A_{pp} in (17.5) is diagonal.
2. Show that the submatrix A_{uu} in (17.5) is symmetric.
3. Show that $A_{up} = A_{pu}^t$ in (17.5).
4. Conclude that the stiffness matrix for the linear elasticity (and Stokes) equations is symmetric.
5. Show that, for $\lambda < \infty$, the diagonal submatrix A_{pp} in (17.5) has negative main-diagonal elements.
6. Conclude that $-A_{up}A_{pp}^{-1}A_{pu}$ is SPD.

7. Show that the submatrix A_{uu} is positive semidefinite.
8. Conclude that the Schur complement submatrix $S(A; u)$ in (17.8) is SPD as well.
9. Use (17.7) to show that the stiffness matrix for the linear elasticity equations $(\lambda < \infty)$ is indefinite.
10. Show that the reduced linear elasticity system has a well-posed weak formulation, in the sense that it has a unique solution with derivatives in $L_2(\Omega)$. The solution can be found in Chapters 11 and 20 in [103].
11. Show that the stiffness matrix for the reduced linear elasticity equations in Section 17.8 is symmetric.
12. Show that it is also positive semidefinite.
13. Show that it is also nonsingular.
14. Conclude that it is also SPD.
15. Write the computer code that implements the semialgebraic multilevel method for the reduced linear elasticity equations in Section 17.8. The solution can be found in Chapter 20 in [103].
16. Show that the submatrix A_{uu} in (17.5) is nonsingular.
17. Conclude that it is SPD.
18. Conclude that the Schur complement corresponding to p, defined by

$$S(A; p) = A_{pp} - A_{pu} A_{uu}^{-1} A_{up}$$

is symmetric and negative definite.
19. Use the reversed block-LU decomposition

$$A = \begin{pmatrix} I & A_{pu}A_{uu}^{-1} \\ & I \end{pmatrix} \begin{pmatrix} S(A;p) & \\ & A_{uu} \end{pmatrix} \begin{pmatrix} I & \\ A_{uu}^{-1}A_{up} & I \end{pmatrix} \tag{17.13}$$

to show that A is indefinite not only for the linear elasticity equations but also for the Stokes equations.
20. Use your above code (with $\nu = 0$) to solve problems of the form

$$A_{uu}e = r, \tag{17.14}$$

where r is a given $2n$-dimensional vector and e is a $2n$-dimensional vector of unknowns.
21. Use your above code in (17.13) to obtain a direct solver for the Stokes equations. [Solve the Schur-complement subsystem in (17.13) by a conjugate gradient iteration, with an inner PCG iteration with the semialgebraic multilevel preconditioner to solve (17.14)].
22. In the model case in which the Stokes equations are defined in a square with periodic boundary conditions and discretized on a uniform mesh as in Figure 4.13, expand the discrete pressure function p using the 2-D Fourier transform as in the exercise at the end of Chapter 2:

$$p_{j,m}^{(k,l)} = \exp\left(2\pi\sqrt{-1}(kj + lm)h\right),$$

bound the corresponding eigenvalues of $S(A; p)$ from above and below, and conclude that

$$\kappa(S(A; p)) = O(1)$$

as $h \to 0$. Conclude that the above conjugate gradient iteration for solving the Schur-complement subsystem in (17.13) should converge rapidly.

23. Write the computer code that uses (17.7) as a preconditioner for the Stokes equations. The solution can be found in Chapter 21 in [103].

24. Which of the above two solvers for the Stokes equations is more efficient?

25. Let w be a coarse-grid function with u-values defined at the subdomain corners and constant p-values defined in the subdomains in Figure 17.3. Let P be the prolongation operator indicated in Figures 17.4 and 17.5. Show that Pw is indeed rather small in terms of the norm induced by A (energy norm). Conclude that the A-induced norm of the operator PJ_c is moderate.

26. Show that the ith column in Q in the algorithm in Section 17.11 can be calculated by

$$Qe^{(i)} = R(A(Pe^{(i)})),$$

where $e^{(i)}$ is the ith column of the identity matrix of the same order as Q.

27. Show that all the columns of Q can be calculated simultaneously in parallel.

28. Show that Q in Section 17.11 is indeed sparse in the sense that coarse-grid u and p values are coupled in Q only with other coarse-grid values at subdomains that are at most two subdomains away (neighbor of neighbor subdomains). Conclude that the dimension of $e^{(i)}$ above could actually be much smaller than the order of Q.

29. Explain the advantage of the domain decomposition two-level method in Section 17.11 in terms of cache access.

Part VII

Appendices

In the following appendices, we consider two kinds of problems: time-dependent problems and nonlinear problems. Each problem of either of these kinds requires the solution of many linear systems of algebraic equations. Each of these linear systems can be solved efficiently by multigrid. Furthermore, when complicated domains and variable coefficients are also used, the advantage of matrix-based multigrid is apparent.

Time-Dependent Parabolic PDEs

The (semi-) implicit discretization of a time-dependent parabolic PDE produces a long sequence of linear systems of algebraic equations. Each linear system can by itself be solved efficiently by matrix-based multigrid.

18.1 Parabolic PDEs

So far in this book, we have considered elliptic PDEs in a spatial domain Ω, with boundary conditions given on $\partial\Omega$. Here we consider parabolic PDEs, with an extra dimension t, which represents the time. In this kind of problem, boundary conditions are given on $\partial\Omega$ and initial conditions are given at the initial time $t = 0$. For this reason, the problem is referred to as an initial boundary-value problem.

As we will see below, the numerical solution of this problem is obtained from a sequence of elliptic boundary-value problems at discrete time steps. Each of these elliptic problems can be solved iteratively by multigrid.

18.2 The Parabolic Diffusion Equation

The parabolic diffusion equation is obtained from the diffusion equation in (3.2) by also adding the time derivative

$$u_t - \nabla \cdot (\mathcal{D}\nabla u) = \mathcal{F}$$

in $\Omega \times [0, \mathbf{T}]$, where $\mathbf{T}$ is a positive number indicating the maximal time, $0 \le t \le T$ is the time variable, the '∇' operator is as in Section 17.2, $\mathcal{D}$ is a given 2×2 symmetric and uniformly positive definite matrix function in Ω (as in Section 3.2), and the unknown solution u and the right-hand side $\mathcal{F}$ are functions in $\Omega \times [0, \mathbf{T}]$.

The boundary conditions are of Dirichlet type in the subset $\Gamma_D \subset \Omega$:

$$u(x, y, t) = \mathcal{F}_1(x, y, t) \quad (x, y) \in \Gamma_D, \ 0 \le t \le T,$$

where $\mathcal{F}_1$ is a given function in $\Gamma_D \times [0, \mathbf{T}]$. On the rest of $\partial\Omega$, that is, on $\Gamma = \partial\Omega \backslash \Gamma_D$, the boundary conditions are of the mixed type:

$$(\mathcal{D}\nabla u) \cdot \mathbf{n} + \mathcal{G}_1 u = \mathcal{G}_2$$

in $\Gamma \times [0, \mathbf{T}]$, where $\mathbf{n} = (n_1(x, y), n_2(x, y))$ is the outer normal vector to Γ in R^2, $\mathcal{G}_1$ is a given nonnegative function in Γ, and $\mathcal{G}_2$ is a given function in $\Gamma \times [0, \mathbf{T}]$.

In order to complete the definition of the problem, one must also impose initial conditions at the initial time $t = 0$:

$$u(x, y, 0) = u_0(x, y), \quad (x, y) \in \Omega,$$

where u_0 is a given function in Ω that is compatible with the boundary conditions. This completes the strong formulation of the initial boundary-value problem.

18.3 The Weak Formulation

In the above, the parabolic PDE is given in its strong formulation. A better-posed formulation, however, is the weak formulation, which is obtained from the strong formulation by multiplying the above PDE by a function $v(x, y)$ that vanishes in Γ_D, integrating over Ω, and using Green's formula and the mixed boundary conditions in Γ:

$$\int_\Omega u_t v \, dx dy + a(u, v) = \int_\Omega \mathcal{F} v \, dx dy + \int_\Gamma \mathcal{G}_2 v \, d\Gamma,$$

where

$$a(u, v) = (\mathcal{D}\nabla u) \cdot \nabla v \, dx dy + \int_\Gamma \mathcal{G}_1 u v \, d\Gamma.$$

This weak formulation will be used in the finite-element discretization below.

18.4 The Semi-Implicit Time Discretization

Let us first discretize the above equation in the time direction t. For this purpose, we use the semi-implicit scheme, which is of second-order accuracy in both space and time.

Let M be an integer number denoting the number of discrete time steps used to discretize the time interval $[0, \mathbf{T}]$. This way, if $\Delta t = \mathbf{T}/M$, then the discrete time points are:

$$t = 0, \Delta t, 2\Delta t, 3\Delta t, \ldots, M\Delta t.$$

For $1 \le k \le M$, let $\mathcal{F}^{(k)}$ be the function in Ω that coincides with $\mathcal{F}$ at the half time step $t = (k - 1/2)\Delta t$:

$$\mathcal{F}^{(k)}(x, y) = \mathcal{F}(x, y, (k - 1/2)\Delta t), \quad (x, y) \in \Omega.$$

Similarly, we also define

$$\mathcal{G}_2^{(k)}(x, y) = \mathcal{G}_2(x, y, (k - 1/2)\Delta t), \quad (x, y) \in \Gamma$$
$$\mathcal{F}_1^{(k)}(x, y) = \mathcal{F}_1(x, y, (k - 1/2)\Delta t), \quad (x, y) \in \Gamma_D$$
$$\mathcal{F}_{1\,t}^{(k)}(x, y) = \mathcal{F}_{1\,t}(x, y, (k - 1/2)\Delta t), \quad (x, y) \in \Gamma_D$$

(derivative of $\mathcal{F}_1$ with respect to t at the half time step).

Furthermore, for $0 \le k \le M$, let $u^{(k)}$ be the function in Ω that approximates the solution at the kth time steps:

$$u^{(k)}(x, y) \doteq u(x, y, k\triangle t), \quad (x, y) \in \Omega.$$

In particular, $u^{(0)}$ is obtained from the initial conditions:

$$u^{(0)}(x, y) = u_0(x, y), \quad (x, y) \in \Omega.$$

The semi-implicit time discretization of the above weak formulation is obtained by approximating the term u_t by symmetric finite difference around the half time step:

$$(\triangle t)^{-1} \int_\Omega \left(u^{(k)} - u^{(k-1)} \right) v dx dy + \frac{1}{2} a \left(u^{(k)} + u^{(k-1)}, v \right)$$
$$= \int_\Omega \mathcal{F}^{(k)} v dx dy + \int_\Gamma \mathcal{G}_2^{(k)} v d\Gamma.$$

This completes the semi-implicit time discretization. In the next section, we also discretize in space, using a finite-element triangulation of Ω.

18.5 The Finite-Element Discretization

Here we discretize the above equation in space as well. For this purpose, we use a finite-element triangulation T of Ω. We assume that there are n nodes in $T \setminus \Gamma_D$.

Let ϕ_j be the nodal basis function that is continuous in T, linear in each individual triangle, and takes the value 1 at node j and 0 at all the other nodes in T. For every two nodes i and j, define

$$K_{i,j} = \int_T \phi_j \phi_i dx dy \quad \text{and} \quad a_{i,j} = a(\phi_j, \phi_i).$$

These definitions induce the definitions of the so-called mass matrix

$$K = (K_{i,j})_{i,j \in T \setminus \Gamma_D}$$

and stiffness matrix

$$A = (a_{i,j})_{i,j \in T \setminus \Gamma_D},$$

both of which are SPD matrices of order n.

Now, let us use the finite-element mesh to approximate the functions $u^{(k)}$ in Section 18.4 above by

$$u^{(k)} \doteq \sum_{j \in T \setminus \Gamma_D} x_j^{(k)} \phi_J + \sum_{j \in \Gamma_D} \mathcal{F}_1(j) \phi_j,$$

where the $x_j^{(k)}$ are the unknown components in the vector

$$x^{(k)} = \left(x_1^{(k)}, x_2^{(k)}, \ldots, x_n^{(k)} \right)^t.$$

[Note that at the initial time step $k = 0$ these components are known: $x_j^{(0)} = u_0(j)$.]

Finally, define the components

$$b_i^{(k)} = \int_T \mathcal{F}^{(k)} \phi_i \, dx \, dy + \int_\Gamma \mathcal{G}_2^{(k)} \phi_i \, d\Gamma - \sum_{j \in \Gamma_D} K_{i,j} \mathcal{F}_1^{(k)}(j) + a_{i,j} \mathcal{F}_{1\,t}^{(k)}(j)$$

$(1 \leq i \leq n)$ of the vector

$$b^{(k)} = \left(b_1^{(k)}, b_2^{(k)}, \ldots, b_n^{(k)} \right).$$

By making the substitutions

$$u^{(k)} \leftarrow \sum_{j \in T \backslash \Gamma_D} x_j^{(k)} \phi_J + \sum_{j \in \Gamma_D} \mathcal{F}_1(j) \phi_j$$

and

$$v \leftarrow \phi_i$$

in the equation at the end of Section 18.4 above, we have the numerical scheme

$$(\Delta t)^{-1} K \left(x^{(k)} - x^{(k-1)} \right) + \frac{1}{2} A \left(x^{(k)} + x^{(k-1)} \right) = b^{(k)}.$$

By marching across the time steps indexed by $k = 1, 2, 3, \ldots, M$ and calculating the unknown vectors $x^{(k)}$, we obtain the numerical solution to the original initial boundary-value problem. In the next section, we show that this scheme is indeed stable and, hence, accurate.

18.6 Stability Analysis

The above equations can be written more compactly as

$$\mathcal{A}x = b,$$

where x is the nM-dimensional vector of unknowns whose first n-dimensional subvector is $x^{(1)}$, its second subvector is $x^{(2)}$, and so on, until the Mth subvector $x^{(M)}$, b is the nM-dimensional vector whose first n-dimensional subvector is $b^{(1)} + (\Delta t)^{-1} Kx^{(0)} - Ax^{(0)}/2$, its second subvector is $b^{(2)}$, its third subvector is $b^{(3)}$, and so on, until the Mth subvector $b^{(M)}$, and $\mathcal{A}$ is the $nM \times nM$ block-bidiagonal matrix

$$\mathcal{A} = \begin{pmatrix} \frac{K}{\Delta t} + \frac{A}{2} & & & & \\ \frac{A}{2} - \frac{K}{\Delta t} & \frac{K}{\Delta t} + \frac{A}{2} & & & \\ & \frac{A}{2} - \frac{K}{\Delta t} & \frac{K}{\Delta t} + \frac{A}{2} & & \\ & & \ddots & \ddots & \\ & & & \frac{A}{2} - \frac{K}{\Delta t} & \frac{K}{\Delta t} + \frac{A}{2} \end{pmatrix}.$$

Actually, $\mathcal{A}$ can be written as the product

$$\mathcal{A} = \mathcal{B}(I - \mathcal{Q}),$$

where $\mathcal{B}$ is the block-diagonal matrix

$$\mathcal{B} = \begin{pmatrix} \frac{K}{\Delta t} + \frac{A}{2} & & & & \\ & \frac{K}{\Delta t} + \frac{A}{2} & & & \\ & & \frac{K}{\Delta t} + \frac{A}{2} & & \\ & & & \ddots & \\ & & & & \frac{K}{\Delta t} + \frac{A}{2} \end{pmatrix}$$

and $\mathcal{Q}$ is the matrix with a single nonzero block diagonal:

$$\mathcal{Q} = \begin{pmatrix} \left(\frac{K}{\Delta t} + \frac{A}{2}\right)^{-1}\left(\frac{K}{\Delta t} - \frac{A}{2}\right) \overset{(0)}{} & & & \\ & \left(\frac{K}{\Delta t} + \frac{A}{2}\right)^{-1}\left(\frac{K}{\Delta t} - \frac{A}{2}\right) \overset{(0)}{} & & \\ & & \ddots & \\ & & & \left(\frac{K}{\Delta t} + \frac{A}{2}\right)^{-1}\left(\frac{K}{\Delta t} - \frac{A}{2}\right) \overset{(0)}{} \end{pmatrix}.$$

Because both K and A are SPD, $\mathcal{B}$ is SPD as well. We say that the numerical scheme is stable if $\mathcal{A}^{-1}$ is well bounded in terms of the norm induced by $\mathcal{B}$. In this context, the above decomposition of $\mathcal{A}$ proves to be useful. Indeed, from Lemma 2.5, we have

$$\|\mathcal{A}^{-1}\|_{\mathcal{B}} = \|(I - \mathcal{Q})^{-1}\mathcal{B}^{-1}\|_{\mathcal{B}} \leq \|(I - \mathcal{Q})^{-1}\|_{\mathcal{B}}\|\mathcal{B}^{-1}\|_{\mathcal{B}}.$$

In order to use this estimate, let us first bound $\mathcal{Q}$ in terms of the norm induced by $\mathcal{B}$. For this purpose, note that the maximum used in the norm of $\mathcal{Q}$ is obtained at an nM-dimensional vector v whose mth time step $v^{(M)}$ vanishes and whose former time steps are the same, namely, $v^{(1)} = v^{(2)} = \cdots = v^{(M-1)}$:

$$\|\mathcal{Q}\|_{\mathcal{B}}^2$$

$$= \max_{v \in R^{nM}, v \neq 0} \frac{\|\mathcal{Q}v\|_{\mathcal{B}}^2}{\|v\|_{\mathcal{B}}^2}$$

$$= \max_{v^{(1)}, v^{(2)}, \ldots, v^{(M-1)} \in R^n} \frac{\sum_{k=1}^{M-1} \left\|((\Delta t)^{-1}K + A/2)^{-1}((\Delta t)^{-1}K - A/2)v^{(k)}\right\|_{(\Delta t)^{-1}K + A/2}^2}{\sum_{k=1}^{M-1}\|v^{(k)}\|_{(\Delta t)^{-1}K + A/2}^2}$$

$$= \max_{v \in R^n, v \neq 0} \frac{(M-1)\left\|((\Delta t)^{-1}K + A/2)^{-1}((\Delta t)^{-1}K - A/2)v\right\|_{(\Delta t)^{-1}K + A/2}^2}{(M-1)\|v\|_{(\Delta t)^{-1}K + A/2}^2}$$

$$= \left\|\left((\Delta t)^{-1}K + A/2\right)^{-1}\left((\Delta t)^{-1}K - A/2\right)\right\|_{(\Delta t)^{-1}K + A/2}^2.$$

Furthermore, since $((\triangle t)^{-1}K + A/2)^{-1}((\triangle t)^{-1}K - A/2)$ is symmetric with respect to the inner product induced by $(\triangle t)^{-1}K + A/2$, we have from Lemmas 2.13 and 2.14 that

$$
\begin{aligned}
&\left\|\left((\triangle t)^{-1}K + A/2\right)^{-1}\left((\triangle t)^{-1}K - A/2\right)\right\|_{(\triangle t)^{-1}K+A/2} \\
&= \max_{v \in R^n, v \neq 0} \frac{\left|\left(v, \left((\triangle t)^{-1}K - A/2\right)v\right)_2\right|}{\left(v, \left((\triangle t)^{-1}K + A/2\right)v\right)_2} \\
&= \max_{v \in R^n, v \neq 0} \frac{\left|\left(v, (\triangle t)^{-1}Kv\right)_2 - (v, Av/2)_2\right|}{\left(v, (\triangle t)^{-1}Kv\right)_2 + (v, Av/2)_2} \\
&\leq 1.
\end{aligned}
$$

In summary, we have

$$
\|\mathcal{Q}\|_{\mathcal{B}} \leq 1.
$$

Using this result and Lemma 2.5, we have

$$
\begin{aligned}
\|(I - \mathcal{Q})^{-1}\|_{\mathcal{B}} = \|\sum_{k=0}^{M-1} \mathcal{Q}^k\|_{\mathcal{B}} \\
\leq \sum_{k=0}^{M-1} \|\mathcal{Q}\|_{\mathcal{B}}^k \\
\leq \sum_{k=0}^{M-1} 1 \\
= M.
\end{aligned}
$$

Let us now estimate $\mathcal{B}^{-1}$ in terms of the norm induced by $\mathcal{B}$. From Lemma 2.17, we have

$$
\|\mathcal{B}^{-1}\|_{\mathcal{B}} = \|\mathcal{B}^{-1}\|_2.
$$

Let h be the typical meshsize in the triangulation T. Because both K and A contain the factor h^2 that follows from the integration over each individual triangle, we have that

$$
\|\mathcal{B}^{-1}\|_{\mathcal{B}} = \|\mathcal{B}^{-1}\|_2 = O(\triangle t h^{-2})
$$

as both h and $\triangle t$ approach zero at the same time. As a conclusion, we have

$$
\|\mathcal{A}^{-1}\|_{\mathcal{B}} \leq \|(I - \mathcal{Q})^{-1}\|_{\mathcal{B}} \|\mathcal{B}^{-1}\|_{\mathcal{B}} = O(M\triangle t h^{-2}) = O(\mathbf{T} h^{-2}).
$$

Because $\mathcal{A}^{-1}$ is bounded independently of the number of time steps M, the numerical scheme is considered as stable. Below we will see that the above estimate implies that the scheme is also accurate.

18.7 Accuracy of the Numerical Scheme

The above stability analysis is helpful in estimating the discretization error in terms of the norm induced by $\mathcal{B}$. This is done as follows. Let $\tilde{u}$ be the nM-dimensional vector obtained from restricting the solution $u(x, y, t)$ to the discrete time-space grid. The discretization error can then be written as

$$\tilde{u} - x = \mathcal{A}^{-1}\left(\mathcal{A}\tilde{u} - b\right).$$

Using the norm induced by $\mathcal{B}$, we have

$$\|\tilde{u} - x\|_\mathcal{B} = \|\mathcal{A}^{-1}\left(\mathcal{A}\tilde{u} - b\right)\|_\mathcal{B} \le \|\mathcal{A}^{-1}\|_\mathcal{B}\,\|\mathcal{A}\tilde{u} - b\|_\mathcal{B} = O\left(\mathbf{T}h^{-2}\,\|\mathcal{A}\tilde{u} - b\|_\mathcal{B}\right).$$

All that is left to do is to estimate the truncation error $\mathcal{A}\tilde{u} - b$ in terms of the norm induced by $\mathcal{B}$. For this purpose, one should bear in mind that the spatial discretization is in the undivided form; that is, both A and K contain an extra factor of $O(h^2)$.

Let us carry out this estimate in the model PDE

$$u_t - (D_1 u_x)_x - (D_2 u_y)_y = \mathcal{F}, \quad 0 < t < \mathbf{T},\ 0 < x, y < 1,$$

discretized on the uniform finite-element mesh in Figure 4.13, where D_1 and D_2 are piecewise constant functions in the unit square, with discontinuity lines that align with the mesh. In this case, we have

$$\mathcal{A}\tilde{u} - b = O(h^2((\triangle t)^2 + h^2))$$

pointwise in the time-space grid. Since the entire time-space grid contains $O(M/h^2)$ points, and since $\|\mathcal{B}\|_2 = O(h^2/\triangle t)$,

$$\|\mathcal{A}\tilde{u} - b\|_\mathcal{B}^2 = (\mathcal{A}\tilde{u} - b, \mathcal{B}(\mathcal{A}\tilde{u} - b))_2 = O\left(\frac{M}{h^2} \cdot \frac{h^2}{\triangle t}h^4\left((\triangle t)^2 + h^2\right)^2\right),$$

or

$$\|\mathcal{A}\tilde{u} - b\|_\mathcal{B} = O\left(\frac{\sqrt{\mathbf{T}}}{\triangle t}h^2\left((\triangle t)^2 + h^2\right)\right).$$

Using also the stability estimate from the previous section, we have the following accuracy estimate:

$$\|\tilde{u} - x\|_\mathcal{B} = O\left(\mathbf{T}^{3/2}(\triangle t + h^2/\triangle t)\right).$$

The best accuracy is therefore obtained when $\triangle t$ is of the same order as h.

18.8 The Algebraic Multilevel Preconditioner

The above numerical scheme requires the solution of a linear system with the coefficient matrix $(\triangle t)^{-1}K + A/2$ at each time step. It is thus most important to have

Table 18.1. Number of PCG iterations used in each refinement level in the adaptive-refinement algorithm (with threshold of 0.01) applied to a particular time step in the semi-implicit scheme for the parabolic diffusion problem in the complicated domain in Figure 16.2. (It is assumed that $\triangle t = 2$, so the coefficient matrix is $K + A$, where A is the stiffness matrix and K is the mass matrix.)

Level	Nodes	PCG-AML	PCG-AMG	PCG-ILU	PCG-SGS
1	11	5	5	3	5
2	34	4	4	5	11
3	107	8	8	8	21
4	342	12	12	17	40
5	1069	18	18	30	61
6	3095	22	23	63	120
7	8331	36	34	122	216
8	18608	54	52	420	386

an efficient iterative solver for this system. For this purpose, it is recommended to use PCG with the algebraic multilevel preconditioner.

To illustrate how efficient this preconditioner is, we carry out the numerical experiments in Section 16.7 above, only this time we use the coefficient matrix $K + A$ rather than A, where A is the stiffness matrix in Section 16.7 and K is the mass matrix corresponding to the same finite-element mesh.

The details of the algorithms are as in Section 16.7. The iteration counts reported in Table 18.1 illustrate the advantage of the algebraic multilevel methods.

19

Nonlinear Equations

Here we discuss another kind of problem in which matrix-based multigrid proves to be most useful. Indeed, the Newton iteration for solving a nonlinear boundary-value problem requires the solution of a long sequence of linear systems of algebraic equations. Each linear system can by itself be solved efficiently by an inner iteration of matrix-based multigrid.

19.1 Nonlinear PDEs

So far in this book, we have considered only linear PDEs, whose discretization produces the algebraic system (3.5). Here, however, we consider also the more difficult case of a nonlinear PDE that produces a nonlinear algebraic system of the form

$$A(x) = b, \tag{19.1}$$

where b is a given N-dimensional vector, x is the N-dimensional vector of unknowns, and $A() : R^N \rightarrow R^N$ is a nonlinear vector function. In this case, the usual residual equation is no longer relevant, as discussed below. A more general formulation is necessary.

19.2 The Residual Equation

Let $x^{(0)}$ be an approximation for the solution x. Let $e = x - x^{(0)}$ denote the error. Then e can be viewed as the correction that should be added to $x^{(0)}$ to have the solution $x = x^{(0)} + e$. In other words, since $x^{(0)}$ is mapped by $A()$ to $A(x^{(0)})$ rather than b, it should be corrected by adding e to it, which produces a vector that is indeed mapped onto b, as required. In fact, in terms of mapped vectors, e satisfies the residual equation

$$A(x^{(0)} + e) - A(x^{(0)}) = b - A(x^{(0)}), \tag{19.2}$$

which tells us what vector e should be added to $x^{(0)}$ in order to add $b - A(x^{(0)})$ to the mapped value, and change it from $A(x^{(0)})$ to the required vector b.

Indeed, when (19.2) is added to the trivial equation

$$A(x^{(0)}) = A(x^{(0)}),$$

the sum of these two equations produces the required equation

$$A(x^{(0)} + e) = b,$$

which shows that $x^{(0)} + e$ is indeed the required solution.

Thus, (19.2) is the correct form of the residual equation, which determines the required correction e.

Of course, when $A()$ is linear, (19.2) takes the simpler and more usual form

$$A(e) = b - A(x^{(0)}).$$

In the present nonlinear case, however, $A()$ may produce an interaction between $x^{(0)}$ and e, so such a simplification is not allowed. One must stick to (19.2), and use it also in the multigrid algorithm [24].

19.3 Defect Correction

The operator that approximates $A()$ on a coarser grid can be viewed as a special case of the following formulation. Let $A_1()$ be an approximation to $A()$. For example, if $A()$ is a second-order approximation to the original nonlinear differential operator $T()$ (see the exercises at the end of Chapter 8):

$$A() = (1 + O(h^2))T(),$$

then $A_1()$ could be a first-order approximation to the same operator:

$$A_1() = (1 + O(h))T().$$

Usually, $A_1()$ is much easier to invert than $A()$. Therefore, one would like to solve (19.1) without inverting $A()$. This is indeed done in the defect-correction method as follows. First, replace $A()$ in the left-hand side in (19.2) by $A_1()$:

$$A_1(x^{(0)} + e^{(0)}) - A_1(x^{(0)}) = b - A(x^{(0)}). \tag{19.3}$$

Clearly, $e^{(0)}$ is not necessarily the same as the exact error $e = x - x^{(0)}$ used in (19.2). Still, it can approximate it rather well, and produce an improved approximation to x, denoted by $x^{(1)}$:

$$x^{(1)} = x^{(0)} + e^{(0)}.$$

This procedure can repeat iteratively, producing the sequence of vectors $x^{(i)}$, satisfying

$$A_1(x^{(i+1)}) - A(x^{(i)}) = b - A(x^{(i)}). \tag{19.4}$$

If the sequence $x^{(i)}$ converges in some norm, then the right-hand side of (19.4), must tend to zero, so $x^{(i)}$ must converge to x. Thus, (19.1) has been solved using inversions of $A_1()$ only.

In the special case in which $A_1()$ is linear, one can apply $A_1^{-1}()$ to both sides of (19.4) to obtain the iteration

$$x^{(i+1)} = x^{(i)} + A_1^{-1}\left(b - A(x^{(i)})\right),$$

which means that A_1 actually serves as a preconditioner for the iterative solution of (19.1).

In the following, we will see how (19.3) can be used in a multigrid setting, with $A_1()$ being interpreted as a coarse-grid approximation to $A()$.

19.4 Geometric Multigrid

The multigrid algorithm in [24] for the nonlinear equation (19.1) indeed uses (19.3) with $A_1()$ being a coarse-grid approximation to $A()$. More specifically, in the geometric multigrid approach, $A_1()$ is obtained from a rediscretization of the original differential operator [$T()$ in the exercises at the end of Chapter 8] on the coarse grid. In the following, the multigrid iteration is described in more detail.

Let P be the prolongation operator from the coarse grid, R the restriction operator to the coarse grid, and $Q()$ a coarse-grid nonlinear vector function that approximates $A()$ in some sense [e.g., both $A()$ and $Q()$ are obtained from discretization of the same nonlinear differential operator $T()$]. Then (19.2) can be approximated on the coarse grid by

$$Q(Rx^{(0)} + E) - Q(Rx^{(0)}) = R\left(b - A(x^{(0)})\right)$$
$$e = PE.$$

Here E is a coarse-grid approximation of the fine-grid error e. First, E is calculated by solving the coarse-grid version of (19.2); then, it is prolonged back to the fine grid to produce an approximation to e.

In order to be approximated well on the coarse grid, both $x^{(0)}$ and e must be sufficiently "smooth" (have low variation). This can be achieved by performing pre and postrelaxations before and after the coarse-grid correction, provided that the underlying PDE has a continuous solution. For PDEs with discontinuous (shock-wave) solutions, on the other hand, the situation may be more complicated: even though one may assume that e has been smoothed by the prerelaxations, there is no guarantee that $x^{(0)}$ is smooth as well, so its coarse-grid restriction $Rx^{(0)}$ may be completely smeared and spoiled.

Another drawback in the above algorithm is the poor suitability of $Q()$. As discussed earlier in the book, when the underlying PDE has discontinuous coefficients, rediscretization no longer produces a suitable coarse-grid approximation for $A()$. In such cases, matrix-based multigrid should be used rather than geometric multigrid; unfortunately, here no matrix is available, because $A()$ is a nonlinear function. The only cure is linearization.

19.5 The Newton Iteration

The Newton iteration (see the exercises at the end of Chapter 8) is based on linearization. For $i = 0, 1, 2, \ldots$, it produces better and better approximations $x^{(i)}$ for x, the solution of (19.1).

In order to compute the next iteration $x^{(i+1)}$, the Newton method uses the Jacobian of $A()$ at $x^{(i)}$, denoted by $A_{x^{(i)}}$:

$$x^{(i+1)} = x^{(i)} + A_{x^{(i)}}^{-1} \left(b - A(x^{(i)}) \right).$$

(Compare with the exercises at the end of Chapter 8, in which a slightly different linearization, based on the underlying differential operator, is used.)

Each implicit inversion of $A_{x^{(i)}}$ in this iteration can be done efficiently by matrix-based multigrid.

References

1. Adams, L. M.; and Jordan, H. F.: Is SOR Color-Blind? *SIAM J. Sci. Stat. Comput.* 7 (1986), pp. 490–506.

2. Alcouffe, R.; Brandt, A.; Dendy, J. E.; and Painter J.: The multigrid method for the diffusion equation with strongly discontinuous coefficients. *SIAM J. Sci. Statist. Comput.* 2 (1981), pp. 430–454.

3. Aruliah, D. A.; and Ascher, U. M.: Multigrid preconditioning for time-harmonic Maxwell's equations in 3D. On www.mgnet.org/mgnet/papers/Aruliah-Ascher/mgmax.ps.gz.

4. Axelsson, O.; and Padiy, A.: On the additive version of the algebraic multilevel iteration method for anisotropic elliptic problems. *SIAM J. Sci. Comput.* 20 (1999), pp. 1807–1830.

5. Bank, R. E.: A comparison of two multilevel iterative methods for nonsymmetric and indefinite finite element equations. *SIAM J. Num. Anal.* 18 (1981), pp. 724–743.

6. Bank, R. E.; and Douglas, C. C.: Sharp estimates for multigrid rates of convergence with general smoothing and acceleration. *SIAM J. Numer. Anal.* 22 (1985), pp. 617–633.

7. Bank, R. E.; and Dupont, T. F.: An optimal order process for solving elliptic finite element equations. *Math. Comp.* 36 (1981), pp 35–51.

8. Bank, R. E.; Dupont, T. F.; and Yserentant, Y.: The hierarchical basis multigrid method. *Numer. Math.* 52 (1988), pp. 427–458.

9. Bank, R. E.; and Gutsch, S.: Hierarchical basis for the convection-diffusion equation on unstructured meshes. In *Ninth International Conference on Domain Decomposition Methods*, Bjorstad, P., Espedal, M., and Keyes, D. (eds.), Bergen Univ. Press, Norway (1998).

10. Black, M. J.; Sapiro, G.; Marimont, D.; and Heeger, D.: Robust anisotropic diffusion: Connections between robust statistics, line processing, and anisotropic diffusion. In *Scale-Space Theory in Computer Vision*, Lecture Notes in Computer Science 1252, Springer, New York, pp. 323–326 (1997).

11. Black, M. J.; Sapiro, G.; Marimont, D. H.; and Heeger, D.: Robust anisotropic diffusion. *IEEE T. Image Process.* 7 (1998), pp. 421–432.

12. Botta, E. E. F.; and van der Ploeg, A.: Preconditioning techniques for matrices with arbitrary sparsity patterns. In *Proceedings of the 9th International Conference on Finite Elements in Fluid Dynamics, New Trends and Applications* (1995), pp. 989–998.

13. Braess, D.: Towards algebraic multigrid for elliptic problems of second order. *Computing* 55 (1995), pp. 379–393.

14. Bramble, J. H.; Ewing, R. E.; Pasciak, J. E.; and Schatz, A. H.: A preconditioning technique for the efficient solution of problems with local grid refinement. *Comput. Methods Appl. Mech. Engrg.* 67 (1988), pp. 149–159.

15. Bramble, J. H.; Leyk, Z.; and Pasciak, J. E.: Iterative schemes for non-symmetric and indefinite elliptic boundary value problems. *Math. Comp.* 60 (1993), pp. 1–22.

16. Bramble, J. H.; and Pasciak, J. E.: New convergence estimates for multigrid algorithms. *Math. Comp.* 49 (1987), pp. 311–329.

17. Bramble, J. H.; and Pasciak, J. E.: New estimates for multigrid algorithms including the V-cycle. *Math. Comp.* 60 (1993), pp. 447–471.

18. Bramble, J. H.; Pasciak, J. E.; and Schatz, A. H.: The constructuring of preconditioners for elliptic problems on regions partitioned into substructures I. *Math. Comp.* 46 (1986), pp. 361–369.

19. Bramble, J. H.; Pasciak, J. E.; and Schatz, A. H.: The constructuring of preconditioners for elliptic problems on regions partitioned into substructures II. *Math. Comp.* 47 (1986), pp. 103–134.

20. Bramble, J. H.; Pasciak, J. E.; Wang, J.; and Xu, J.: Convergence estimates for multigrid algorithms without regularity assumptions. *Math. Comp.* 57, 1991, pp. 23–45.

21. Bramble, J. H.; Pasciak, J. E.; and Xu, J.: Parallel multilevel preconditioners. *Math. Comp.* 55 (1990), pp. 1–22.

22. Brand, C. W.: An incomplete factorization preconditioning using repeated red black ordering. *Numer. Math.* 61 (1992), pp. 433–454.

23. Brandt, A.: Multi-level adaptive solutions to boundary-value problems. *Math. Comp.* 31 (1977), pp. 333–390.

24. Brandt, A.: Guide to multigrid development. In *Multigrid Methods*, Hackbusch, W. and Trottenberg, U. (eds.), Lecture Notes in Mathematics 960, Springer-Verlag, Berlin, Heidelberg (1982).

25. Brandt, A.: Algebraic multigrid theory: The symmetric case. *Appl. Math. Comp.* 19 (1986), pp. 23–56.

26. Brandt, A.; and Ta'asan, S.: Multigrid Methods for Nearly Singular and Slightly Indefinite Problems. In *Multigrid Methods II*, Hackbusch, W. and Trottenberg, U. (eds.), Lecture Notes in Mathematics 1228, Springer-Verlag, New York (1985), pp. 100–122.

27. Brandt, A.; and Yavneh, I.: Inadequacy of first-order upwind difference schemes for some recirculating flows. *J. Comp. Phys.* 93 (1991), pp. 128–143.

28. Brandt, A.; and Yavneh, I.: On multigrid solution of high reynolds incompressible entering flows. *J. Comp. Phys.* 101 (1992), pp. 151–164.

29. Brandt, A.; and Yavneh, I.: Accelerated multigrid convergence and high reynolds recirculating flows. *SIAM J. Sci. Statist. Comput.* 14 (1993), pp. 607–626.

30. Brackenridge, K.: Multigrid and cyclic reduction applied to the helmholtz equation. In *Sixth Copper Mountain Conference on Multigrid Methods*, Melson, N. D., McCormick, S. F., and Manteuffel, T. A. (eds.), NASA, Langley Research Center, Hampton, VA (1993), pp. 31–42.

31. Brenner, S. C.; and Scott, L. R.: *The Mathematical Theory of Finite Element Methods*. Texts in Applied Mathematics 15, Springer-Verlag, New York, (2002).

32. Cai, X. C.; and Widlund, O. B.: Domain decomposition algorithms for indefinite elliptic problems. *SIAM J. Sci. Statist. Comput.* 13 (1992), pp. 243–258.

33. Caselles, V.; Kimmel, R.; and Sapiro, G.: Geodesic Active Contours. *IJCV* 22 (1997), pp. 61–79.

34. Chan, T. F.; and Goovaert, D.: A note on the efficiency of domain decomposed incomplete factorizations. *SIAM J. Sci. Statist. Comput.* 11 (1990), pp. 794–803.

35. Chan, T. F.; and Resasco, C.: A domain decompositioned fast poisson solver. *SIAM J. Sci. Statist. Comput.* 8 (1987), 514–527.

36. Chan, T. F.; and Vanek, P.: Detection of strong coupling in algebraic multigrid solvers. In *Multigrid Methods VI*, Vol. 14, Springer-Verlag, Berlin (2000), pp. 11–23.

37. Chang, O.; and Huang, Z.: Efficient algebraic multigrid algorithms and their convergence. *SIAM J. Sci. Comput.* 24 (2002), pp. 597–618.

38. D'Azevedo, E. F.; Romine, C. H.; and Donato, J. H.: Coefficient adaptive triangulation for strongly anisotropic problems. In *Preproceedings of the 5th Copper Mountain Conference on Iterative Methods*, Manteuffel, T. A. and McCormick, S. F. (eds.) (1998).

39. Dendy, J. E.: Black box multigrid. *J. Comp. Phys.* 48 (1982), pp. 366–386.

40. Dendy, J. E.: Black box multigrid for nonsymmetric problems. *Appl. Math. Comp.*, 13 (1983), pp. 261–283.

41. Dendy, J. E.: Two multigrid methods for the three-dimensional problems with discontinuous and anisotropic coefficients. *SIAM J. Sci. Statist. Comput.* 8 (1987), pp. 673–685.

42. Dendy, J. E.: Semicoarsening multigrid for systems. *ETNA* 6 (1997), pp. 97–105.

43. Dendy, J. E.; Ida, M. P.; and Rutledge, J. M.: A semicoarsening multigrid algorithm for SIMD machines. *SIAM J. Sci. Statist. Comput.* 13 (1992), pp. 1460–1469.

44. Dendy, J. E.; and Tazartes C. C.: Grandchild of the frequency decomposition multigrid method. *SIAM J. Sci. Comput.* 16 (1995), pp. 307–319.

45. Dendy, J. E.; and Tchelepi, H.: Multigrid applied to implicit wells problems. Report n98/6, Institut fuer Computeranwendungen der Universitaet Stuttgart (1988), pp. 18–34.

46. Diskin, B.: M.Sc. Thesis, Weismann Institute of Science, Rehovot, Israel (1993).

47. Douglas, C. C.: Cache based multigrid algorithms. In *MGnet Virtual Proceedings of the 7th Copper Mountain Conference on Multigrid Methods* (1997).

48. Dutto, L. C.; Habashi, W. G.; and Fortin, M.: An algebraic multilevel parallelizable preconditioner for large-scale CFD problems. In *MGnet Virtual Proceedings of the 7th Copper Mountain Conference on Multigrid Methods* (1997).

49. Elman, H. C.; Ernst, O. G.; and O'Leary, D. P.: A multigrid method enhanced by Krylov subspace iteration for discrete Helmholtz equations. *SIAM J. Sci. Comput.* 23 (2001), pp. 1291–1315.

50. Elman, H. C.; and Golub, G. H.: Line iterative methods for cyclically reduced discrete convection-diffusion problems. *SIAM J. Sci. Statist. Comput.* 13 (1992), pp. 339–363.

51. Evans, D. J.: *Preconditioning Methods*. Gordon and Breach, New York (1983).

52. Frederickson, P. O.; and McBryan, O. A.: Parallel superconvergent multigrid. In *Multigrid Methods*, McCormick, S.F. (ed.), Lecture Notes in Pure and Applied Mathematics 110, Marcel Dekker, New York (1988).

53. Freund, R. W.: Conjugate gradients type methods for linear systems with complex symmetric coefficient matrices. *SIAM J. Sci. Stat. Comput.* 13 (1992), pp. 425–448.

54. Freund, R. W.: Transpose free quasi-minimal residual algorithm for non-Hermitian linear systems. *SIAM J. Sci. Statist. Comput.* 14 (1993), pp. 470–482.

55. Goldenberg, R.; Kimmel, R.; Rivlin, E.; and Rudzsky, M.: Fast geodesic active contours. In *Scale-Space Theories in Computer Vision*, Nielsen, H., Johansen, P., Olsen, O. F., and Weickert, J. (eds.), Lecture Notes in Computer Science 1682, Springer, Berlin (1999).

56. Gustafsson, S.: On modified incomplete factorization methods. In *Numerical Intergration of Differential Equations and Large Linear Systems*. Hinze, J., (ed.), Lecture Notes in Mathematics 968, Springer, Berlin (1982) pp. 334–351.

57. Hackbusch, W.: *Multigrid Methods and Applications*. Springer-Verlag, Berlin, Heidelberg (1985).

58. Hackbusch, W.: The frequency decomposition multigrid algorithm. In *Robust Multigrid Methods*, Hackbusch, W. (ed.), Proceedings of the fourth GAMM seminar, Kiel (1988).

59. Hackbusch, W.: The frequency decomposition multigrid method, Part 1: Application to anisotropic equations. *Numer. Math.* 56 (1989), pp. 229–245.

60. Hart, L.; and McCormick, S. F.: Asynchronous multilevel adaptive methods for solving partial differential equations: Basic ideas. *Parallel Comput.* 12 (1990), pp. 131–144.

61. Henson, V. H.: Towards a fully-parallelizable algebraic multigrid. Presented in the *7th Copper Mountain Conference on Multigrid Methods*, April 6–11, 1997.

62. Hestenes, M.; and Stiefel, E.: Methods of conjugate gradients for solving linear systems. *J. Res. Nat. Bar. Stand.* 49 (1952), pp. 409–425.

63. Kettler, R.: Analysis and comparison of relaxation schemes in robust multigrid and preconditioned conjugate gradients methods. In *Multigrid Methods*, Hackbusch, W. and Trottenberg, U. (eds.), Lecture Notes in Mathematics 960, Springer-Verlag, Berlin, Heidelberg (1982).

64. Kettler, R.; and Meijerink, J. A.: A multigrid method and a combined multigrid-conjugate gradient method for elliptic problems with strongly discontinuous coefficients in general domains. Shell Publ. 604, KSELP, Rijswijk, The Netherlands (1981).

65. Kou, C. C. J.; and Levy, B. C.: Two-color fourier analysis of the multigrid method with red-black Gauss-Seidel smoothing. *Appl. Math. Comp.* 29 (1989), pp. 69–87.

66. Kraus, J. K.; and Schicho, J.: Algebraic multigrid based on computational molecules, 1: Scalar elliptic problems. RICAM report, March 2005.

67. Lee, B.; Manteuffel, T. A.; McCormick, S. F.; and Ruge, J.: First order system least squares for the Helmholtz equation. *SIAM J. Sci. Comput.* 21 (2000), pp. 1927–1949.

68. Maitre, J. F.; and Musy, F.: Multigrid methods: Convergence theory in a variational framework. *SIAM J. Numer. Anal.* 21 (1984), pp. 657–671.

69. Maitre, J. F.; and Musy, F.: Algebraic formulation of the multigrid method in the symmetric and positive definite case—A convergence estimation for the V-Cycle. In *Multigrid Methods for Integral and Differential Equations*, Paddon, D.J. and Holstein, H. (eds.), Clarendon Press, Oxford (1985).

70. Mavriplis, D. J.: Directional coarsening and smoothing for anisotropic Navier-Stokes problems. In *MGnet Virtual Proceedings of the 7th Copper Mountain Conference on Multigrid Methods* (1997).

71. McCormick, S. F.: An algebraic interpretation of multigrid methods. *SIAM. J. Numer. Anal.* 19 (1982), pp. 548–560.

72. McCormick, S. F.; and Quinlan, D.: Asynchronous multilevel adaptive methods for solving partial differential equations: Performance results. *Parallel Computing* 12 (1090), pp. 145–156.

73. McCormick, S. F.; and Quinlan, D.: Idealized analysis of asynchronous multilevel methods. Internal memo, University of Colorado at Denver.

74. Meijerink, J. A.; and Van der Vorst, H. A.: An iterative solution method for linear systems of which the coefficients matrix is a symmetric M-matrix. *Math. Comp.* 31 (1977), pp. 148–162.

75. Meijerink, J. A.; and Van der Vorst, H. A.: Guidelines for the usage of incomplete decompositions in solving sets of linear equations as they occur in practical problems. *J. Comp. Phys.* 44 (1981), pp. 134–155.

76. Mitchell, W. F.: Optimal multilevel iterative methods for adaptive grids. *SIAM J. Sci. Statist. Comput.* 13 (1992), pp. 146–167.

77. Munson, D. C.: A note on Lena. *IEEE T Image Process.* 5, (1996).

78. Oman, M. E.: Fast multigrid techniques in total variation-based image reconstruction. In *Seventh Copper Mountain Conference on Multigrid Methods*, Melson, N.D., Manteuffel, T.A., McCormick, S.F., and Douglas, C.C. (eds.), NASA CP 3339, Hampton, VA (1996), pp. 649–660.

79. Ortega, J. M.: *Introduction to Parallel and Vector Solutions of Linear Systems.* Plenum Press, New York (1988).

80. Patera, A. T.: A spectral element method for fluid dynamics; laminar flow in a channel expansion. *J. Comp. Phys.* 54 (1984).

81. Perona, P.; and Malik, J.: Scale space and edge detection using anisotropic diffusion. *IEEE T. Pattern Anal.* 12 (1990), pp. 629–639.

82. van-der-Ploeg, A.; Botta, E. F. F.; and Wubs, F. W.: Nested grids ILU-decomposition (NGILU). *J. Comput. Appl. Math.* 66 (1996), pp. 515–526.

83. Quarteroni, A.; and Zampieri, E.: Finite element preconditionering for legendre spectral collocation approximations to elliptic equation and systems. *SIAM J. Numer. Anal.* 29 (1992), pp. 917–936.

84. Reusken, A.: A multigrid method based on incomplete gaussian elimination. *Numer. Linear Algebra Appl.* 3 (1996), pp. 369–390.

85. Ruge, J. W.; and Stuben, K.: Efficient solution of finite difference and finite element equations by algebraic multigrid. In *Multigrid Methods for Integral and Differential Equations*, Paddon, D. J. and Holstein, H. (eds.), Oxford Univ. Press, New York (1985), pp. 169–212.

86. Ruge, J. W.; and Stuben, K.: Algebraic multigrid. In *Multigrid Methods*, McCormick, S. F. (ed.), SIAM, Philadelphia (1987).

87. Saad, Y.: ILUM, a multi-elimination ILU preconditioner for general sparse matrices. *SIAM J. Sci. Comput.* 17 (1996), pp. 830–847.

88. Saad, Y.; and Schultz, M.H.: A generalized minimal residual algorithm for solving nonsymmetric linear systems. *SIAM J. Sci. Stat. Comput.* 7 (1986), pp. 856–869.

89. Sapiro, G.; and Ringach, D.L.: Anisotropic diffusion of multivalued images with applications to color filtering. *IEEE T Image Process.* 5 (1996), pp. 1582–1586.

90. Shapira, Y.: Two-level analysis and multigrid methods for SPD, non-normal and indefinite problems. Technical Report #824 (revised version), Computer Science Department, Technion, Haifa, Israel (1994).

91. Shapira, Y.: Black-box multigrid solver for definite and indefinite problems. In *Algebraic Multi-Level Iteration Methods with Applications*, Axelsson, O. and Polman, B. (eds.), Nijmegen, The Netherlands, (1996), pp. 235–250.

92. Shapira, Y.: Multigrid techniques for highly indefinite equations. In *Seventh Copper Mountain Conference on Multigrid Methods*, Melson, N.D., Manteuffel, T.A., McCormick, S.F., and Douglas, C.C. (eds.), NASA CP 3339, Hampton, VA (1996), pp. 689–705.

93. Shapira, Y.: Algebraic interpretation of continued fractions. *J. Comput. Appl. Math.* 78 (1997), pp. 3–8.

94. Shapira, Y.: Multigrid methods for 3-D definite and indefinite problems. *Appl. Numer. Math.* 26 (1998), pp. 377–398.

95. Shapira, Y.: Coloring update methods. *BIT* 38 (1998), pp. 180–188.

96. Shapira, Y.: Parallelizable approximate solvers for recursions arising in preconditioning. *Linear Algebra Appl.* 274 (1998), pp. 211–237.

97. Shapira, Y.: Analysis of matrix-dependent multigrid algorithms. *Numer. Linear Algebra Appl.* 5 (1998), pp. 165–201.

98. Shapira, Y.: Algebraic domain decomposition method for unstructured grids. In *9th International Conference on Domain Decomposition Methods*, Bjorstad, P., Espedal, M., and Keyes, D. (eds.), Bergen Univ. Press, Norway (1998), pp. 205–214.

99. Shapira, Y.: Multigrid for locally refined meshes. *SIAM J. Sci. Comput.* 21 (1999), pp. 1168–1190.

100. Shapira, Y.: Model-case analysis of an algebraic multilevel method, *Numer. Linear Algebra Appl.* 6 (1999), pp. 655–685.

101. Shapira, Y.: Algebraic multilevel method with application to the Maxwell equations. *J. Comp. Appl. Math.* 137 (2001), pp. 207–211.

102. Shapira, Y.: Adequacy of finite difference schemes for convection-diffusion equations. *Numer. Methods Partial Differential Equations* 18 (2002), pp. 280–295.

103. Shapira, Y.: *Solving PDEs in C++*. SIAM, Philadelphia, 2006.
104. Shapira, Y.; Israeli, M.; and Sidi, A.: An automatic multigrid method for the solution of sparse linear systems. In *Sixth Copper Mountain Conference on Multigrid Methods*, Melson, N.D., McCormick, S.F. Manteuffel, T.A., and Douglas, C.C. (eds.), NASA, Langley Research Center, Hampton, VA (1993), pp. 567–582.
105. Shapira, Y.; Israeli, M.; and Sidi, A.: Towards automatic multigrid algorithms for SPD, nonsymmetric and indefinite problems. *SIAM J. Sci. Comput.* 17 (1996), pp. 439–453.
106. Shapira, Y.; Israeli, M.; Sidi, A.; and Zrahia, U.: Preconditioning spectral element schemes for definite and indefinite problems. *Numer. Methods Partial Differential Equations* 15 (1999), pp. 535–543.
107. Sidi, A.; and Shapira, Y.: Upper bounds for convergence rates of acceleration methods with initial iterations. *Numer. Algorithms* 18 (1998), pp. 113–132.
108. Smith, R. A.; and Weiser, A.: Semicoarsening multigrid on a hypercube. *SIAM J. Sci. Statist. Comput.* 13 (1992), pp. 1314–1329.
109. Sochen, N.; Kimmel, R.; and Malladi, R.: A general framework for low level vision. *IEEE T Image Process.* 7 (1998), pp. 310–318.
110. Sonneveld, P.: CGS, a fast Lanczos-type solver for nonsymmetric linear systems. *SIAM J. Sci. Statist. Comput.* 10 (1989), pp. 36–52.
111. Sonneveld, P.; Wesseling, P.; and de Zeeuw, P.M.: Multigrid and conjugate gradient methods as convergence acceleration techniques. In *Multigrid Methods for Integral and Differential Equations*, Paddon, D.J. and Holstein, H. (eds.), Oxford University Press, Oxford (1985), pp. 117–168.
112. Strang, G.; and Fix, G.: *An Analysis of the Finite Element Method*. Prentice-Hall, Englewood Cliffs, NJ (1973).
113. Stuben, K.; and Trottenberg, U.: Multigrid methods: Fundamental algorithms, model problem analysis and applications. In *Multigrid Methods*, Hackbusch, W., and Trottenberg, U. (eds.), Lecture Notes in Mathematics 960, Springer-Verlag, Berlin, Heidelberg (1982), pp. 1–176.
114. Ta'asan, S.: Multigrid methods for highly oscillatory problems. Ph.D. Thesis, Weismann Institute of Science, Rehovot, Israel (1984).
115. Tanabe, K.: Projection methods for solving a singular system of linear equations and its applications. *Numer. Math.* 17 (1971), pp. 203–214.
116. Vanek, P.; Brezina, M.; and Mandel, J.: Convergence of algebraic multigrid based on smoothed aggregation. *Numer. Math.* 88 (2001), pp. 559–579.
117. Vanek, P.; Mandel, J.; and Brezina, M.: Algebraic multigrid by smooth aggregation for second and fourth order elliptic problems. *Computing* 56 (1996), pp. 179–196.
118. Varga, R.: *Matrix Iterative Analysis*. Prentice-Hall, NJ (1962).
119. van der Vorst, H. A.: Iterative solution methods for certain sparse linear systems with a non-symmetric matrix arising from PDE-problems. *J. Comp. Phys.* 49 (1982), pp. 1–19.
120. Wesseling, P.: A robust and efficient multigrid solver. In *Multigrid Methods*, Hackbusch, W. and Trottenberg, U. (eds.), Lecture Notes in Mathematics 960, Springer-Verlag, Berlin, Heidelberg (1982), pp. 614–630.
121. Wesseling, P.: *An Introduction to Multigrid Methods*. Wiley, Chichester, UK (1992).
122. Yavneh, I.: Multigrid smoothing factors for red black Gauss-Seidel applied to a class of elliptic operators. *SIAM J. Numer. Anal.* 32 (1995), pp. 1126–1138.
123. Yavneh, I.: On red black SOR smoothing in multigrid. *SIAM J. Sci. Comput.* 17 (1996).
124. Yavneh, I.: Coarse grid correction for nonelliptic and singular perturbation problems. *SIAM J. Sci. Comput.* 19 (1998), pp. 1692–1699.
125. Young, D.: *Iterative Solution of Large Linear Systems*. Academic Press, New York (1971).

126. Yserentant, H.: On the multi-level splitting of finite element spaces. *Numer. Math.* 49 (1986), pp. 379–412.
127. Yserentant, H.: Two preconditioners based on the multi-level splitting of finite element spaces. *Numer. Math.* 58 (1990), pp. 163–184.
128. de Zeeuw, P. M.: Matrix-dependent prolongations and restrictions in a blackbox multigrid solver. *J. Comput. Appl. Math.* 33 (1990), pp. 1–27.
129. de Zeeuw, P. M.: Multigrid and advection. In *Numerical Methods for Advection-Diffusion Problems*, Vreugdenhil, C. B. and Koren, B. (eds.), Notes on Numerical Fluid Mechanics 45, Vieweg Verlag, Braunschweig (1993), pp. 335–351.

Index